Econometrics

A practical approach

H. R. Seddighi, K. A. Lawler and A. V. Katos

London and New York

First published 2000
by Routledge
11 New Fetter Lane, London EC4P 4EE

Simultaneously published in the USA and Canada
by Routledge
29 West 35th Street, New York, NY 10001

Routledge is an imprint of the Taylor & Francis Group

Typeset in Times by Wearset, Boldon, Tyne and Wear
Printed and bound in Great Britain by TJ International Ltd,
Padstow, Cornwall

British Library Cataloguing in Publication Data
A catalogue record for this book is available from the British Library

Library of Congress Cataloging in Publication Data

A catalog record for this book has been applied for

ISBN 0-415-15644-0 (hbk)
ISBN 0-415-15645-9 (pbk)

To our families

Contents

Figures

Tables

Acknowledgements

The authors would like to express their sincere appreciation to Professor Eric Fletcher and Mr. P. J. Ripley, University of Sunderland. The Microfit appendix was written by Professor Fletcher. However, the responsibility for any errors remains ours.

The authors are grateful for the assistance with proof reading undertaken by Kin-Pui Lee and Antonis Theocharous, both of whom are PhD candidates at Sunderland Business School.

HAMID SEDDIGHI
KEVIN LAWLER

1 Econometric methods of research

INTRODUCTION

In this chapter we develop some basic econometric models and outline the key specific assumptions underlying the traditional 'Specific to General' (SG) approach to model building. The key features of the meaning of the disturbance term in the classical methodology are discussed as are some limitations of the approach. A more detailed review of methodological issues is held over until Chapter 6 where the significance of the normality assumption is considered in other approaches. This chapter attempts to investigate basic econometric features of an income-consumption model and outlines critical aspects of development. The stress here concerns the direct links of the econometric measurement with economic theory. Hypotheses drawn from theory are modelled in a manner to achieve data consistency. Data limitations and requirements are considered in the light of traditional 'SG' modelling procedures.

Key points

- Economic modelling
- Economic modelling using (SG) approach
- The ordinary least squares estimators

1 Theory and modelling

Economic theories are developed to provide logical and coherent explanations of economic issues. These issues range from the activities of households and firms to the activities of government concerning employment, inflation and economic growth.

Typically, economic theory consists of a set of conceptual definitions about the economic variables under consideration and a set of assumptions about the behaviour of economic agents including households, firms and the government. Economists then follow a process of logical deduction deriving the implications of the assumptions. These implications are the predictions of the economic theory. These are usually reported as conditional statements in the form of testable hypotheses concerning the issues and causal linkages.

The implications of theory when configured as economic hypotheses constitute an economic model. Economic models are, therefore, derived from theoretical arguments and are, essentially a simple representation of the complex real world economic relationships. Economic models can be represented verbally, geometrically or algebraically. In this latter form, the relationship among economic variables implied by theoretical arguments are expressed by mathematical symbols and equations. These three types of presentation of economic models are frequently combined to explain various economic phenomena under examination.

Algebraic presentations of economic models have, however, a number of advantages. These are that they provide: (i) frameworks within which the relationships among economic variables are expressed in consistent and logical sequences; (ii) the bases for the model builder to generalise theoretical arguments and ascertain implications; and (iii) frameworks for the empirical investigation of economic hypotheses.

To illustrate the process of economic model building we develop two economic models, these are: (a) a model of household consumption expenditure; and (b) a model of demand for competitive imports.

2 Economic modelling in practice

A *model of household consumption*

This model is designed to explain factors influencing the planned consumption expenditure of a household. That is the amount of expenditure that a household plans to spend out of income. There are a number of important theoretical contributions in this field. The simplest of these theories is the Keynesian absolute income hypothesis (AIH). In developing this theory, Keynes argued that, of the many factors that influence the level of household consumption expenditure, the most important is the level of household current disposable income. The way in which consumption expenditure is influenced by the disposable income is, according to what Keynes called 'the fundamental psychological law', that 'men are disposed, as a rule and on average, to increase their consumption as their income increases, but not by as much as the increase in their income'. (Keynes, 1936.) The change in consumption expenditure per unit of a change in income is called: Marginal Propensity to Consume (MPC). According to Keynes the MPC is less than unity. Furthermore, Keynes argued that the proportion of income saved would increase as household income increased. The proportion of income consumed is called the average propensity to consume. Given Keynes' view, the average propensity to consume falls as real income rises. Moreover, Keynes argued that the marginal propensity to consume would be less than the average propensity to consume.

To derive an algebraic model from these theoretical arguments, we define two variables: Y = consumption expenditure of household; X = household income.

Economic theory identifies income to be the major determinant of household consumption expenditure Y. This statement may be expressed mathematically as follows: Y = f(X). That is Y depends on X (Y is a function of X) and the direction of causation is from X to Y, so X influences Y, but y does not influence X. In this presentation, Y is called the dependent variable and X the independent variable of the model.

This mathematical relationship is the maintained hypothesis of the theory, from which theoretical reasoning is used to explain the nature of the relationship between Y and X. According to (AIH), household increase consumption as disposable income increases, but not by as much as income. In other words, as income changes, consumption also changes, but the corresponding change in consumption expenditure is assumed to be less than the change in income. Mathematically, we may define: ΔY = a change in income; ΔX = a change in consumption expenditure; $\Delta Y/\Delta X$ is assumed to be less than one. Or equivalently: $\Delta Y/\Delta X \leq 1$, where the ratio $\Delta Y/\Delta X$ is the marginal propensity to consume (MPC). This shows the change in consumption per unit of a change in income. Hence, the average propensity to consume can be written as: APC = Y/X where apc > mpc and apc, falls as real income rises.

The algebraic form of the model may now be summarised as:

$Y = f(X)$

$0 \leq MPC = \Delta Y/\Delta X \leq 1$

$mpc < apc; apc = Y/X$

In general, economic theorists do not specify the exact functional forms likely to exist among economic variables. The specification of functional forms is left to the model builder. It is, however, customary to consider initially linear relationship; because of the ease of presentation and analysis. Following this tradition, the AIH in a linear form may be presented as: $Y = a + bX; a \geq 0; 0 \leq b \leq 1$.

Where a is the intercept term and b is the slope of the linear function. Both a and b are unknown constants and are called parameters of the model. An important task for econometrics is to provide estimates of the unknown parameters of the economic models on the basis of economic data on economic variables. This model can be presented graphically by assuming certain hypothetical values for the unknown parameters, a and b.

This is done in Figure 1.1, where the dependent variable is measured along the vertical axis and the independent variable along the horizontal axis.

According to Figure 1.1, there appears to exist a one-to-one relationship between Y and X. That is given a value for X, such as X_1, there is a unique value for Y_1. We return to this point in subsequent sections.

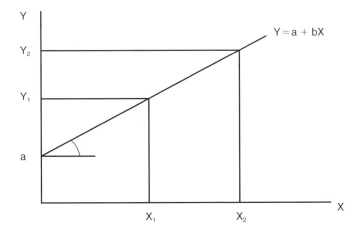

Figure 1.1 A one-to-one deterministic relationship for consumption and income

Modelling import demand

Theory identifies three major factors determining a country's demand for competitive imports, that is, those imports for which there are domestic substitutes.

The first factor is level of final expenditure, that is; total expenditure in a given period in an economy. The composition of total expenditure is also important given the degree of the import content of the different components of total expenditure (consumption expenditure, investment expenditure and expenditure on exports). In developing a model of competitive imports, we allow for a general case and distinguish between the three broad categories of final expenditure, namely total consumption expenditure by the private and public sector, investment expenditure and expenditure on exports. The underlying assumption being that each aggregate component of final expenditure has a different impact on imports.

The second factor developed by theory is the price of imports relative to the price of domestic substitutes. Where the price of imports is measured in the units of domestic currency and expressed in index form. The domestic price level is also expressed in index form, for example, in the form of wholesale price indices or GDP price deflators. A rise in the relative prices would normally lead to a fall in demand for competitive imports. In most cases to explain how prices are determined, it is usually assumed that the supply elasticities are infinite which implies that, whatever the level of domestic demand for imports these are supplied and therefore that import prices are determined outside the model, through the interactions of demand and supply. The domestic price index is also usually taken as given and is assumed to be flexible thereby eliminating excess demand at home. Theory also identifies the capacity of the import substitution sector to produce and supply the goods as an important factor too in determining the demand for imports. However, the capacity variable is essentially a short-run phenomenon and is relevant only if excess demand at home cannot be eliminated by a change in domestic prices. To generate an algebraic model we begin by defining the economic variables considered to be important in determining the demand for competitive imports. These variables are defined as follows:

M = the volume of imports (imports in constant price);

CG = the sum of private and public sector consumption expenditure in real terms (constant prices);

INVT = expenditure on investment goods including gross domestic fixed capital formation and stockbuilding in real terms (constant prices);

EXPT = expenditure on exports in real terms (constant prices);

PM = an index price of imports in units of domestic currency;

PD = an index price of domestically produced goods;

PM and PD may be measured by price deflators as follows:

$$PM = \frac{\text{imports in current prices}}{\text{imports in constant prices}}$$

$$PD^* = \frac{\text{GDP in current prices}}{\text{GDP in constant prices}}$$

Having defined these variables, we are now in the position to present the general form of the algebraic model as follows:

$$M = F(CG, INVT, EXP, PM/PD) \tag{1.1}$$

$$(+ve) \quad (+ve) \quad (+ve) \quad (-ve)$$

Hence, according to theoretical arguments, the demand for competitive imports depends on ('is a function of') the main components of aggregate demand, CG, INVT and EXP. Each of these having a separate impact on import demand. Moreover, one expects that the relationship between each one of these variables and imports to be positive. That is, a rise/fall in any of these variables would result in a rise/fall in the level of demand for competitive imports. With regard to the relative price term, we expect that a rise/fall in this term to lead to a fall/rise in the level of imports. We indicate this negative relationship by using a negative sign below the relative price variable in the previous equation. In this particular presentation of the model, M is the dependent variable, and CG, INVT, EXP, PM/PD are the independent variables.

To analyse in more detail the nature of the interactions between the dependent variable and each one of the independent variables for any economic model, we need to assume a certain functional form for the general economic model such as that one given by Equation (1.1). This will be done as follows.

3 The specification of models

Generally, the exact specifications for economic models are best left to model builders. Ease of presentation and interpretation, in most applications means that model builders tend to use either linear or log-linear specifications. We consider each of these specifications in turn.

A linear model of import demand

In this presentation, the dependent and independent variables of the model are linked together via a linear equation in which the parameters of the model appear in linear form. In applying this type of specification to the model of demand for imports, a linear model may be obtained as follows:

$$M = \alpha_1 + \alpha_2 CG + \alpha_3 INVT + \alpha_4 EXP + \alpha_5\, PM/PD \tag{1.2}$$

where α_1, α_2, α_3, α_4, α_5 are unknown constants and are called the parameters of the model α_1 is the intercept term of the linear equation and α_2, α_3, α_4, α_5 are slope parameters.

In this linear presentation of the model each slope parameter shows the impact of a marginal change (a one unit change) in a particular independent variable – while other independent variables are constant – on the average value of the dependent variable. For example α_2 shows the impact of one unit change in CG, while INVT, EXP, PM/PD are kept constant, on the average value of import demand. Symbolically:

$$\alpha_2 = \frac{\Delta M}{\Delta CG} \,|\, INVT,\, EXP,\, PM/PD (cons\,tant) \tag{1.3}$$

α_2 may be termed marginal propensity to import out of aggregate consumption. It is expected that its value lies between 0 and 1.

Similarly, α_5 shows the impact of a one unit change in the relative price variable, while all components of aggregate demand are kept constant, on the average level of demand for competitive imports. That is:

$$\alpha_5 = \frac{\Delta M}{\Delta^{PM}/_{PD}} | INVT, EXP, CG(cons\, tant) \qquad (1.4)$$

According to theory, one expects α_5 to be negative.

A log-linear specification

In log linear specifications, the logs of the dependent and independent variables are linked via a linear equation in which all parameters appear in a linear form. Equation (1.5) demonstrates a log-linear specification of the model of demand for imports:

$$\log M = \beta_1 + \beta_2 \log CG + \beta_3 \log INVT + \beta_4 \log EXP + \beta_5 \log PM/PD \qquad (1.5)$$

where β_1 is the intercept term and β_2–β_5 are the slope parameters.

The slope parameters β_2, β_3, β_4 and β_5 are partial elasticities each showing the percentage change in the dependent variable with respect to a percentage change in any one of the independent variables under consideration. For example, the parameter β_2 shows the percentage change in the dependent variable per unit of a percentage change in the independent variable CG, ceteris paribus. Therefore,

$$\beta_2 = \frac{\%\Delta M}{\%\Delta CG} \text{ (all other independent variables constant)}$$

Notice that in a log-linear specification each slope parameter shows the percentage change in the dependent variable (and not its log) per unit of a percentage change in any one of the independent variables. In economic analysis such a parameter is called an 'elasticity'. Here, β_2 is the partial elasticity of import demand with respect to consumption expenditure, showing the percentage change in the level of imports for every one percentage change in the level of CG. For example, if β_2 is found to be 5%, it implies that, for every single percentage change in CG (all other independent variables are kept constant), demand for import changes by 5%. This implies that the demand is highly elastic. Log-linear economic models are extremely popular in applied economic/econometric analysis due to the ease of specification and interpretation of results.

A summary of key steps required for modelling

Step 1   Give a clear statement/explanation of the economic theory underlying the economic phenomena under consideration. Pay particular attention to the assumptions noting implications and limitations.

Step 2   Use simple linear relationships to present the economic model in algebraic form. Use linear relationships to model/link economic variables to measure the impact of marginal changes on the dependent variables.

Step 3   Use log linear specifications, when you want to measure elasticities. Make sure you understand how the variables are linked through the log linear specification. Remember that a slope coefficient in a log linear specification shows the percentage change in the dependent variable per unit of a percentage change in an independent variable (to which the slope coefficient is attached).

Chapter 1: Key terms

Econometrics   A branch of economics dealing with empirical measurement and confirmation of economic models

Traditional/conventional approach (S–G)   An empirical methodology based on specific economic theory/model.

Disturbance term/error term   A random variable included in economic models to change a deterministic model into a stochastic framework compatible with economic data.

Multicollinearity   A high degree of linear association between two or more of variables.

OLS method   The Ordinary Least Squares (OLS) method-based on minimisation of the residual sum of squares.

BLUE   Best Linear Unbiased Estimator

Monte-Carlo study   A controlled experiment designed to generate/determine properties of estimators.

ECONOMETRICS

Economic models, presented in an algebraic form, provide the starting point for an econometric enquiry. An econometric enquiry basically aims to (a) quantify the relationship among economic variables providing a numerical estimate for the unknown parameters of the model, and (b) test to see whether the economic hypotheses derived from theory are consistent with facts in the form of economic data.

For example, in the previous examples the objective of the study is to determine estimates of the parameters in the models. Specifically referring to the household consumption and income/model, it is interesting to find the value of the household marginal propensity to consume, so as to predict the impact of a marginal change in household income on household average consumption expenditure. Similarly, with respect to the model of competitive imports, researchers are interested to estimate the model to establish the value of various partial elasticities. For example, it is of research interest to know the partial elasticity of demand for imports with respect to each component of final expenditure. Robust estimates of the partial elasticities would have solid policy implications in terms of inflation and/or balance of payment issues.

In addition to parameter estimation, researchers are also interested to know how well deductive/theory-based algebraic models explain real world economic relationships. Thus, if we cannot find evidence against a particular model, we can say that the theory/model conforms to the available facts. If this is the case, the model may then be used for policy analysis/economic forecasting by practitioners. Alternatively, if we find empirical evidence against a particular model, we can conclude that the model does not conform to the reality and therefore, the economic theory upon which the model is based, should either be discarded, or modified. Moreover, we need a procedure for choosing between competing economic theories. Often there are a number of competing economic theories designed to explain the same economic phenomena. For example, in order to explain price inflation, there are demand-pull, cost-push and monetary theories of inflation. The key question is which one of these, if indeed any, explains price inflation best in a given country over a period of time. The only way to answer this type of question is to resort to empirical investigation.

For these reasons there is a vital need to conduct empirical analysis in economics providing empirical content to abstract economic models. To this end, we need a framework/mechanism for confronting economic models with economic data. This framework is provided by econometrics. Thus a working definition of econometrics would be:

Definition
Econometrics is a branch of economics dealing with empirical measurement of economic relationships. The term 'metrics' signifies measurement. It aims to provide empirical content to abstract economic models with the purpose of verifying or refuting them.

Given this definition, the question which now arises is how does one carry out econometrics work/research in practice? In other words what is the methodology of econometric/empirical analysis in economics? There are three major rival methodologies for empirical work/analysis in economics: 'Specific to General' (SG); 'General to Specific' (GS); and cointegration strategies. These three alternative strategies are most popular in practice. Here we consider briefly each of these in turn. In this chapter we start with the traditional approach to econometric modelling, often termed the Specific to General (SG) approach to econometric modelling.

1 The traditional approach – specific to general modelling strategies

The Specific to General (SG) approach to econometric modelling is the oldest methodology used to carry out empirical research in economics. It dates back to the 1930s when the Econometric Society was established to encourage and promote empirical work/research in economics. From the 1930s, the SG approach to econometric analysis dominated research work in applied econometrics until the late 1970s.

The SG approach begins procedures with economic theory and applies this to the issues under investigation. In this situation, the practitioner using this strategy, must be careful in explaining the theory to be utilised in the analysis. However, there are occasions when relevant economic theory/model concerning a particular economic issue has not yet been developed. In this situation, care has to be taken to ensure that the economic model/hypothesis underlying the empirical analysis is explained. Once an economic hypothesis is selected for the purposes of empirical investigation, its algebraic form must be specified. There are no concrete guidelines as to how to specify the algebraic forms. In practice, it is customary to use linearities in parameter specification, particularly when one is interested to measure the impact of marginal variations in the dependent variable with respect to small changes in the independent variables. Log linear specifications are also extremely popular, for measuring various partial elasticities, in both time series and cross-section analysis. However, mathematical models are exact relationships and are not compatible with inexact economic data. To reflect the inexact nature of the data, economic models are transformed into 'econometric models'. Subsequently, we discuss in detail the nature of such modifications. Once an econometric model is developed, the assumptions are explained. Nowadays, there are many specially designed software packages for estimation of econometric models (e.g. Microfit, Give, SPSS). In subsequent chapters procedures governing the estimation and confirmation of econometric models are explained. If the model is found to be adequate, passing all key diagnostic tests/checks designed for confirmation, it may then be used for policy simulation. If, however, the model is found to be inadequate, it should either be modified or discarded. Figure 1.2 schematically shows key stages in a traditional econometric investigation based on the SG modelling strategy.

2 An econometric model

To investigate the relationships among economic variables captured by an algebraic model, and to confirm or otherwise, the underlying economic theory, the algebraic model has to be confronted with appropriate economic data. However, there is a problem associated with this process which has to be resolved before the process of measurement and confirmation can begin. The problem arises from the fact that, whereas an algebraic economic model intrinsically portrays an exact relationship among a number of variables, the data seldom exhibits such exact relationships. The economic data, however, has to be taken as given and cannot be modified. The solution therefore lies in the algebraic economic model which needs to be modified to reflect the nature of economic data. To undertake this modification to the algebraic model, we need to establish the likely role of the economic model in generating the economic data. In order words we need to explain the particular feature of the relevant data which the model purports to show. This is a deductive process. To perform the necessary modifications, we need a set of assumptions concerning a feasible data generation process in the underlying

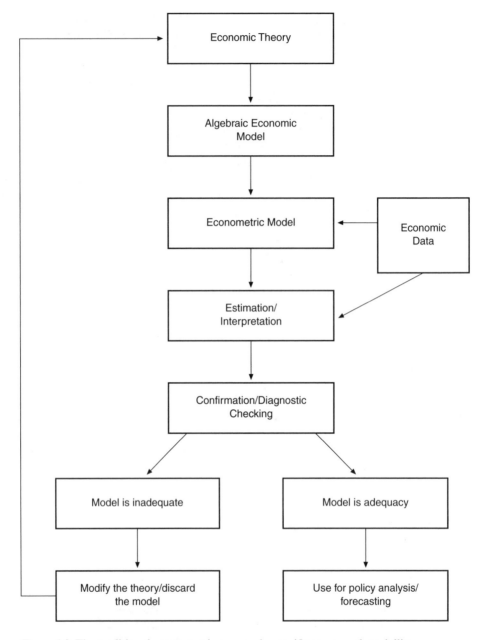

Figure 1.2 The traditional econometric approach: specific to general modelling

economic model. To fix these ideas we consider the simple household consumption/income model again. Thus, when:

$$Y = a + bX$$

This algebraic relationship is an exact one, showing a one-to-one relationship between income X and consumption Y. So given a value for X (income) the model generates a unique value for Y (household consumption). In other words for a particular level of income, according to this relationship, there can be only one level of household consumption expenditure. This is inconsistent with the underlying data, which shows that households with identical income levels seldom have identical consumption levels over specific periods of time. What could be a true picture is that at any specific level of income, there is a distribution or, a range of unknown consumption expenditure levels. This type of relationship is depicted in Figure 1.3, which shows a hypothetical distribution of monthly consumption levels at different levels of monthly income.

For each level of income, it is reasonable to assume that there would be a probability distribution of associated consumption levels for households. Notice that, we talk about likely probability distributions because we do not know the exact level of monthly consumption expenditure of each household. The proposition here is that, households with the same level of income say, X_1, are likely to have different levels of consumption. Consumption levels are likely to be different because there are many factors other than income (interest rates, size of family, location, habits and savings behaviour) which influence monthly expenditure. It is difficult to measure the net influence of all these economic and behavioural factors, but they do exist leading to a probability of distributions of consumption expenditure at each level of income. We forage to go deeper, making assumptions concerning how each of these probability of distributions could have been generated. A probability distribution is identified by three basic characteristics: the shape; the central value; and a measure of dispersion of values around the centre of the distribution. We start by considering the standard assumptions of the classical linear regression model.

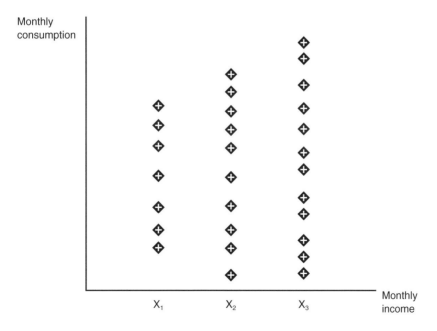

Figure 1.3 Consumption and income

The classical model of regression

(a)   The normality assumption

In the absence of any specific information concerning the shape of the distribution, it is customary to assume that for each level of the independent variable (income levels), the dependent variable (consumption expenditure) is normally and independently distributed. This assumption is depicted in Figure 1.4:

A consideration of the basic features of a normal distribution allows further exploration of the normality assumption. A normal distribution is bell-shaped and symmetrical around its central value (Figure 1.5).

The central value of the distribution is called the expected value and is usually denoted by E(X) (X here is any one variable). Half of the distribution lies to the left of E(X) with the other half to its right. A measure of average dispersion of values around the central value of the distribution is given by the variance of the distribution and is usually denoted by var(X). The variance of the distribution is defined as:

$$\text{Var}(X) = E[X - E(X)]^2 \tag{1.6}$$

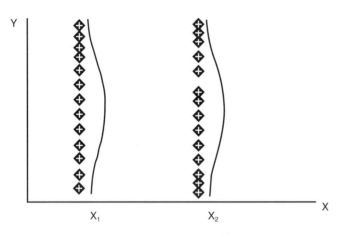

Figure 1.4 Normally distributed consumption levels

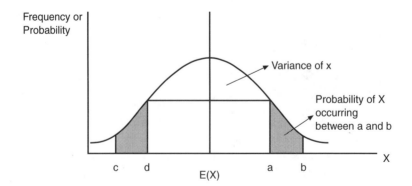

Figure 1.5 A normally distributed variable

The variance of the distribution has no meaningful unit of measurement so to generate a meaningful measure of the dispersion of values around the central value, the square root of variance is used in practice. This is called the standard deviation of the distribution denoted by SD(X) and the unit of measurement is the same as X, the variable under consideration. The area under the curve between any two points shows the probability of X occurring between those two points (Figure 1.5). Hence, $\Pr(a \leq X \leq b)$ = area under the normal curve between a and b.

An important property of the normal distribution is that the probability of X occurring between a and b would be the same as X occurring between c and d, provided the distance of these points from the centre of the distribution is the same (measured in units of standard deviations). That is the probability of X occurring between c and d to the left of E(X) is the same as the probability of X occurring between a and b to the right of X, as long as (c–d) and (a–b) are the same distance from the centre of distribution. What are the implications of these characteristics of the normal distribution for the data generation process concerning the household consumption/income model? First, it is assumed that most of the consumption expenditure is concentrated around the average consumption expenditure or expected expenditure, at any particular income level. Second, the consumption expenditure of households occur 'above average' or 'below the average' depending on each household's circumstances and behaviour. However, the probability of consumption levels occurring 'above average' is the same as the probability of consumption levels occurring below the expected consumption level for each class of income. Third, at each level of income, the probability that household consumption expenditure lies between a certain range, say above expected consumption is the same as the probability of that household consumption expenditure is between a certain range below the expected consumption expenditure, provided that the two ranges of consumption expenditure have the same distance/difference from expected consumption. These are the main implications of the normality assumption. They can only be considered valid, if the normality assumption is confirmed through diagnostic testing.

(b) The expected value of each distribution is determined by the algebraic economic model

This is an important assumption and it is being made to bring the economic model to the forefront of analysis. If the model is correctly specified and is consistent with the data generation process, then its role is to show the central value/expected value for each distribution. Given this framework, the algebraic economic model represents the average value or the expected value of the dependent variable, at any specific value of the independent variable. Using the consumption/income model, the expected level of consumption for any given level of income is assumed to be determined by this simple relationship. Given X_1 is a level of income, the expected level of consumption expenditure $E(Y/X_1)$ (reads expected consumption given X_1 level of income) is determined by algebraic model as follows:

$E(Y/X_1) = a + bX_1$ [at the centre of distribution of Y given X_1] similarly the expected level of consumption expenditure given the level of income of X_2 is:

$$E(Y/X_2) = a + b X_2 \tag{1.7}$$

The straight line $E(Y/X) = a + bX$ then goes through centre of each distribution, connecting the expected levels of consumption expenditure at various income levels. This is depicted in Figure 1.6. The line $E(Y) = a + bX$ is called the population regression line or function. This line is unknown and it is to be estimated on the basis of a set of observations/data on Y and X. Notice that this assumption has to be tested and confirmed later using appropriate diagnostic tests. All that is being said is that, provided the algebraic model is correct or is based on a valid economic theory, then it shows the average level for the dependent variable, at various levels for the independent variable. This assumption also enables us to represent each observation in terms of the expected value of each distribution and the deviation of the data points from the expected level of distribution. For example, the consumption expenditure of the ith household, corresponding to the income class X_i can be represented as follows:

$$Y_i = E(Y_i/X_i) + u_i \qquad (1.8)$$

where u_i shows the deviations of the ith household's consumption expenditure from the expected/average expenditure of household with the income of X_i. u_i can be positive, in which case, households consume more than average or, it can be negative implying that the ith household consumes less than expected consumption for the income group X_i. We can substitute for $E(Y_i/X_i)$ in terms of the algebraic model to obtain:

$$Y_i = a + bX_i + u_i \qquad (1.9)$$

$a + bX_i$ = the algebraic economic model showing expected consumption at income level X_i; u_i = deviation of ith household consumption expenditure from the expected consumption.

In this simple model, u_i represents the net influence of all factors/variables other than income on the ith household consumption expenditures. Some of these factors are economic factors influencing consumption such as interest rates, wealth and possibly rates of inflation. For household consumption expenditure, family size is also an important factor. In a more sophisticated model, these factors should be included. In such cases, there are a number of independent variables (income, interest rates and wealth), each influencing consumption expenditure independently. Other factors that are captured by u_i, however, cannot be measured quantitatively, these include the behavioural patterns of households and tastes. These factors essentially influence the consumption expenditure in unsystematic random ways. In other words we cannot precisely measure all these net influences. Given this, the econometric relationship is indeterministic or stochastic. Given a level of income such as X_i we get a range of values for consumption expenditure, because, the influence of random factors, as captured by the term u_i, are different for different households. The term u_i is called a random disturbance term. Its value occurs randomly and its inclusion into analysis disturbs an otherwise deterministic algebraic economic model.

Using this idea of the disturbance term, we can now present the consumption expenditure of each household (the data points) as follows:

$$Y_i = a + bX_i + u_i \qquad (1.10)$$

$$i = 1, 2, 3, \ldots, n$$

Hence each observation obtained from household consumption expenditure population can be decomposed as a sum of two elements. The expected consumption for the income class (i.e. $E(Y_i/X_i)$) and the deviation from the expected value which reflects the special characteristics/behaviour for that household. This model is called an econometric model. It is a modified economic model. The economic model is modified through inclusion of the disturbance term u_i, to reflect the stochastic/random nature of economic data. This econometric model is compatible with the data and although it is based on a deterministic algebraic economic model, it can be used with stochastic economic data to quantify the underlying economic relationships.

Notice that the two data generation assumptions can also be made in terms of the distribution of the disturbance term. In particular, the probability distribution of each consumption expenditure is generated through the probability distribution of the disturbance term. In other words, if we assume that consumption expenditure is normally distributed, then by implication, the disturbance terms associated with each value of the independent variable, are also normally distributed. With regard to the second data generation assumption, this implies that the expected value for each distribution of the disturbance term must be zero. To see this, consider the simple econometric model again:

$$Y_i = E(Y_i/X_i) + u_i \tag{1.11}$$

Now take average/expected value of both sides, we have:

$$E(Y_i/X_i) = E(Y_i/X_i) + E(u_i/X_i) \tag{1.12}$$

Notice that average of Y_i, given X_i is $E(Y_i/X_i)$ and that average/expected value of $E(Y_i/X_i)$ is in fact itself, since there is only one expected value! Therefore, by implication, $E(u_i/X_i)$ must be zero. This is called the zero mean assumption. So the net influence of all factors other than independent variable (income), when averaged out are zero, provided that the underlying model is sound.

(c) Independence assumption

How are the data points/observations of the consumption expenditure generated? The standard assumption here is that dependent variable (consumption) and by implication the disturbance term are independently distributed. So that the consumption expenditure of different households is independent of each other. So factors captured by the disturbance terms for different households are independent of each other. For example, consumption expenditure of the *i*th household is independent of consumption expenditure of the *j*th household, symbolically, the assumption of independence is usually written as follows:

$$Cov(u_i, u_j) = 0 \quad i \neq j \tag{1.13}$$

Cov is short for covariance. Covariance measures the degree of association between two random variables. This assumption is often termed the non-autocorrelation assumption, which implies no association between the values of the same variable (disturbance term).

(d) The constant variance assumption

Now we need an assumption concerning the dispersion of the values around the central value of each distribution. In the absence of any information concerning the variance of each distribution and the way it might be determined, the standard assumption usually made is that the variance of the distribution of dependent variable/disturbance term is constant and does not change across distributions. In terms of the consumption/income model, this assumption implies that, the spread/variance of distribution of the consumption expenditure does not change across the different income groups. In other words, the variance of the distribution does not change with the level of independent variable, income. This is a strong assumption, and in practice it often breaks down particularly, using cross-section data. This is because, in this type of data, independent variables tend to change significantly from one distribution/group to the next. Significant variations in the levels or scales of the independent variables might well, in turn, influence the variance/spread of each distribution generating changes in the variance across distributions. As with the other assumptions about the data generation process, we therefore need to check/test the appropriateness of this assumption when analysing the empirical model.

We are now in position to create an econometric model developed on the basis of these assumptions. We use the same notation as the consumption/income model, representing the dependent variable by Y and the independent variable by X. The model may be presented as follows:

$$Y_i = a + bX_i + u_i \qquad i = 1, 2, 3, \ldots, n \text{ (n observations)} \tag{1.14}$$

Y_i is distributed normally and independently around the expected value of $E(Y_i/X_i) = a + bX_i$, with a constant variance. We use some standard notation to present the standard assumptions as follows:

$$Y_i \sim \text{nid}(a + bX_i, \sigma^2)$$

where n = normal

   i = independently

   d = distributed

$E(Y_i/X_i) = a + bX_i$, the expected value of each distribution

   σ^2 = notation for a constant variance

Alternatively, we may state the assumption in terms of the distribution of the disturbance term u_i:

$$u_i \sim \text{nid}(0, \sigma^2)$$

Notice that the expected value of the disturbance term associated with each of the independent variables is assumed to be zero. This model is depicted in Figure 1.6.

The straight line $E(Y_i/X_i) = a + bX_i$ is called the population regression line or, sometimes the population regression function. The aim of the analysis is to estimate

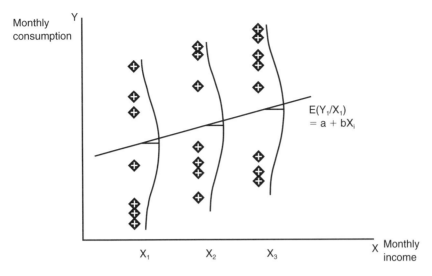

Figure 1.6 A linear consumption-income model

this population regression line, that is, to estimate a and b, on the basis of a set of observations on Y and X. Before we consider this issue, we first expand the theoretical model to include more independent variables. In practice, it is seldom the case that only one independent variable influences the dependent variable.

ECONOMETRIC METHODS

1 The multiple linear regression model and its assumptions

The two variable linear regression model is essentially a theoretical model with little practical use. It is used to explain the basic features of an econometric model and assumptions. Because there are only two variables involved, one can use a simple two-dimensional diagram to present the model. In practical applications, it is seldom the case that a particular economic variable under consideration (dependent variable) is determined on the basis of only one other economic variable (independent variable). The relationship between economic variables are much more involved and complex and one needs to make use of all the information provided by economic theory to specify an econometric model. In most applications, the model under consideration is determined by a number of other variables, based upon logical reasoning and economic theory.

For practical purposes, we must therefore abandon the restrictive features of the two variable model in favour of models where there are two or more variables that appear as regressors. These models form the basis for almost all empirical investigations in economics. The main feature of these models is that there are a number of variables, each exerting a systematic influence on the dependent variable under consideration, for example:

A basic demand model: homogeneous product

According to the economic theory the demand for a product depends upon a number of factors, each exerting a partial influence on demand. In particular, according to traditional utility maximisation theory, the demand for a product depends on the price of the product, on the prices of related products and on consumer income. Hence:

$$Q^D = f(P, P*, Y)$$

where:

Q^D = the quantity demanded of a product,

P   = the price of the product,

P* = the average price of related products and,

Y   = money income.

Each of the independent variables (P, P* and Y) have a particular influence on quantity demanded. In particular, for a normal product, a rise in P is expected to lead to a fall in quantity demand (Q^D) ceteris paribus. Hence, the relationship between P and Q^D is expected to be inverse or negative. A rise in the price of other related products (P*) will increase the demand for the product, if these are close substitutes, and reduce demand if these goods are complementary. Finally, a rise in income is expected to lead to an increase in quantity demanded (a positive relationship).

We now modify this model to generate a simple econometric model for the purpose of empirical analysis. The necessary steps in this modification are as follows:

(a) Specification of the economic model

We specify the functional form of the economic model. There are many options; however, the two most popular forms are linear and log linear models of demand.

A linear model of demand may be specified as follows:

$$Q^D = \beta_1 + \beta_2 P + \beta_3 P* + \beta_4 Y \qquad (1.15)$$

Notice that the model is linear in terms of its parameters β_1, β_2, β_3 and β_4.

In a log linear specification each variable is expressed in a natural logarithmic form, as follows:

$$\log Q^D = \beta_1 + \beta_2 \log P + \beta_3 \log P* + \beta_4 \log Y \qquad (1.16)$$

Notice that the model is linear in its parameters since there are no powers.

The 'correct' functional form is not known at this stage and we therefore need to conduct a number of diagnostic tests to see whether or not the chosen functional form is consistent with the data.

(b) Data generation assumptions

The next step in developing an econometric model is to specify data generating assumptions. That is we need to explain how each observation on quantity demanded might have been generated. The basic idea is the same as in the case of the two variable model. We assume that for each set of values of the regressors (P, P* and Y), we get a probability distribution for quantity demanded. These probability distributions are generated as a result of influences of random factors on quantity demanded. For example, random factors such as changes in tastes, fashion and the behaviour of buyers. The net influence of these factors is captured by the disturbance term u. The data generation assumptions are then concerned with the shape of each conditional probability distribution, the expected value of each distribution and the variance of each distribution.

Following the standard assumptions developed previously, we assume that for each set of values of the regressors (P, P*, Y), the associated disturbance term to be normally and independently distributed with a zero mean and a constant variance, say σ^2. Given these assumptions, an econometric model of demand, based on a log linear specification may be written as follows.

$$\log Q_t^D = \beta_1 + \beta_2 \log P_t + \beta_3 \log P_t^* + \beta_4 \log Y_t + u_t$$

$$\text{With } u_t \sim \text{nid}(0, \sigma^2); \text{ for } t = 1, 2, \ldots, n \tag{1.17}$$

Thus each observation on quantity demand (or its log) can be written as the sum of two components: a systematic component showing the expected level of demand for the product at any given level of prices and incomes, so:

$$E(\log Q_t^D/\log P_t, \log P_t^*, \log Y_t) = \beta_1 + \beta_2 \log P_t + \beta_3 \log P_t^* + \beta_4 \log Y_t \tag{1.18}$$

and a stochastic component, u_t, showing the net influence of factors other than prices and incomes on quantity demanded. These random factors, for example, changes in the behaviour of buyers, might generate deviations from the expected level of demand, causing the level of demand to be less/more than expected levels.

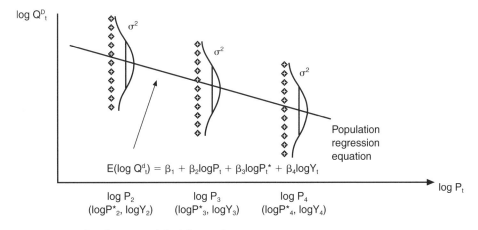

Figure 1.7 A log-linear model of demand

The disturbance term u_t is added to the expected level of demand to account for these random deviations. We can present the model graphically, using a two-dimensional diagram, by allowing only one regressor (e.g. P_t) to change and keeping the other two regressors constant.

2 A general multiple linear regression model

We are now in the position to generalise the presentation of the multiple linear regression model. It is customary to present the dependent variable by 'Y', the regressors by 'X' and the parameters of the model by β.

Suppose according to economic theory the dependent variable Y is a function of 'k' regressors i.e. $Y = f(X_1, X_2, X_3, \ldots, X_k)$.

To present a general linear econometric equation, we assume that there exists a linear relationship between the dependent and independent variables (linearity in parameters). Hence:

$$Y = \beta_1 X_1 + \beta_2 X_2 + \beta_3 X_3 + \ldots + \beta_k X_k \tag{1.19}$$

With this specification, each observation on the dependent variable can be written as the sum of two components as follows:

$$Y_t = \beta_1 X_{1t} + \beta_2 X_{2t} + \beta_3 X_{3t} + \ldots + \beta_k X_{kt} + u_t \tag{1.20}$$

where subscript 't' indicates the relevant observation at time 't'. To allow for an intercept term, we set the value of X_{1t} to be unity i.e.:

$$Y_t = \beta_1 + \beta_2 X_{2t} + \beta_3 X_{3t} + \ldots + \beta_k X_{kt} + u_t \qquad (t = 1, 2, 3, \ldots) \tag{1.21}$$

Alternatively, we can use the subscript of 'i' instead of 't' i.e. (cross-section data):

$$Y_i = \beta_1 + \beta_2 X_{2i} + \beta_3 X_{3i} + \ldots + \beta_k X_{ki} + u_i \qquad (i = 1, 2, 3, \ldots) \tag{1.22}$$

A summary of key assumptions of multiple linear regression models

1   *The specification assumption:* Each observation on the dependent variable is assumed to be a linear function of 'k' independent variables and a disturbance term (u) (linearity in parameters). This assumption may breakdown under any one or more of the following circumstances, producing specification errors:

1.1 Omission of the relevant regressors or inclusion of irrelevant regressors.
1.2 Incorrect functional form.
1.3 Changing Parameters.
1.4 Simultaneous relationships between dependent and independent variables.

2   *Zero mean assumption:* The expected value of each distribution of the disturbance term is zero, i.e. $E(u_t | X_{1t}, X_{2t}, \ldots) = 0$ for all 't' which implies that the expected value of the dependent variable is determined by the linear economic model, i.e.:

$$E(Y_t | X_{2t}, X_{3t}, \ldots) = \beta_1 + \beta_2 X_{2i} + \beta_3 X_{3i} + \ldots \tag{1.23}$$

The breakdown of this assumption occurs with incorrect specification of the economic model. Such as the omission of a relevant regressor.

3   *The homoscedasticity (constant variance) assumption:* The variance of each conditional distribution of the disturbance term (or the dependent variable) is the same and equal to an unknown constant variance (σ^2) i.e.

$$Var(u_t | X_{2t}, X_{3t}, \ldots) = \sigma^2 \text{ for all 't'} \tag{1.24}$$

also

$$Var(Y_t | X_{2t}, X_{3t}, \ldots) = \sigma^2 \text{ for all 't'} \tag{1.25}$$

The failure of this assumption is known as heteroscedasticity. Heteroscedasticity usually occurs in cross-sectional data analysis where changes in the magnitude of one or more regressors causes variations in the variance of the dependent variable/disturbance term.

4   *The non-autocorrelation assumption:* The disturbance terms are not correlated. So there is no association between any pair of disturbance terms 'u_t & u_s', hence

$$cov(u_t, u_s) = 0 \text{ for all } t \neq s \tag{1.26}$$

This assumption implies that the values of the dependent variables are independent of each other. The breakdown of this assumption is known as autocorrelation. It usually occurs in time series analysis due to prolonged influence of random shocks which are captured by the disturbance terms. Autocorrelation could also be due to the specification errors known as dynamic misspecification.

5   Independent variables are non-random and are fixed in repeated sampling. In practice, this assumption can fail due to:

i   One or more of the independent variables being measured with random errors. This phenomenon frequently occurs in empirical analysis and there is a need to check for random errors by statistical tests.

ii   Some or all of the independent variable contained in the model may not be independent of the dependent variable. This phenomenon occurs when dependent and independent variables are jointly determined. In this situation, a single equation model is not an adequate presentation of the data generation process and the independent variables may not be fixed in repeated sampling.

6   *Lack of perfect multicollinearity:* Multicollinearity occurs when there exists a near or, perfect linear association between any two or more of the explanatory variables in the model. In this situation, it is not possible to measure the separate impact of each independent variable on the dependent variable. Multicollinearity essentially reflects problems with data sets rather than the model. It usually occurs when

there are little variations in the values of the regressors over the sample period. To reduce the extent of the multicollinearity we must ensure that there are sufficient variations in the values of the regressors.

There is, however, always a certain degree of multicollinearity existing in econometric models due to the nature of economic data. The stated assumption implies that it is possible to measure the separate impact of each regressor on the dependent variable.

7   For each set of values of independent variables/regressors, the dependent variable/the disturbance terms are normally distributed. In certain situations specially where the data contains a significant number of outliers, the assumption of normality fails and the normal procedures of model evaluation are no longer valid.

The model and the assumptions may be written as follows:

Time series data: $Y_t = \beta_1 + \beta_2 X_{2t} + \beta_3 X_{3t} + \ldots + \beta_k X_{kt} + u_t, u_t \sim nid(0, \sigma^2)$ (1.27)

Cross section data: $Y_i = \beta_1 + \beta_2 X_{2i} + \beta_3 X_{3i} + \ldots + \beta_k X_{ki} + u_i,$

$u_i \sim nid(0, \sigma^2)$ for i = 1, 2, ...   (1.28)

3 Estimation and falsification

We now consider the following general linear regression model introduced previously.

$$Y_t = \beta_1 + \beta_2 X_{2t} + \beta_3 X_{3t} + \ldots + \beta_k X_{kt} + u_t, (t = 1, 2, 3, \ldots), u_t \sim nid(0, \sigma^2) \qquad (1.29)$$

The model implies that each observation on the dependent variable Y, can be written as a linear sum of 'k' independent variables (linear in parameters) and a disturbance term (u).

Moreover, the expected value of Y, for any given values of the independent variables is determined by the deterministic part of the model:

$$E(Y_t | X_{2t}, X_{3t}, \ldots) = \beta_1 + \beta_2 X_{2t} + \beta_3 X_{3t} + \ldots + \beta_k X_{kt} \qquad (1.30)$$

Each slope parameter, $\beta_2, \beta_3, \ldots, \beta_k$, measures the impact of a marginal change in the corresponding independent variable on the expected value of Y, while the other regressors are held constant. For example, β_2 shows the impact of one unit (a marginal change) in X_2 on the expected value of Y, while $X_3 \ldots X_k$ are held constant i.e.

$$\beta_2 = \frac{\Delta E(Y_t | \ldots)}{\Delta X_{2t}} \quad \bigg| \quad X_3 \ldots X_k \text{ (constant)}$$

or in general...

$$\beta_i = \frac{\Delta E(Y_t | X_{2t} \ldots X_{kt})}{\Delta X_{it}} \quad \bigg| \quad \text{all other regressors, for i = 2, \ldots k held constant}$$

Given this interpretation, the slope parameters are sometimes called the partial regression coefficients. These parameters, along with the intercept parameter are, unknown. The first task in the empirical analysis is to quantify or, estimate each of the unknown parameters, on the basis of a set of observations on the dependent and independent variables. For example, the estimated parameters for β_1, β_2, β_3 and β_k can be denoted as β_1^*, β_2^*, β_3^* and β_k^*.

These are point estimates of the unknown parameters obtained from a sample of observations on the dependent variable and the independent variables. The actual process by which the numerical values of these parameters are obtained is discussed subsequently. At this stage it is suffice to say that our aim is to generate estimates for the unknown parameters from a sample of observations on the dependent and independent variables. Notice that the expected value of the dependent variable is given by the expression:

$$E(Y_t \mid X_{1t}, X_{2t}, \ldots) = \beta_1 + \beta_2 X_{2t} + \beta_3 X_{3t} + \ldots + \beta_k X_{kt} \tag{1.33}$$

Once the intercept term (β_1) and partial regression coefficients ($\beta_2 \ldots \beta_k$) are estimated, we substitute for these estimates into Equation (1.33) to generate the expected value of the dependent variable for any given values of the independent variables. The estimated expected value of the dependent variable is usually denoted by Y_t^*, and can be expanded as follows:

$$Y_t^* = \beta_1^* + \beta_2^* X_{2t} + \beta_3^* X_{3t} + \ldots + \beta_k^* X_{kt}; \qquad (t = 1, 2, 3, \ldots) \tag{1.34}$$

In practice, β_1^*, β_k^* are numerical values and to generate the estimated expected value of the dependent variable we need to substitute into Equation (1.34) the values of independent variable. For example, let us consider the following hypothetical example. Suppose there are four parameters and on the basis of a set of observations on the dependent and independent variables, we obtain: $\beta_1^* = 1$, $\beta_2^* = 2$, $\beta_3^* = 3$ and $\beta_4^* = 4$.

Thus the estimated expected values for Y are:

$$Y_t^* = 1 + 2X_{2t} + 3X_{3t} + 4X_{kt} \tag{1.35}$$

To generate a specific value for the Y_t^* variable, we substitute a specific set of values for the independent variable in Equation (1.35). For example, the estimated expected value of the dependent variable (Y_t^*) when $X_2 = 100$, $X_3 = 300$ and $X_4 = 400$ can be calculated as follows:

$$Y_t^* = 1 + 2(100) + 3(300) + 4(400) = 1 + 200 + 900 + 1600 = 2701 \tag{1.36}$$

In addition, knowledge of the parameter estimates enables us to provide information on the impact of a marginal change (a one unit change) in the value of any one of the regressor on the estimated expected value of the dependent variable, while all other regressors are held constant. In the hypothetical example, $\beta_2^* = 2$

$$\beta_2^* = \left. \frac{\Delta Y_t^*}{\Delta X_{2t}} \right| \begin{array}{l} X_3 = 2 \\ X_4 \end{array}$$

Therefore when X_2 changes by one unit, Y_t^* changes by two units in the same direction as X_2. Similarly:

$$\beta_3^* = \frac{\Delta Y_t^*}{\Delta X_{3t}} \bigg| = 3$$

$$\beta_4^* = \frac{\Delta Y_t^*}{\Delta X_{4t}} \bigg| = 4$$

The reliability of the parameter estimates depends crucially upon whether or not the econometric model is consistent with the data. In particular, only when each of the assumptions are found to be consistent with the observed data, we can consider the point estimate to be reliable. The second aim of an empirical investigation is therefore to test to see whether each assumption is consistent with the data. These assumptions include those concerning model specification as well as data generation assumptions. The assumption concerning specification of the model is based upon our understanding of the economic theory underlying the data. We are testing to see if the assumptions of economic theory are data consistent.

The testing procedures are based on a series of diagnostic tests designed to confirm the model and its assumptions. Based upon the results of the diagnostic tests, two situations can arise:

1   We may fail to falsify each of the assumptions. In this case, the model is consistent with the data and we proceed to test economic hypotheses implied by the model and use the model.
2   The second situation that occurs more frequently in empirical economic analysis, is when according to the results of the diagnostic tests, one or more assumptions are found to be inconsistent with the data. In this situation, on the basis of the empirical results obtained, our task is to improve the economic model. There is therefore a link between economic theory and empirical analysis. Econometric analysis not only provides empirical content to abstract economic models but also helps us to understand complex economic relationships. This process means that we modify and improve economic theories to achieve a better understanding of economic issues.

We now consider the estimation of econometric models.

Given a multiple linear regression model of the form:

$$Y_t = \beta_1 + \beta_2 X_{2t} + \beta_3 X_{3t} + \ldots + \beta_k X_{kt} + u_t, \; u_t \sim \text{nid}(0, \sigma^2) \quad \text{for } t = 1, 2, \ldots, n \tag{1.37}$$

The task is to estimate the unknown parameters $\beta_1, \beta_2, \beta_3, \ldots, \beta_k$ on the basis of a set of observations on dependent and independent variables (i.e. Y_t and $X_{2t}, \ldots X_{kt}$). A general data set is presented as follows:

Each column in Table 1.1 depicts the observations obtained on the dependent and independent variables (for example the first row is the first set of observations on the dependent and independent variables).

Table 1.1 A general data set

Y	X_2	X_3	X_4	...	X_k
Y_1	X_{21}	X_{31}	X_{41}	...	X_{k1}
Y_2	X_{22}	X_{32}	X_{42}	...	X_{k2}
...	...	...	...	...	...
...	...	...	...	...	...
Y_n	X_{2n}	X_{3n}	X_{4n}	...	X_{kn}

Estimators

We need now to convert the observations on the dependent and independent variables into numerical values for the unknown parameters β_1, β_2, β_3, ..., β_k. A mechanism useful for this purpose is called an estimator. In particular, an estimator is a mathematical formula which is used to convert the observations on the dependent and independent variables into numerical values of the unknown parameters. The resulting numerical values are called 'point estimates'. Each parameter has its own estimator which can be used to generate point estimates. Conventionally, estimators are shown by a hat (ˆ) on top of the parameter.

$\hat{\beta}_1 = f_1$ (observation on dependent and independent variable) = point estimate of unknown β_1

$\hat{\beta}_2 = f_2$ (observation on dependent and independent variable) = point estimate of unknown β_2

...

$\hat{\beta}_k = f_k$ (observation on dependent and independent variable) = point estimate of unknown β_k

where $f_i (i = 1 \dots k)$ denote a particular mathematical formula.

How do econometricians determine these estimators? Econometricians seek reliable point estimates of the unknown parameters. The reliability of an estimator cannot be judged on the basis of only one point estimate since the actual value of the parameter is unknown. Under this condition, the only way to judge an estimator is to look at the performance of the estimator under repeated estimation/sampling procedures.

This is a hypothetical procedure used to generate sampling distributions of each estimator. Econometricians will then select an estimator whose sampling distribution has a number of desirable/optimum properties.

Sampling distribution of estimators

To fix these ideas, let us investigate how to generate sampling distribution of an estimator by hypothetical repeated sampling procedures.

The basic idea is to retain the value of the regressors fixed from sample to sample, obtaining data on the dependent variable. In general, because of the influence of factors other than regressors on the dependent variable (the disturbance term), we are likely to get different values for the dependent variable from sample to sample, although the values of the regressors are fixed. If we repeat this procedure a large number of times, we obtain large sample data containing the same values of the regressors but different values for the dependent variable as in the following table.

In Table 1.2, the values of the regressors are fixed at certain levels. Each column shows the observations on the dependent and independent variables in the hypothetical sampling procedure. The first row shows the first set of observations, the second row the second set of observations and the final row the mth set of observations. The interesting point here is that although the values of the regressors are kept fixed from sample to sample, we are likely to obtain different values for the dependent variable from sample to sample due to the fact that the dependent variable contains the influence of the random disturbance term. These values of the dependent variable are presented by Y_{ij} where 'i' is the number of observations and 'j' is the number in the sample. For example, Y_{21} is the second observation on Y in the first sample, and so on.

In this hypothetical experiment, if we substitute for each sample data in an estimator (say $\hat{\beta}_k$ – the estimator of β_k) we get a large number of point estimates for β_k. More specifically, we can write:

$$\hat{\beta}_k = f_k(Y_{11}, \ldots, Y_{n1}; X_{21}^*, X_{22}^*, \ldots, X_{2n}^* \ldots) \to \hat{\beta}_{k1} \text{ point estimate obtained}$$
from sample 1 \hfill (1.38)

$$\hat{\beta}_k = f_k(Y_{12}, \ldots, Y_{n2}; X_{21}^*, X_{22}^*, \ldots, X_{2n}^* \ldots) \to \hat{\beta}_{k2} \text{ point estimate obtained}$$
from sample 2 \hfill (1.39)

. . .
. . .
. . .

$$\hat{\beta}_k = f_k(Y_{1m}, \ldots, Y_{nm}; X_{21}^*, X_{22}^*, \ldots, X_{2n}^* \ldots) \to \hat{\beta}_{km} \text{ point estimate obtained}$$
from sample mm \hfill (1.40)

By arranging the point estimates into class intervals, it is possible to generate the sampling distribution of estimator of β_k as follows:

Class Interval	Frequency
$\beta^*_{k2} < \hat{\beta}_k < \beta^*_{k1}$	f_1
. . .	. . .
. . .	. . .
. . .	. . .

Table 1.2 Sample values of regressors and independent variables

Sample 1				Sample 2				Sample m			
Y	X_2	X_3	X_4	Y	X_2	X_3	X_4	Y	X_2	. . .	X_m
Y_{11}	X_{21}^*	X_{31}^*	X_{41}^*	Y_{12}	X_{21}^*	X_{31}^*	X_{41}^*	Y_{1m}	X_{21}^*	. . .	X_{m1}^*
Y_{21}	X_{22}^*	X_{32}^*	X_{42}^*	Y_{22}	X_{22}^*	X_{32}^*	X_{42}^*	Y_{2m}	X_{22}^*	. . .	X_{m2}^*
. . .	. . .	. . .	. . .	. . .	. . .	. . .	. . .	. . .	. . .	. . .	. . .
. . .	. . .	. . .	. . .	. . .	. . .	. . .	. . .	. . .	. . .	. . .	. . .
. . .	. . .	. . .	. . .	. . .	. . .	. . .	. . .	. . .	. . .	. . .	. . .
Y_{n1}	X_{2n}^*	X_{3n}^*	X_{4n}^*	Y_{n2}	X_{2n}^*	X_{3n}^*	X_{4n}^*	Y_{nm}	X_{2n}^*	. . .	X_{mn}^*

By measuring the relative frequency on the vertical axis and the corresponding class interval on the horizontal axis, we can obtain the histogram corresponding to the hypothetical sampling distribution and thus, generate the sampling distribution of the estimator. Figure 1.8 represents one such sampling distribution (note: the mid points of each bar of the histogram are connected together to form the sampling distribution).

An estimator whose sampling distribution has a number of desirable character-istics is then considered to be 'optimal' to be used for the purpose of estimating parameters of the econometric models. These ideal characteristics are now con-sidered:

Unbiasedness property of estimators

The expected value of a distribution represents the central value of the distribution. It is considered desirable for the expected value of the distribution to coincide with the true value of the unknown parameter (i.e. $E(\hat{\beta}_k) = \beta_k$). In other words, if we were to repeat the process of sampling a large number of times, the estimator has the property that its average/expected value (i.e. the expected value of its sampling distribution) is the same value as the unknown parameter.

The bias of the estimator in this case will be zero and the estimator is said to be unbiased. Hence: $E(\hat{\beta}_k) - \beta_k = 0$ (where $E(\hat{\beta}_k) - \beta_k$ is defined as the bias of the estimator).

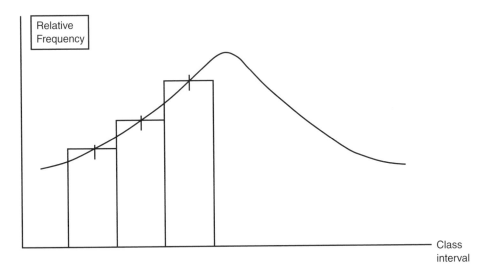

Figure 1.8 The sampling distribution of an estimator

Sampling distribution of $\hat{\beta}_k$

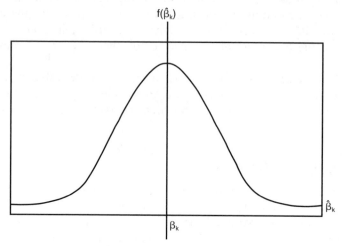

Figure 1.9a Sampling distribution for an unbiased estimator

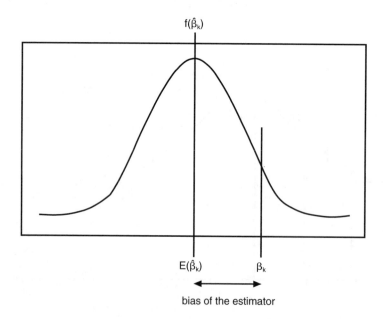

Figure 1.9b Sampling distribution for a biased estimator

(a) The minimum variance property of the estimator

The variance of an estimator provides a measure of average dispersions of values of the estimator around expected/central values.

It is desirable for the variance of an estimator to be as small as possible. The smaller the variance, the narrower is the sampling distribution. Neither the properties of unbiasedness nor minimum variance alone, are sufficient to enable selection amongst estimators. However, when they are combined, they provide a powerful criteria for selection. In particular it is desirable for an estimator to be unbiased as well as possessing the minimum variance property. So:

$$E(\hat{\beta}_k) = \beta_k \text{ (unbiased property)}$$

$$Var(\hat{\beta}_k) = E[\beta_k - E(\hat{\beta}_k)]^2 \text{ (as small as possible)}$$

Such an estimator is called an 'efficient' estimator. The standard deviation (i.e. the error associated with the sampling distribution) of the minimum variance estimator is less than any other unbiased estimator. Hence, most of the points generated by such an estimator would cluster around the expected value of the distribution (Figure 1.10).

(b) The sufficiency property of estimators

In addition to the efficiency property it is desirable for an estimator to make use of all sample data to generate point estimates. An estimator which uses all observations on the dependent and independent variables to generate point estimates is known as a 'sufficient' estimator. All observations including extreme values are used to generate point estimates and no information is discarded.

A 'good' estimator is therefore defined to be a sufficient estimator whose sampling distribution is centred on the parameter to be estimated and the dispersion of values about the central value is as small as possible. In many cases however, it is difficult to ascertain amongst many unbiased estimators, which estimator has the smallest variance. To overcome this problem, another requirement is added to that of efficiency. In particular, it is desirable for an estimator to be a linear function of observations on the value of the dependent variable. Through linear manipulation

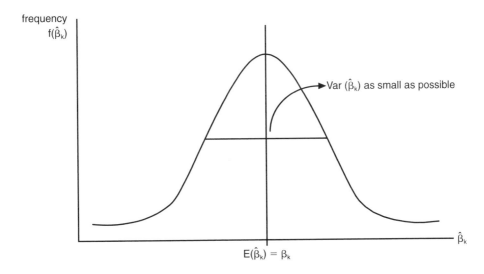

Figure 1.10 The sampling distribution of an efficient estimator

of the data on the dependent variable, a point estimate can be generated. Such an estimator is called Best Linear Unbiased Estimator (BLUE).

The asymptotic property of estimators

The above properties of the sampling distribution of estimators provide powerful criteria for selection of a 'good' estimator. In particular amongst all linear, unbiased estimators, we select an estimator whose sampling distribution has the smallest variance. Such an estimator is called an efficient estimator and given the assumptions of the econometric model, we need a method of estimation which generates an efficient estimator.

In many econometric applications, however, it is not possible to identify the properties of the sampling distribution of an estimator when the sample size is small and consequently, these criteria of selection no longer apply. However, we can determine what happens to the sampling distribution as the sample size becomes larger and larger. In practice, we need to know if the estimator becomes unbiased as the sample size increases and whether the variance of the estimator has a tendency to become smaller and smaller as the sample size becomes larger and larger. The properties of the sampling distribution of an estimator obtainable only when one allows the sample size to become extremely large are called asymptotic properties. These properties provide the criteria for the selection of a 'good' estimator when the sampling distribution of the estimator and other properties cannot be determined due to the small size of the sample. In particular, the following asymptotic properties are desirable:

a   *The Consistency Property.* The estimator is such that its asymptotic distribution becomes concentrated on the true value of the parameter as the sample size becomes extremely large. So as the sample size becomes larger and larger, the centre of the sampling distribution of the estimator shifts towards the true value of the parameter. Moreover, the spread of the distribution becomes smaller and smaller as sample size increases. In the limit, the sampling distribution simply becomes a vertical line concentrated on the true value of the parameter. This is shown in Figure 1.11.

b   *Asymptotically Efficient Estimators.* An asymptotically efficient estimator is such that its variance approaches zero faster than any other consistent estimator. In situations where small sample properties of the sampling distribution of the estimator cannot be determined, the usual practice is to consider asymptotic properties, and select an estimator which is asymptotically efficient. In practice, the properties of the sampling distribution of the estimator and their asymptotic characteristics are determined using Monte-Carlo simulation/studies.

ESTIMATION METHODS

We are now in a position to develop a method for estimating the parameters of the multivariate linear regression model. We seek estimators which are efficient, given the model's assumptions. Amongst all methods of estimation, there are only two types which satisfy the efficiency criterion given the assumptions, these are:

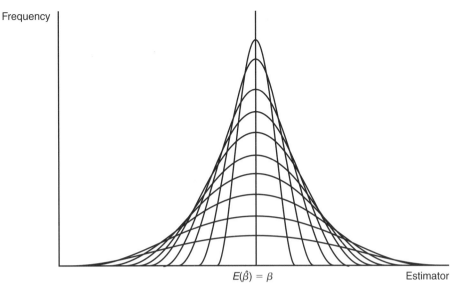

Frequency

$E(\hat{\beta}) = \beta$ Estimator

Figure 1.11 Consistency property

1   Ordinary Least Squares (OLS) estimation; and
2   Maximum Likelihood (ML) methods

OLS estimation can be obtained without reference to the normality assumption whilst ML estimation on the other hand, requires the assumption of normality to hold. Both methods generate identical estimators of parameters.

(a) The OLS method

Let us begin by considering a k-variable linear regression model such as:

$$Y_t = \beta_1 + \beta_2 X_{2t} + \beta_3 X_{3t} + \ldots + \beta_k X_{kt} + u_t, \quad u_t \sim \text{nid}(0, \sigma^2) \text{ for } t = 1, 2, \ldots \quad (1.41)$$

We aim here to generate efficient estimators of the unknown parameters β_1 to β_k so that each parameter can be estimated given a set of observations on the dependent and independent variables. We denote the estimators as follows:

$\hat{\beta}_1 \rightarrow$ estimator of $\beta_1 \rightarrow$ a set of observations on Y and $x_1 \ldots x_k \rightarrow$

$\hat{\beta}_1$ (a point estimate)

$\hat{\beta}_2 \rightarrow$ estimator of $\beta_2 \rightarrow$ a set of observations on Y and $x_1 \ldots x_k \rightarrow$

$\hat{\beta}_2$ (a point estimate)

$\ldots$

$\hat{\beta}_k \rightarrow$ estimator of $\beta_k \rightarrow$ a set of observations on Y and $x_1 \ldots x_k \rightarrow$

$\hat{\beta}_k$ (a point estimate)

The expected value of the dependent variable is given by the following expression:

$$E(Y_t) = \beta_1 + \beta_2 X_{2t} + \beta_3 X_{3t} + \ldots + \beta_k X_{kt} \tag{1.42}$$

and the estimated expected value is obtained by substitution of the point estimates into Equation (1.42):

$$\hat{Y}_t = \hat{\beta}_1 + \hat{\beta}_2 X_{2t} + \hat{\beta}_3 X_{3t} + \ldots + \hat{\beta}_k X_{kt}; \quad (t = 1, 2, 3, \ldots) \tag{1.43}$$

where $\hat{Y}_t$ is the estimated expected value of Y or simply fitted values. In the theoretical model the difference between each individual value of the dependent variable and the expected value of the dependent variable is known as the disturbance term i.e.:

$$Y_t - E(Y_t) = u_t \text{ for all } t = 1, 2, 3, \ldots \tag{1.44}$$

The empirical counterpart of the disturbance term (u_t) is known as the residual and is usually denoted as e_t. It can be obtained as follows:

$$e_t = Y_t - \hat{Y}_t = Y_t - (\hat{\beta}_1 + \hat{\beta}_2 X_{2t} + \hat{\beta}_3 X_{3t} + \ldots + \hat{\beta}_k X_{kt}), \text{ for all } t = 1, 2, 3, \ldots \tag{1.45}$$

Each residual shows the difference between an observed value of Y and the estimated expected value of Y and it is in this sense, that a residual may be considered to be the empirical counterpart/estimate of a disturbance term.

The OLS method makes use of the concept of the residuals to obtain parameter estimates. In particular, under the OLS method, the parameter estimates are obtained such that the sum of the squared residuals is minimised. Note that a residual can be positive or negative, depending upon whether or not the estimated expected value of the dependent variable is greater than or less than a particular value of Y. We can square each residual and then sum up the squared residuals to obtain the residual sum of squares (RSS) as follows:

$$\text{Residual sum of squares} = \text{RSS} = e_1^2 + e_2^2 + e_3^2 + \ldots e_k^2$$

$$\text{or simply} \ldots \sum_{i=1}^{n} e_i^2 \tag{1.46}$$

where $\sum$ is the summation operator. We can substitute for e_t into the above expression and we arrive at the following equation:

$$\text{RSS} = \sum e_i^2 = \sum_{i=1}^{n} [Y_t - \hat{\beta}_1 - \hat{\beta}_2 X_{2t} - \hat{\beta}_3 X_{3t} - \ldots - \hat{\beta}_k X_{kt}]^2 \tag{1.47}$$

Under the OLS method, $\hat{\beta}_1 \ldots \hat{\beta}_k$ are obtained in terms of observations of the dependent and independent variables $Y_t, X_{1t}, \ldots, X_{kt}$, such that the residual sum of squares is minimised i.e. we find the OLS estimators such that $\sum e_i^2 = \sum [Y_t \ldots]^2$ is as small as possible. This procedure yields OLS estimators $\hat{\beta}_{1ols} \ldots \hat{\beta}_{kols}$ etc. Each estimator is a particular function of the dependent and independent variables, yielding point estimates for each set of observations on Y and $X_1 \ldots X_k$, i.e.

$$\hat{\beta}_{iOLS} = f_i(Y_t, X_{1t}, \ldots, X_{kt}) \qquad (1.48)$$

The exact mathematical formula of each estimator can be obtained by matrix algebra and minimisation procedures. In practice, one uses software where the OLS method is routinely used to generate point estimates. The main reason for the popularity of the OLS method is that under repeated sampling procedures, and given the assumptions of the model, the sampling distribution of each OLS estimator has a number of desirable/optimal properties. In particular, it can be shown that if all assumptions of the model are consistent with the data then:

(1) OLS estimators are unbiased i.e. the expected value of the sampling distribution of an OLS estimator is equal to the unknown parameter. Symbolically:

$$E(\hat{\beta}_k) = \beta_k, \text{ for } k = 1, \ldots, \qquad (1.49)$$

This is a conditional property dependent upon the validity of the zero mean assumption which in turn depends upon the model to have been specified correctly.

(2) OLS estimators have minimum variance. This property is conditional on the validity of the assumption of non-autocorrelation and homoscedasticity. Symbolically:

$$Var(\hat{\beta}_k) = E[\beta_k - E(\hat{\beta}_k)]^2 \qquad (1.50)$$

If all assumptions can be maintained, then it can be shown that among all unbiased estimators, the OLS estimators have the smallest variance. The mathematical proof of this statement is given by the Gauss-Markov Theorem, which provides formal justification for the use of OLS methods of estimation.

Given the properties of (1) and (2), OLS estimators are said to be efficient. However, it should be noted that the efficiency depends crucially on the validity of the model's assumptions. If any of the assumptions fail, then the OLS method is no longer efficient and another method of estimation would need to be found. It is therefore important that we verify the assumptions at stages in the empirical analysis.

These properties identify two important features of the sampling distribution of the OLS. The shape of the distribution depends upon the assumption concerning the shape distribution of the dependent variable. If we maintain the assumption of normality, the sampling distribution should also be normal. This is because the sampling distribution is obtained from the values of the dependent variable which are consistent with the normal distribution. The sampling distribution of an OLS estimator ($\hat{\beta}_k$) is presented below:

Symbolically, $\hat{\beta}_k \sim N(\beta_k, Var(\hat{\beta}_k))$.

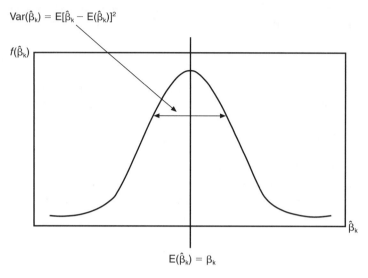

$$\text{Var}(\hat{\beta}_k) = E[\hat{\beta}_k - E(\hat{\beta}_k)]^2$$

$f(\hat{\beta}_k)$

$\hat{\beta}_k$

$$E(\hat{\beta}_k) = \beta_k$$

Figure 1.12 Sampling distribution of an OLS estimator

(b) Monte-Carlo studies

Monte-Carlo experiment: a practical method for checking properties of the estimators

Estimators are selected on the basis of performance under the repeated sampling/estimation procedures. This hypothetical sampling procedure generates sampling distributions of the estimators. A 'good' estimator is then identified as where the sampling distribution has a mix of desirable properties including unbiasedness, efficiency and consistency. Given the assumptions in the classical linear regression model, the Gauss Markov theorem provides a theoretical foundation for the use of OLS estimators. In particular the theorem demonstrates that amongst all linear unbiased estimators OLS estimators have the smallest variance and in this sense they are the Best Linear Unbiased Estimators (BLUE). In addition to theoretical reasoning, econometricians also employ Monte-Carlo simulation experiments to understand the sampling properties of estimators. A Monte-Carlo experiment provides a framework within which econometricians have complete control of the way the data is generated. Within this controlled experimental environment, a practitioner sets the values of the unknown parameters from the outset to see how well a particular method of estimation can track values. This is in fact the reverse of the estimation procedure. Within the framework of a Monte-Carlo study, the value of the parameters are known from the start. However, what is not known is how close a particular method of estimation generates these known values.

In what follows we demonstrate the steps undertaken in a Monte-Carlo study with reference to the OLS estimators.

(1) Set specific values for the parameters of the model to begin the process. For example, in a simple two-variables regression model

$$Y_t = \beta_1 + \beta_2 X_t + u_t \qquad (1.51)$$

Set:

$\beta_1 = 0.5$
$\beta_2 = 0.8$

(2) Collect data on independent variable X_t. For example
$$\begin{bmatrix} X_1 \\ X_2 \\ \dots \\ X_n \end{bmatrix}$$

(3) Generate some random numbers for u_t from a normal distribution with a mean of zero and a variance of unity. In practice random numbers are usually generated by computer software. For example:

$$\begin{bmatrix} u_1 \\ u_2 \\ \dots \\ u_n \end{bmatrix}$$

Random numbers for u_t.

(4) Use values of X_t and random numbers for u_t to generate a set of observations for Y_t, the dependent variable of the model.

For example:

$Y_1 = 0.5 + 0.8 X_1 + u_1$

$Y_2 = 0.5 + 0.8 X_2 + u_2$

$\dots$

$Y_n = 0.5 + 0.8 X_n + u_n$

(5) Choose a particular estimation method, for example, OLS. Use the value of X_t and Y_t (generated in step 4) to generate point estimates for the parameters of the model.

For example, using the OLS method, the point estimates for β_1 and β_2 can be generated as:

$$\hat{\beta}_2 = \frac{\sum_{i=1}^{n} (Y_i - \bar{Y})(X_i - \bar{X})}{\sum_{i=1}^{n} (X_i - \bar{X})^2} \dots [\text{point estimate}] \qquad (1.52)$$

$$\hat{\beta}_1 = \bar{Y} - \hat{\beta}_2 \bar{X} \dots [\text{point estimate}] \qquad (1.53)$$

where $\bar{X}$ is sample mean of observations on X and $\bar{Y}$ is sample mean of generated data for Y.

(6) Compare the point estimates with the preset parameter values. Are these closely matched?

(7) Repeat this controlled experiment from step (3) many times to generate 'sampling' distributions for the OLS estimators (i.e. $\hat{\beta}_1$ and $\hat{\beta}_2$).

(8) Check the following properties of the 'sampling' distributions:

i Shape: Are the distributions symmetrical around their central values?

ii Expected value: Are the expected value/central values close or the same as the pre set parameter value? So

$$E(\hat{\beta}_1) = 0.5 \text{ and } E(\hat{\beta}_2) = 0.8$$

iii Variance: Is the variance of each distribution 'relatively' small?

iv Consistency: Are the 'sampling' distributions becoming narrower and narrower as the sample size of the controlled experiment increases?

Using a Monte-Carlo study such as this can be shown that OLS estimators are: unbiased, possess minimum variance, and are efficient and consistent. However, these properties depend crucially on the generation of random data from a normal distribution with a mean of zero and a constant variance. Through a Monte-Carlo simulation experiment, econometricians have been able to investigate the impact of a change in any one of the data generation conditional assumptions on the behaviour of the estimators. This type of analysis enables the investigators to select 'best' estimation method to fit the type of data available in different situations.

(c) Maximum likelihood estimators

The maximum likelihood estimators are frequently employed by the practitioners for the estimation of single equation and particularly, systems of equations. The intuitive idea underlying the ML methodology is quite appealing: thus estimators are derived such that the likelihood/probability of obtaining all data points on the dependent variable is maximum.

For example:

$$Y_t = \beta_1 + \beta_2 X_t + \beta_3 Z_t + u_t \tag{1.54}$$

$$\text{observations on } y = \begin{bmatrix} Y_1 \\ Y_2 \\ \cdots \\ \cdots \\ Y_n \end{bmatrix}$$

We find β_1, β_2 and β_3 such that the probability/likelihood of obtaining $\begin{bmatrix} Y_1 \\ Y_2 \\ \cdots \\ \cdots \\ Y_n \end{bmatrix}$ is maximised. To derive ML estimators we need the normality assumption to hold. Given this assumption the ML estimators are derived from the joint probability density function of the dependent variable in the model (the mathematical form of the model will not be attempted here and interested readers are directed to texts such as D. N. Gujerati (1997)). The mathematical form of the ML estimators are the same as OLS estimators. In this respect, the OLS estimators are also ML estimators. That is, provided, the normality assumption holds, the OLS estimators generate the highest probability of obtaining any sample observations on the dependent variable. This is yet another theoretical justification for the popularity of the OLS estimators.

DISCUSSION QUESTIONS

1   Suggest an econometric model for the estimation of demand for tourism in a particular country. Explain how you would measure each variable included in your model. Explain the role of any disturbance term included in your model and outline and discuss the standard assumptions concerning this term. Why are these assumptions necessary?
2   Discussion the role of empirical analysis in economics.
3   Explain what you understand by the sampling distribution of an estimator. How are the sampling distributions used in the search for a 'good' estimator?
4   Explain the ordinary least squares (OLS) method. What do you understand by the statement that OLS estimators are 'BLUE'?
5   Explain what you understand by a Monte-Carlo study.

ADDITIONAL READING

See D. N. Gujarati 'Basic Econometrics' McGraw Hill, 1997. For a good discussion of some of the basic issues. See also, Peter Kennedy 'A guide to Econometrics', Blackwell, 1999, for a lucid account of econometric methods.

2 Econometric modelling and diagnostic testing (specific to general approach)

INTRODUCTION

The aim of this chapter is to develop a model of the demand for imports in the UK. A step-by-step approach is utilised which allows expertise to be developed in forming an econometric model from basic theory, developing this model against data limitations and rigorously testing the specification. A comprehensive testing procedure is laid down whence assumptions, definitions and data consistency are investigated. Eventually the full range of misspecification tests are discussed and applied to the import demand model. The misspecification tests used are those commonly found in regression software. The appropriate cautions required in evaluating parameter stability, consistency and reliability are included. The interpretation of the results yielded by the model is accorded special significance. All in all, working through this chapter enables readers to gain confidence in econometric modelling providing that they have a working knowledge of fundamental theory relating to import substitution demand. In the previous chapter we developed the theoretical arguments underlying linear regression models. In this chapter we put these ideas into practice by estimating a multiple linear regression model. The computations are carried out by software packages which are widely available. These packages generate various diagnostic test results which are used to evaluate the regression model. We consider a number of diagnostic tests. The computer package used is Microfit 4.

Key points

- An econometric model of demand for imports
- Estimation and diagnostic testing in practice

1 IMPORT DEMAND MODEL

The first step in developing an econometric model is consideration of theoretical relationship about the variable. Thus, on the basis of assumptions and deductive reasoning, economic theory provides a framework within which the main variables influencing economic phenomenon are linked in a causal way. We therefore begin by considering the economic theory concerning the demand for imports. We have discussed features of this model in the previous chapter. In what follows, some of the basic assumptions of this model are reviewed again.

Economic theory focuses on three major variables determining the demand for competitive imports. For example, imports of manufactured goods to the UK are competitive imports, since the UK produces all types of manufactured goods. The economic factors influencing imports of competitive goods are first, the level of final expenditure, that is total expenditure in a given period in the economy. Since total expenditure on goods and services in a given period reflects the level of gross domestic product (GDP), this first variable is usually proxied by the level of GDP. It should, however, be noted that the composition of total expenditure might also be important to the extent that the import content of different components of total expenditure/GDP differs. We therefore have two options concerning the treatment of total expenditure: (1) To use GDP as proxy for final expenditure; or (2) to allow for each component of GDP (i.e. consumption, investment, exports), to appear separately under the assumption that each macro component of final expenditure may have a different effect on imports (H. R. Seddighi, 1996). In this chapter we follow the simpler model to assist exposition.

A second variable influencing the demand for imports, ceteris paribus is the price of close substitutes. In particular, a rise in relative price levels is expected to lead to a fall in demand for imports. In most studies of import functions it is assumed that supply elasticities are infinite and therefore import prices are taken as being determined outside the theoretical model. Moreover, it is usually assumed that the domestic prices are flexible and change to eliminate excess demand. Under these conditions import prices and the domestic price levels are determined outside the model through the interaction of world demand and supply. Moreover, infinite supply elasticities assumed means that income distribution is unaffected. The third factor is the capacity of the country to produce and supply the goods domestically. However, the capacity factor is essentially a short-run phenomenon and is relevant only if excess demand at home cannot be eliminated by a change in domestic prices (A. P. Thirwall, 1991).

Given this reasoning, the import demand function may be written symbolically as follows:

$$M = f\left(\begin{matrix} GDP, & PM/PD \\ (+) & (-) \end{matrix} \right) \tag{2.1}$$

where M = aggregate imports in units of domestic currency/constant prices:

GDP = Gross Domestic Product at constant prices;
PM = an index price of imports expressed in domestic currency;
PD = an index price of domestically produced rival goods.

On the theoretical grounds, we expect a positive relationship between M and GDP. So that a rise/fall in the level of GDP, with PM/PD unchanged, is expected to lead to a rise/fall in the level of aggregate imports. On the other hand, a rise/fall in relative price term, PD/PM (with GDP held constant), is expected to lead to a fall/rise in the level of aggregate imports. More specifically, we expect to see a positive relationship between aggregate imports and GDP and a negative relationship between aggregate imports and the price of imports relative to the price of domestic goods, where the prices of imports are expressed in domestic currency.

2 AIMS OF THE EMPIRICAL STUDIES

Before we go any further it is useful to state the aims of the empirical study. In most empirical studies of economic phenomena based on the traditional (SG) approach, the aims may be stated as:

1   To provide empirical content to the abstract economic theories/models. To this end we aim to quantify/estimate the relationship between aggregate competitive imports, M, and main determinants.
2   To evaluate the theoretical explanation of the data generation process.
3   In the light of results enhance the economic model as an explicator of the observed phenomena.
4   To use the econometric model for basic forecasting.

Modelling requirements

Given these aims, the basic tools are:

1   An econometric model, designed to explain observations concerning aggregate competitive imports.
2   A set of observations (a data set) of the variables included in the model. In this example, these are M, GDP, PM and PD.
3   A method of estimation which can be used to convert the data set into point estimates of the parameters of the model. Given the data generation assumptions, the estimators require desirable properties, these include: efficiency, sufficiency and consistency features.
4   A set of criteria for the evaluation of the results and a set of diagnostic tests designed to test assumptions. These assumptions include specification and data generation assumptions.
5   A regression package which generates the desired diagnostic tests/information. These basic requirements are now discussed.

Specification of the econometric model

The econometric model is derived from an economic model to explain observations. To this end, the economic model must be modified in two respects: (a) we must suggest a specific functional form for the model, in other words, we must state a specific mathematical relationship between aggregate imports, M and its main determinants; (b) the model must be data admissible. This requirement implies that a deterministic economic model must be modified to reflect the stochastic nature of the data.

The model $M = f(GDP, PM/PD)$ is a general mathematical relationship, saying that, based on the theoretical reasoning aggregate imports depends mainly on total expenditure and relative prices. The economic model must be given a specific functional form. Unfortunately, theory does not help in specifying an exact mathematical relationship.

In this situation, the practitioner has no option but to assume a certain functional form from the outset. The assumption of the specific functional form chosen can then be tested. In studies of the demand for aggregate imports, most practitioners have either used a linear specification or, alternatively, log linear specification. In a linear specification variables under consideration are connected to each other by a set of linear parameters. The model is said to be linear in parameters.

$$M = \alpha_1 + \alpha_2 GDP + \alpha_3(PM/PD) \tag{2.2}$$

In this specification, α_1 is intercept term α_2 and α_3 are respective slope parameters.

In a log linear specification the log values (expressed as natural log) are connected to each other through a set of linear parameters as follows: thus:

$$\log M = \beta_1 + \beta_2 \log GDP + \beta_3 \log(PM/PD) \tag{2.3}$$

Both linear and log-linear models are frequently used in practice.

The choice between the two is essentially an empirical one and can be resolved by diagnostic testing. However, in the analysis of import functions, practitioners tend to prefer the log-linear specifications for some or all, of the following reasons:

i   In log linear specifications the slope parameters. β_2 and β_3, represent the partial elasticity of imports with respect to GDP and the relative prices terms, respectively. The income elasticity and partial elasticity of imports with respect to relative prices are thus estimated directly.

ii It is easy to interpret estimated slope parameters as these show percentage changes. Moreover, the units by which the variables are measured do not influence the magnitudes of estimated coefficients, unlike linear specifications.

iii Log transformations of variables reduce the variability in data. This potentially reduces the likelihood of heteroscedasticity.

We also make use of a log linear model to estimate partial elasticities of demand for imports with respect to income and relative prices as follows:

$$\log M_t = \beta_1 + \beta_2 \log GDP_t + \beta_3 \log(P_M/P_D)_t + u_t; \qquad \text{for } t = 1, \ldots, n \tag{2.4}$$

where u_t is a random disturbance term added to the equation to capture the impact of all other variables. These influences include random variables. The equation suggests that each observation obtained on logs of imports are made up of two components; namely: (i) the 'average' expected value of imports at any specific level of GDP and PM/PD, given by:

$$E[\log Mt] = \beta_1 + \beta_2 \log GDP_t + \beta_3 \log(P_M/P_D)_t \tag{2.5}$$

and (ii) a term showing the deviations from the expected value given by u_t. Moreover, the expected value of imports is determined by variables suggested by economic theory. Where the deviations from the average are essentially due to random factors. We start the process by making some standard assumptions concerning how data are generated. These assumptions concern the distribution of the disturbance term. This may be presented symbolically:

$$u_t \sim nid(0, \sigma^2); \text{ for } t = 1, \ldots, n. \tag{2.6}$$

Hence, the disturbance term is normally and independently distributed around a mean of zero with a constant variance. The econometric model may therefore be written as follows:

$$\log M_t = \beta_1 + \beta_2 \log GDP_t + \beta_3 \log (P_M/P_D)_t + u_t$$

$$u_t \sim nid(0, \sigma^2); \text{ for } t = 1, \ldots, n$$

(2.7)

3 THE DEFINITION OF VARIABLES, ESTIMATION AND DIAGNOSTIC TESTING

The way in which each variable is measured and used in the model has important consequences for the results. Occasionally economic theory provides insights regarding the appropriate definition and the use of economic variables. Practitioners must be fully aware of these insights and wherever possible use them. Regarding the model of imports, the variables may be defined as:

Mt = Volume of imports. This yields the total value of imports at constant prices. Imports in a given year are measured in a particular base year.

GDP = Gross Domestic Product at constant prices (1985) prices.

P_M = Any price index of imported goods with prices measured in domestic currency. This is calculated as:

$$P_M = \frac{\text{imports in current prices}}{\text{imports in constant prices}}$$

P_D = A price index of domestic goods including GDP deflator/wholesales price index.

In this example we have defined P_D to be the GDP deflator as:

$$P_D = \frac{\text{GDP in current prices}}{\text{GDP in constant prices}}$$

For the purpose of analysis there are 68 observations of quarterly data taken from the Economic trends Annual Supplement (1998). The period covered is quarter 1 of 1980 to quarter 4 of 1996.

Estimation: The model was estimated by the ordinary least squares method using Microfit 4. Under the OLS method, the parameters are estimated so that the residual sum of squares is minimised.

The Microfit package produces a set of diagnostic tests designed to test the adequacy of the model. In what follows the microfit output is discussed.

Presentation of the regression results

$$\text{Log}\,\hat{M}_t = -13.4337 + 2.025\,\text{log}\,\text{GDP}_t - 0.905\,\text{log}\,(P_M/P_D)_t \tag{2.8}$$

$$(0.78558) \quad (0.066527) \qquad\qquad (0.74096)$$

$R^2 = 0.98109$

F-statistic $(2, 65) = 1683.3$

DW $= 0.55343$

SE of Regression $= 0.035263$

This is one way of presenting regression results, with estimated parameters being the coefficients attached to each corresponding variable, estimated standard errors written below corresponding coefficients followed by R^2 (coefficient of determination), the F statistic, the Durbin and Watson test statistic and finally the standard error of regression. Each of these computer generated numbers have implications for the adequacy of the model and are used in the evaluation of the regression results. Before we consider these in detail, we need, to develop criteria for the evaluation purposes.

Evaluation criteria

There are three criteria used to evaluate regression results:

(a) The economic criteria

This is a simple but a powerful criterion for the evaluation. This criterion is concerned with the sign and size of the estimated parameters (coefficients). One needs to check that the sign and size of coefficients are consistent with economic theory. It is important to point out that we are quantifying economic relationships linked by some linear parameters. All parameters with the exception of the intercept term which is usually added to the econometric model for computational convenience, have economic meanings. Given this observation, if the economic model is well defined, it is expected that the sign and size of the estimated parameters follow consistent theoretical arguments.

Applying this criterion to the estimated model of imports we observe that: the partial elasticity of imports with respect to GDP/income is estimated as 2.025. The sign is positive and consistent with economic theory. The magnitude suggests that for each 1% change in the level of GDP, with relative prices constant, we expect the volume of imports, on average to change by about 2%. In other words, import demand is fairly elastic with respect to income. Economic theory does not provide us with guidance concerning the magnitude of this parameter. It is advisable to check the magnitude of estimated parameters against other published work. The price elasticity of demand is estimated at -0.905. The sign is negative and consistent with the predictions of economic theory. The size of the coefficient suggests that, for each 1% change in the relative prices of imported goods compared to domestically produced goods, it is expected that imports fall by only 0.9 of 1%. The value suggests that the demand for imports could be unit elastic. We need to bear this in mind in further analysis. The economic criteria appear to be satisfied. This is only one yardstick of the evaluation and we need further analysis. If on the other hand, the sign and the magnitude of coefficients were found to be contrary to what is suggested by the economic theory, we should resolve this problem before going further. This could be done by checking the adequacy of the data set, for example.

(b) Statistical criteria for evaluation

Statistical criteria are used to provide underpinnings for the econometric model. In particular we test to see if there exists statistical evidence against or/in favour of inclusion of each regressor. These individual/single tests are called tests of significance and are routinely carried out in applied work. We also test to see if there exists any evidence in favour of all regressors. This is a joint test of significance and in this case we are testing for the *overall* significance of regression.

In addition to tests of significance it is also useful to know how close the estimated line (plane) is to the scatter of observations which have been used to estimate it. This is a measure of 'goodness of fit' of the estimated line and is usually called the coefficient of determination or R^2. We begin by considering R^2.

Coefficient of determination R^2

The coefficient of determination R^2 provides a measure of 'goodness of fit' of the estimated line to the sample data. In particular it shows the percentage/proportion of total sample variation in the dependent variable which is due to sample variations in all independent variables of the model. The higher the value of R^2 the better would be the fit of the estimated linear relationship to the sample data points. To see these issues, we make use of simple two-dimensional diagrams corresponding to import demand. For illustrative purposes we treat relative prices as a constant in the equation; so that the import equation can be drawn in a straight line in a two-dimensional diagram.

Figure 2.1 shows the population regression line corresponding to the log linear model of the import function. At each level of GDP we expect to find a range of demand levels for imports, due to the influence of factors other than GDP on demand. Some of these could be random. Each distribution of import demand levels is symmetrical around an expected value given by the model. Note that each observation has two components, a deviation from the expected value captured by the disturbance term and the expected value of imports, given by the model. The aim is to estimate the population regression line. To do this we select a sample of data values from the population data set and use the OLS method to estimate the line. The sample data set is shown in Figure 2.2.

Remember that the population regression line and the disturbance terms are non-observable, all we have is a set of observations on the dependent and independent variables of the model. These observations are first converted into the OLS estimates of the unknown parameters, giving rise to a sample repression line, such as the one depicted in Figure 2.2. The vertical distances between observed values and the estimated line are the OLS residuals. These may be thought of as being the empirical counterpart of the associated disturbance terms. The smaller the residuals, the better would be the fit of the estimated line to the data points and the

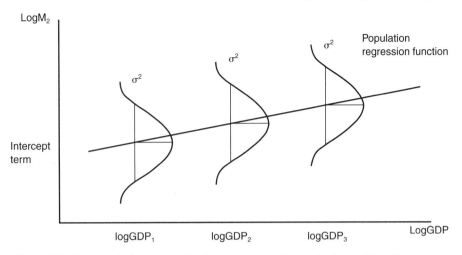

Figure 2.1 The population regression line corresponding to the import function

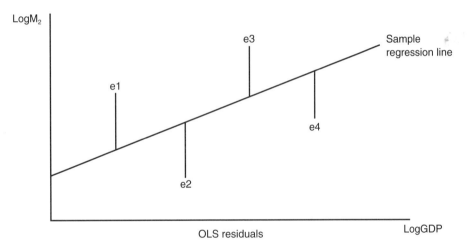

Figure 2.2 Sample regression line

sample regression line would be a better 'fit' of the unknown population regression line.

To find a measure of the 'goodness of fit' based on these observations, we notice that variations in the value of the dependent variable from one observation to the next is due to changes in either the values of the regressor/regressors or, the influence of the disturbance term or, both. If we can separate these two influences, finding the percentage of the sample variations in the imports (or logs of imports) which can be explained by variation in the regressors, then the higher this percentage is, the better should be the fit of the sample regression line to sample data. Therefore, closer it would be to the population regression line. The total sample variation in the dependent variable is defined as the sum of the squares of deviation of each observation from the sample mean value of the dependent variable. That is:

Total sum of squares (TSS) = sum of (observation on the dependent
variable − sample mean of the dependent variable)^2

or, symbolically:

$$TSS = \sum_{t=1}^{68} (\log M_t - \log \overline{M}_t)^2 \qquad (2.9)$$

where $\log \overline{M}_t$ = sample mean of observations on imports expressed in logarithms.

It can be shown that this sum can be divided into sum of two squared components as follows:

Total sum of squares = sample variation due + sample variation due
to changes in the   to changes in factors
regressors   other than regressors.

= Explained sum of   + Residual sum of
squares (ESS)   squares (RSS)

The coefficient of determination, R^2, is then:

$$R^2 = \frac{\text{Explained sum of squares}}{\text{Total sum of squares}} = \frac{\text{ESS}}{\text{TSS}} \qquad (2.10)$$

or, alternatively:

$$R^2 = 1 - \frac{\text{Residual sum of squares}}{\text{Total sum of squares}} = 1 - \frac{\text{RSS}}{\text{TSS}} \qquad (2.11)$$

Multiplying R^2 value by 100% then gives the percentage of sample variations in the dependent variable which can be explained by sample variations in the independent variables of the model. R^2 takes only positive values between zero and one. It shows the extent of linear association between the dependent and all the independent variables of the model. Note that the coefficient of determination is only a measure of the 'goodness of fit' of the linear relationship and in the case of non-linear relationships, R^2 value could be zero, even if there is a perfect non-linear relationship between variables. Moreover, the value of R^2 is highly sensitive to validity of assumptions. In particular when autocorrelation/heteroscedasticity are present, R^2 values are highly unreliable. In practice one has to be sure that all assumptions are consistent with the data before seriously considering the value of R^2. In the example, the value of R^2 is calculated as 0.98109%. That is it appears that over 98% of sample variations in the dependent variable over the period 1980 Q1 to 1996 Q4 can be explained by the regressors, log GDP and the log PM/PD. Only 2% of total variation appears to be due to change in the factor captured by the disturbance term. The model appears to fit the data well. We must not put too much emphasis on this value at this stage since the validity of the assumptions have to be tested. Before we consider these tests, we first look at the tests of significance concerning the parameters.

Significance test on parameters

Let us consider the regression model again:

$$\log M_t = \beta_1 + \beta_2 \log GDP_t + \beta_3 \log (P_M/P_D)_t + u_t \qquad (2.12)$$

The dependent variable ($\log M_t$) is assumed, on theoretical grounds, to be dependent on two independent variables/regressors log GDP and log PM/PD, respectively. The question is: is it possible to provide some statistical support for the inclusion of each, variable in the model? If answer to this question is positive, and provided that all assumptions are valid, statistical tests provide underpinning for the model.

The first step in performing statistical tests of hypotheses is to state the null and alternative hypotheses. In the tests of significance, the null hypothesis is the statement that contrary to the suggestion of economic theory, there is no relationship between the dependent and each one of the independent variables. In other words, the economic model is false. The model is treated as being false, unless contradictory evidence is found through statistical procedures. This is very much like trial by jury. The person is treated as innocent unless evidence is found to 'contradict' the person's presumed innocence. This type of procedure may be called the process of falsification. The test is designed to falsify the economic relationship.

We now state the null and alternative hypotheses concerning the economic model in hand. The null hypothesis is usually denoted as H_0 and the alternative H_1, and may be stated as:

H_0: There is no relationship between imports and GDP

H_1: H_0 is not true

Notice that the object of the test is to test the null hypothesis against the alternative H_1. There is an alternative way of presenting the null hypothesis; in particular, it can be presented in terms of parameters of the model. More specifically, we notice that parameter β_2 links the dependent variable of the model to the independent variable log GPD. Moreover, the value of this parameter is unknown. If the value of this parameter is zero, it is then be implied that there is no relationship between the dependent variable and the log GDP.

The null hypothesis may therefore be written as:

H_0: $\beta_2 = 0$ [no relationship between M and GDP]

The alternative hypothesis can be written more specifically. In particular, in forming the alternative hypothesis we should make use of information that we have concerning the value of this parameter. According to theory, the relationship between imports and the GDP is positive in nature. Hence movements in imports are in the same direction as those of GDP. Making use of this information, the alternative hypothesis may be written as:

H_1: $\beta_2 > 0$ [imports and GDP are positively related]

We therefore have:

H_0: $\beta_2 = 0$

H_1: $\beta_2 > 0$

or, alternatively

H_0: $\beta_2 \leq 0$

H_1: $\beta_2 > 0$

The logic of hypothesis testing

The question that now arises is how to test the null hypothesis? The only information that we have concerning β_2 is its point estimate obtained from an OLS estimator $\hat{\beta}_2$. Moreover, given the assumptions we know that $\hat{\beta}_2$ is distributed normally around a mean of β_2 and an unknown variance, var($\hat{\beta}_2$). Or, in short, $\hat{\beta}_2 \sim N(\beta_2, \text{Var}(\hat{\beta}_2))$.

This distribution can be used for finding the probability that the estimator lies between any two specific values. For example, if one wishes to find the probability of the estimator to lie between β_5 and β_6 (any two specific values) we need to find the area under the normal curve between these two values as shown in Figure 2.3. Mathematically the area is found by finding the integral of the normal distribution, between β_5 and β_6. An important property of the normal distribution is that the value of the area under the curve between any two points depends only on the distance of each point from the mean of the distribution where each distance is expressed per unit of standard error, rather than the mean and the variance of the

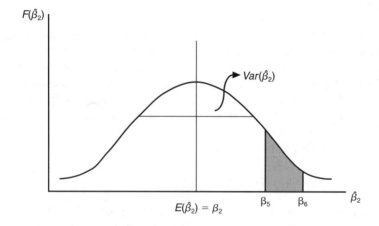

Figure 2.3 The sampling distribution of $\hat{\beta}_2$

distribution. In other words, for all values of mean and variance of a normal distribution, as long as the deviations of the points from the mean per unit of standard error are the same, the probabilities are also the same. Making use of this property to calculate the probability that an estimator, or a random variable lies between any two points, we use the standard normal distribution. The standard normal distribution is centred around a mean of zero and has a variance of unity. Statisticians have calculated the probabilities for various values of random variables and tabulated these in the table of standard normal distribution. Any normal distribution can be converted to the standard normal distribution, by expressing the values of the random variable in terms of deviation from the mean of the distribution per unit of the standard error. For example one can convert the distribution of $\hat{\beta}_2$ into a standard normal one (Z distributions) as follows:

$$Z = \frac{\hat{\beta}_2 - \beta_2}{SE(\hat{\beta}_2)} \sim N(0, 1) \tag{2.13}$$

Or, diagrammatically

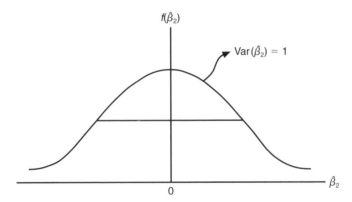

Figure 2.4 A standard normal distribution

The probability that $\hat{\beta}_2$ (OLS estimator) lies between any two specific values can be calculated by converting the units of distribution into the units of standard normal distribution and then finding the area under the standard normal curve. How does this discussion help in testing the null hypothesis? Remember that the process of hypothesis testing is essentially a process of falsification. Therefore, we can argue that, if H_0 is not true the probability that the Z score would be more than, or equal to the value that Z takes under $H_0(\beta_2 \leq 0)$ would be low. To see this, consider the null and alternative hypothesis and the way Z scores are calculated:

H_0: $\beta_2 \leq 0$

H_1: $\beta_2 > 0$

$$Z = \frac{\hat{\beta}_2 - \beta_2}{SE(\hat{\beta}_2)'} \tag{2.14}$$

Z score under $H_0(B_2 = 0)$ is:

$$Z_1 = \frac{\hat{\beta}_2}{SE(\hat{\beta}_2)} \tag{2.15}$$

If H_0 is false then β_2 would take positive values and Z scores would tend to be less than Z_1 quite frequently. Therefore, if H_0 is false, the probability that Z scores would be greater than or, equal to Z_1 under repeated sampling would be small, say less than 5%, as in Figure 2.5.

In Figure 2.5, the shaded area shows $\Pr(Z \geq Z_1)$. If H_0 is false, this probability would be small. The convention is to set this probability from the outset at some predetermined values, usually as 1, 5, or 10%. These are called the level of the significance of the test. The practitioner then calculates the $\Pr(Z \geq Z_1)$ if, this probability is less than 5%, one would be inclined to reject the null hypothesis.

This approach to hypothesis testing is called the ρ-value approach/method. The ρ-values are usually reported by regression packages.

In practice we do not know the standard error of the estimator, $SE(\hat{\beta}_2)$. We need therefore to replace $SE(\hat{\beta}_2)$ with an estimator $S\hat{E}(\hat{\beta}_2)$, in the Z score formula. Doing this, however, changes the shape of the normal distribution.

The resulting distribution is called the student 't' distribution. It is centred on a mean of zero, however, it is slightly flatter than the standard normal distribution, as the variance of the distribution is slightly more than unity, for small number of observations, e.g. $n \leq 30$. However as the number of observations increases to more than 30 and beyond, 't' distribution converges to the normal distribution and the two distributions are not distinguishable for large samples. The 't' distribution is therefore a small sample distribution closely related to the Z distribution.

In practice we look at the ρ-values. If any one of these is less than the 5% level of significance, we can reject H_0, concluding that the independent variable in question is statistically significant. To illustrate these techniques, we consider a computer print out concerning the econometric model under consideration.

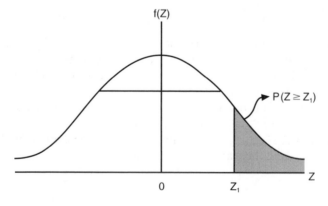

Figure 2.5 Hypothesis testing using the standard normal curve

A test of significance for the GDP

H_0: $\beta_2 \leq 0$

H_1: $\beta_2 > 0$

The ρ-value reported by the regression package is 0.032. In other words if H_0 is true, there is only 3% chance that the value of 't' statistic i.e;

$$t = \frac{\hat{\beta}_2}{S\hat{E}(\hat{\beta}_2)} = \frac{2.025}{0.06652} = 30.44$$

will occur under repeated sampling. Using the conventional 5% level of significance, since the value is less than the chosen level of significance, we reject H_0. This result is, however, only reliable if all assumptions including those concerning the disturbance term are consistent with the data. We therefore need to confirm all assumptions before arriving at an appropriate conclusion.

There is an alternative way of conducting a test of significance using a table of the 't' distribution. Under this method, once the level of significance of the test is set, say at 5%. Practitioners then use tables of the 't' distribution to find the so called critical value of the test. This is a particular value of the 't' statistic for which the probability that the 't' ratio is equal or, greater than the value is 5%:

$$\Pr(t \geq c.v) = 0.05$$

In Figure 2.6 the shaded area is called the critical region of the test, because if the t-ratio falls in the region, we reject H_0 at 5% level of significance. Notice that the method is very similar to the ρ-value approach. Under ρ-value approach, we use the reported probability for rejecting/accepting H_0, whereas under this method, we use the critical value and critical region of the test to arrive at a decision. Both imply the same conclusions. We now perform the 't' test under the alternative method.

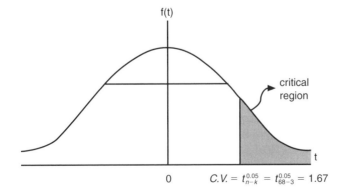

Figure 2.6 Significance testing and the critical region

Under H_0

$$t = \frac{\hat{\beta}_2}{S\hat{E}(\hat{\beta}_2)} = \frac{2.025}{0.06652} = 30.44 \tag{2.15}$$

With the critical value of the test being 1.67, we reject H_0, concluding that the log GDP is a significant regressor.

Significance test for the $\log P_M/P_D$

The null and alternative hypothesis may be stated:

H_0: $\beta_3 \geqq 0$ [no relationship between $\log M$ and $\log P_M/P_D$]

H_1: $\beta_3 < 0$ [a negative/inverse relationship between $\log M$ and $\log P_M/P_D$]

(i) The ρ-value approach

The ρ-value for the test is calculated at 0.0923 which is greater than 0.05. Therefore, we do not reject H_0, concluding that there is no statistical support for the inclusion of P_M/P_D as a regressor in the model. This is an unexpected result. However, as pointed out, the 't' test is only reliable, if all assumptions are valid. We cannot therefore express a firm opinion on the results, at this stage.

(ii) The conventional approach

Under H_0, the 't' value is calculated as:

$$Z = \frac{\hat{\beta}_3}{S\hat{E}C\hat{\beta}_3} = \frac{-0.905}{0.74096} = -1.2213 \tag{2.16}$$

The c.v, the test of 5% level of significance is 1.67.

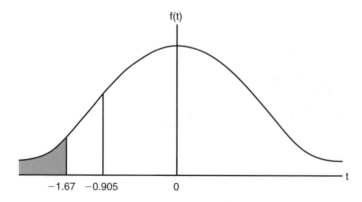

Figure 2.7 Conventional significance testing

$$\lg \hat{M}_t = -13.4337 + 2.025 \lg GDP_1 - 0.905 \lg \left(\frac{P_M}{P_D}\right)$$
$(0.785) \qquad (0.066) \qquad (0.74096)$

Econometric modelling and diagnostic testing 53

Decision rule criteria

Since

$$t = -0.905 > t_{65}^{0.05} = -1.67, \; |t| = 0.905 < t_{65}^{0.05} = 1.67$$

We can not reject H_0 at 5% level of significance, concluding that the relative price term is not statistically significant.

Notice that under H_1, β_2 takes only negative values, the appropriate critical region of the test is the shaded area, as is shown in the Figure 2.7.

Significance test for the intercept parameter

We can perform a test of significance on the intercept term to see whether or not it is statistically significant. In practice, it is advisable to carry out this test, as inclusion of an intercept term is essentially indicated by the type of functional form used. To perform this test we can state the null and alternative hypothesis as follows:

H_0: $\beta_1 = 0$

H_2: $\beta_1 \neq 0$

Note that in the absence of any information concerning the value of the intercept term. The alternative hypothesis is stated such that β_1 can take either positive or negative values.

Under H_0, the t-ratio is calculated as:

$$t_2 = \frac{\hat{\beta}_1}{S\hat{E}(\hat{\beta}_1)} = \frac{-13.4337}{0.78558} = -17.1 \qquad (2.17)$$

Given the way H_1 is formulated, we need to consider both positive and negative values of the t-distribution using 5%* level of significance and dividing it equally between the two tails of distribution, the critical values of the test are 2.0, as shown in Figure 2.8 below.

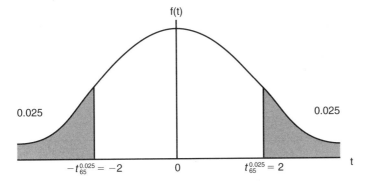

Figure 2.8 Two tailed significance testing using the t-distribution

A decision rule

Because $|t = 17.1| > |t_{65}^{0.025} = 2|$, reject H_0 at the 5% level of significance, concluding that intercept term is statistically significant.

Significance test for parameters restrictions

Before we test the joint significance of regressors, it is useful to consider first a general test procedure which can be used to test the validity or otherwise of linear restrictions on parameters. For example, in the model under consideration, we may be interested to test the hypothesis that the income and price elasticities of imports are of the same magnitude. We can state this hypothesis as follows:

H_0: Income elasticity of demand for imports = Price elasticity of demand for imports

H_1: H_0 is not true

or, in terms of parameters of the model:

H_0: $\beta_2 = \beta_3$

H_1: H_0 is not true

In this formulation, the null hypothesis imposes a linear restriction on parameters in the model. This type of test where one or, a number of linear restrictions, are imposed on parameters, is very popular in practice. How is such a test conducted? We use the example to give a basic insight into the procedure. In this example we have one linear restriction on the parameters of the model. We now impose this restriction. The restricted model can be written as:

$$\log M_t = \beta_1 + \beta_2 \log GDP_t + \beta_2 \log (P_M/P_D)_t + u_t \tag{2.18}$$

or

$$\log M_t = \beta_1 + \beta_2 [\log GDP_t + \log (P_M/P_D)_t] + u_t \tag{2.19}$$

We estimate the restricted model by OLS to obtain the R^2 and the associated residual sum of the squares (RSS). Now if the restriction is valid, what would be the relation between the R^2/RSS of the restricted and unrestricted (original model)? We expect, in this situation that these two values to be very similar, as both models should fit the data equally well. Any observed differences are due to estimation procedures.

Symbolically:

$$R^2_{H_0}(\text{restricted model}) \cong R^2_{H_1}(\text{unrestricted-model}) \tag{2.20}$$

or that, the observed differences between the restricted and unrestricted residuals sum of squares is small. That is:

$$RSS^R - RSS^U = \text{a 'small' magnitude.}$$

To be able to conduct a statistical test, however, what we need is a test statistic which utilises the difference between the two residual sum of squares. Assuming that the assumptions concerning the disturbance term are met, under repeated sampling procedures, the difference between the two residual sum of squares divided by the number of restrictions on the parameters follows a chi-square distribution. The chi-square distribution starts at the origin and is defined only for positive values of a random variable. The shape of a chi-square distribution changes with sample size. As sample size increases, this distribution becomes more symmetrical, approaching a normal distribution shape as sample size approaches infinity (see Figure 2.9).

Where the vertical axis is the frequency/probability of occurrence of values of $(RSS^R - RSS^U)/d$ – under repeated sampling, and the horizontal axis is the value of $(RSS^R - RSS^U)/d$.

In Figure 2.9 d denotes the number of restrictions on the parameters of the model. In our example, there is only linear restriction i.e. $\beta_2 = \beta_3$ and d is therefore equal to one. The denominator d also represents the degrees of freedom of the chi-square distribution which is the number of independent observations utilised to estimate a parameter. The problem with the use of $(RSS^R - RSS^U)/d$, as a test statistic is that the residual sum of squares are sensitive to the units of measurement of the dependent variables. It therefore does not provide an unambiguous measure when the difference between the residual sum of squares is small/large. Statisticians have overcome this problem by making the value in question a relative magnitude. More specifically, the difference between the two residual sum of squares is measured relative to the residual sum of squares of the unrestricted model per degrees of freedom, so:

$$F = \frac{(RSS^R - RSS^U)/d}{RSS^U/n{-}k} \sim F_{d,\,n-k} \tag{2.21}$$

if the assumptions are valid, under the repeated sampling exercise, $RSS^U/n{-}k$ also follows a chi-square distribution with n–k degrees of freedom.

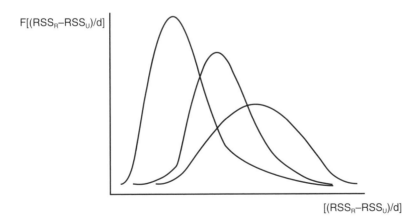

Figure 2.9 A chi-square distribution

It can be shown that the ratio of two independent chi-squares distributions follow an F distribution. The F distribution is skewed and starts from the origin, and ranges to infinity. Since this distribution is related to a pair of chi-square distributions, there are two associated degrees of freedom, d and n–k, degrees of freedom in the numerator and denominator, respectively. Figure 2.10 illustrates an F distribution, like the chi-square distribution from which it is derived, an F distribution becomes symmetrical with increases in the sample size. The F distribution converges to a chi-square distribution with large sample sizes.

There are tables of the F distribution produced for different levels of significance e.g. 1%, 5% and 10%. To find a desired F-value it is necessary to first select the table with the desired probability in the right tail, and locate the entry that matches the appropriate degrees of freedom for the numerator and denominator.

The F statistic may also be written in terms of the R^2 values of the restricted and unrestricted models. This can be done, only if the dependent variable of the model remains the same after the imposition of linear restrictions.

To convert the F statistic into R^2 format, all we need is to divide both the numerator and denominator by the total sum of squares TSS. We have:

$$F = \frac{(RSS^R - RSS^U)/d}{RS^U/n-k} \tag{2.22}$$

$$= \frac{\dfrac{RSS^R}{TSS} - \dfrac{RSS^U}{TSS}}{RSS^U/TSS} \times \frac{n-k}{d} \tag{2.23}$$

Assuming that the dependent variable of the model remains the same after imposition of the linear restrictions, so that TSS for both restricted and unrestricted models are the same, we have:

$$F = 1 - R^2_R - 1 + R^2_U/(1 - R^2_U) \times (n-k)/d$$

$$F = (R^2_U - R^2_R)/(1 - R^2_U) \times (n-k)/d \sim F_{d, n-k}$$

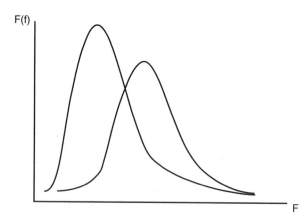

Figure 2.10 An F distribution

To perform tests of linear restrictions, we use the following steps:

1 Estimate the unrestricted model by the OLS and record RSS^U or, R^2_u.
2 Impose the linear restrictions and estimate the restricted model by the OLS to obtain RSS^R or R^2_R.
3 Calculate the F statistic and compare the value with the critical value of the test, obtained from 'F' distribution tables.
4 If the calculated value of F is greater than the critical value of the test we reject H_O at a pre-specified level of significance, concluding that the linear restriction/s are not statistically significant.

To illustrate the use of the F test in what follows, we apply it to test the overall significance of regression.

Testing the overall significance of regression

Given that there are usually more than one regressor in regression models it is necessary to perform a joint test of significance in addition to 't' tests. A joint test of significance is designed to test the significance of all regressors/independent variables of the model, jointly, in contrast to a 't' test which tests the significance of each regressor/independent variable of the model one at a time.

To begin the process we need to state the null and the alternative hypothesis of the test.

H_0: All regressors taken jointly/collectively are not significant.

H_1: H_0 is not true

alternatively:

H_0: $\beta_2 = \beta_3 = 0$ (independent variables GDP, and P_M/P_D, are not significant)

H_1: H_0 is not true

Notice that our aim here is to test the null hypothesis, the H_1 of the test may therefore be stated in a general form which encompasses alternatives.

Under H_0 we have two linear restrictions on the parameter of the model, therefore $d = 2$ which is the same as the number of parameters of the model minus one. So $k - 1 = d$, or, parameters $3 - 1 = 2 = d$.

Since the dependent variable, $\log Mt$, remains the same after imposing the linear restriction, we use the R^2 form of the F test as follows:

$$F = \frac{R_U^2 - R_R^2}{1 - R_U^2} \times \frac{n-k}{d} \sim F_{d,\, n-k} \qquad (2.24)$$

The coefficient of determination of the restricted model is zero, since when we impose the restrictions, the two independent variables are deleted. The test statistic, therefore, takes a simpler form as follows:

$$F = \frac{R_U^2}{1 - R_U^2} \times \frac{n-k}{d} \tag{2.25}$$

substitution of the above variables results in

$$F = \frac{0.98109}{(1 - 0.98109)} \times \frac{68 - 3}{2} = 1683.3 \tag{2.26}$$

In practice, we do not need to calculate the value of F for the test of significance, as it is normally reported by the regression packages.

Using the 5% level of significance, the critical value of $F_{2,65}^{0.05}$ is 3.15.

A decision rule

Because $F = 1683.3 > F_{2,65}^{0.05} = 3.15$, we reject H_0 at the 5% level of significance concluding that independent variables, jointly are highly significant. The reliability of the test depends crucially on the validity of assumptions, which are tested later.

(c) Econometric criteria

The econometric model is constructed on the basis of a number of simplifying assumptions which starts the process of empirical analysis. For example, to conduct an empirical analysis of the import function, we made the following simplifying assumptions:

1   The relationship between imports and its determinants can be adequately presented by a log linear equation (specification assumption).
2   The disturbance term is normally and independently distributed around a mean of zero with a constant variance.

    Notice that there are in fact four assumptions included in this statement. These are:

  i   the normality assumption;
  ii the non-autocorrelation assumption;
  iii the zero mean assumption;
  iv the homoscedasticity assumption.

3   The regressors are non-stochastic and are measured without errors.
4   There is no exact linear relationship among the regressors/independent variables.

The regression results are obtained on the basis that each one of these assumptions is valid. Before considering the regression results as reliable, we need to test the validity of each assumption. The reliability of the regression results depends crucially on the validity of these assumptions. Testing these should be routinely conducted. The econometric criteria deal with testing and evaluation of the assumptions for non-autocorrelation, homoscedasticity, normality and specification.

Although we have divided the criteria for the evaluation of the regression results into: economic; statistical; and econometric criteria, in practice, all these criteria are strongly linked and are used to evaluate results. The regression results can only be considered reliable when all criteria are satisfied. If any one of these criteria are found to be invalid, we need to respecify, and to re-estimate the model.

Diagnostic testing procedures

We are now in the position to consider the econometric criteria for evaluation. As discussed, these criteria are concerned with the data generation assumptions, which include the assumptions of specification of the model, as well as, the assumption of normality, independence, zero mean, and constant variance of the disturbance term. In what follows we will look at a number of diagnostic tests commonly used in practice to test the validity of these. For illustrative purposes we continue to use the model of demand for imports. To begin the process, it is useful to consider the regression model and assumptions:

$$\log M_t = \beta_1 + \beta_2 \log GDP_t + \beta_3 \log (P_M/P_D)_t + u_t$$
$$\text{(specification assumption); } u_t \sim nid(0, \sigma^2_t) \tag{2.40}$$

In the traditional approach of econometric analysis, there are no accepted benchmarks regarding the correct order of diagnostic tests to be used, choice of the tests and the order by which they are performed are left to the practitioner. In what follows, we utilise a Microfit 4 output regarding the types of diagnostic test and the order of testing, distinguishing between specification and misspecification tests.

Chapter 2: Key terms

Specification tests   These involve specific alternative forms of the regression model, for example 't' and 'F' tests of parameter restrictions.

Misspecification tests   These tests do not involve specific alternative forms of the model. They are used to detect inadequate specification. For example, the Durbin Watson Test, the Lagrange multiplier tests for Autocorrelation.

Serial correlation/autocorrelation   The values of the same random variable are correlated over time/space.

The Durbin Watson test   A misspecification test designed to be used with AR(1) process to detect Autocorrelation.

The Lagrange multiplier test   A misspecification test designed to detect higher order Autocorrelation (large sample test).

The Ramsey regression specification error test (RESET)   A misspecification test designed to detect inadequate specification, including omitted variables and incorrect functional forms.

Heteroscedasticity   Breakdown of homoscedasticity assumption. The variance of the disturbance term/dependent variable changes over cross-sectional units/time.

The Koenker-Basset test   A test commonly used to confirm the assumption of homoscedasticity.

The Chow test for a structural break   A test used to detect parameter stability over time.

CUSUM and CUSUMQ tests   Simple diagnostic tests for parameter stability.

Measurement errors   Measurement errors in regressors resulting in inconsistent OLS estimators.

Hauseman test for measurement errors   A general misspecification test for measurement errors/exogeneity.

Instrumental variable (IV) estimation   Estimation method used when the OLS method is inconsistent, for example, when measurement errors are present.

Specification and misspecification tests

The specification tests involve a specific alternative/form of the regression model. For example, significance tests either of the 't' or F variety are specification tests. These tests are designed to confirm or refute a known specification, given by the model under consideration, against a specific alternative. Misspecification tests, on the other hand, do not involve a specific alternative form of the model. They are essentially used to detect an inadequate specification. That is, they tell us that a particular assumption under consideration is inadequate (cannot be confirmed), but how to correct this inadequacy is left to the practitioner. It should also be pointed out that the reliability of a specification test crucially depends on the validity of the assumptions. The practitioner therefore needs to carry out the required misspecification tests first before considering specification tests. In practice, econometric criteria are therefore considered before statistical criteria.

The Durbin Watson diagnostic test

The Durbin Watson test is perhaps the most commonly used misspecification test. It was originally developed to detect serial correlation in the data. Serial correlation is used to describe a situation where the values of the same variable are connected to each other. Another term used in econometrics describing the same phenomenon is autocorrelation. Serial correlation and/autocorrelation are both used to indicate the breakdown of the assumption of independence. The assumption of independence is made from the outset to describe the independence of the distributions of disturbance terms/dependent variable. This means that a disturbance term/dependent variable corresponding to a particular observation does not depend on a disturbance term/dependent variable corresponding to another observation. For example, in a times series study of output, we can use the disturbance term to take into account the influence of random factors such as break-downs of machinery, climatic changes, and labour strikes, on the level of production per period. In this example the assumption of independence implies that a break down of machinery/strikes/or, climatic changes or, for that matter any random changes only affect the level of production in one period. The period during which the random event has actually occurred. In other words, the influence of a random shock does not persist over time. It has only a temporary influence on the dependent variable. Notice that the only way that a prolonged influence of random events can occur in econometric models is when the disturbance terms are correlated over time. So that when a random shock occurs in a particular period, it can also influence the dependent variable in other periods, through interdependency and correlation/connection with the disturbance terms in other times. The assumption of independence or, no autocorrelation/no serial correlation, even when the model is correctly specified, often breaks-down in practice, because of the prolonged influence of random shocks on the economic variables. In all time series applications practitioners should expect the presence of autocorrelation and the break down of the assumption of independence. This is particularly the case when the econometric model is incorrectly specified to begin with. This is usually the case in practice. We explore the connection between autocorrelation and incorrect specification of the model. Recall how the systematic part of an econometric model is developed, use economic theory as a guide in the selection of the independent variables for inclusion in the model.

In practice, there could be other variables, not suggested by the economic theory under consideration, influencing the dependent variable. For example, in developing a model of aggregate consumption expenditure using the Keynesian hypothesis of the Absolute Income Hypothesis AIH, we are guided by the theoretical arguments that real income is the main determinant of short run consumption. In a modern economy there are other variables, equally important. These variables include interest rates, and rates of inflation. Moreover, there are competing theories of aggregate consumption expenditure over time, recommending the inclusion of excluded variables. If a practitioner omits one or some of these variables whose effects are captured by the data, the econometric model would be inconsistent with the data and so is *misspecified*. The cause of misspecification in this case, is omission of relevant economic variables in the econometric model. A particular indication of this type of specification error is the breakdown of the independence assumption and the presence of autocorrelation. This is because one or more of the excluded variables could be autocorrelated variables. For example, if the interest rate is left out of the model of consumption expenditure, we get autocorrelation, because interest rate time series are autocorrelated. What happens in this situation is that the disturbance term picks up the influence of the excluded variable/s which are autocorrelated. We therefore get autocorrelation in the model. A test for autocorrelation is therefore essentially a general misspecification test as the practitioner is unable to determine the exact cause of autocorrelation from test results. Further work would be needed to deal with the problem of autocorrelation.

In the light of this discussion the Durbin Watson test may be regarded as a general misspecification test in the sense that it can detect the autocorrelation, but is incapable of determining its cause. The test is easily carried out in practice and is routinely reported by regression packages. It, therefore, provides an early warning device concerning the specification of the model. Before we consider this test and the way it is used in practice it is useful to look at the consequences of autocorrelation for the OLS methods of estimation.

The consequences of autocorrelation

Recall that we use OLS method of estimation because it generates efficient estimators. So under repeated sampling, the OLS method generates estimators whose distributions have the properties of unbiasedness, minimum variance and consistency. These properties are highly sensitive to assumptions. They will be generated only if assumptions of the model are consistent with the data. In the case of autocorrelation, the consequences for the OLS estimators are serious. While they are still unbiased, they tend to have large variances and become inefficient. In fact there are other estimators (for example GLS) which are efficient and should be used instead. Moreover, the OLS method tends to underestimate the standard errors of the estimators. In this situation, the t-ratios become highly unreliable. In addition, OLS methods underestimate the standard error of regression which in turns leads to unreliable R^2 and F statistics. In short, all the basic tools of analysis including, the t- and F-tests and R^2 become highly unreliable. In practice, it is advisable to test for the presence of autocorrelation as a matter of routine. A simple test designed to detect auto correlation is the Durbin Watson test.

The Durbin Watson test is a simple test designed to deal with first order autocorrelation. First order autocorrelation is a term used to describe a situation where the successive values of the same variable are interrelated/correlated. This is perhaps the simplest form of autocorrelation. We can present first order autocorrelation, using mathematical notation as follows:

$$u_t = \rho u_{t-1} + e_t \qquad e_t \sim \text{nid} (o, \sigma_e^2) \tag{2.41}$$

where, e is a white noise process. Notice that the successive values of the disturbance term are related to each in this simple form. The extent of the linear relationship between these successive values is captured by the term ρ. This term ρ could take on any value between -1 and $+1$. When it is a positive value, autocorrelation is said to be positive. This basically means that the difference between successive values of the disturbance term will be positive or negative for long periods. The disturbance terms do not alternate in sign over a long period of time. This form in turn implies that the nature of the influence of a random shock on the dependent variable remains the same over successive periods. For example, if the disturbance term is added to a model of production to capture the impact of random events on output, positive autocorrelation implies that the impact of a break down on output will be the same over time. It essentially reduces output over a number of time periods. Its sign in this case is negative and will remain so for a period of time. On the other hand if ρ is negative we say that we have negative autocorrelation. In this case the disturbance term alternates in sign, so one period it is negative, next period it is positive, or vice versa. In other words the impact of a random shock on the dependent variable of the model changes from one period to the next. In one period it has a positive impact on the dependent variable in the next period it has a negative impact. Notice that in the case of positive autocorrelation the difference between the successive disturbance terms, for periods of time is either positive or, negative. Whereas in the case of negative autocorrelation the sign of the difference between successive values of the disturbance term would alternate in sign. This simple observation provides the basis for the Durbin Watson test. Given that the disturbance terms are not observable, any test based on this observation, must use the empirical counterpart of a disturbance term, in the case of the OLS estimation the empirical counterparts are the OLS residuals. A simple detection method based on these residuals would be to look at the signs over the sample period. If the residuals do not change sign over a relatively long period, this could indicate positive autocorrelation. Alternatively, if the residuals tend to change signs from one period to the next, this could be a sign of negative autocorrelation. Durbin and Watson extended this idea a bit further to generate a test statistic. They reasoned that in the case of positive autocorrelation the difference between successive OLS residuals tend to be small (because both residuals have the same sign). On the other hand, in the case of a negative autocorrelation the difference between the successive OLS residuals tend to be relatively large (because the residuals alternate in sign). Taking this idea one step further, in order to make all differences positive they square each of residual difference. Moreover, to take into account all differences they add all squared differences. They then reasoned that in the case of a positive autocorrelation this sum is relatively small, whereas in the case of a negative autocorrelation it would be large. So:

$$\sum_{t=2}^{n}(e_t - e_{t-1})^2$$

~ relatively small if positive autocorrelation exists
~ relatively large if negative autocorrelation exists

The pressing question which arises now is what constitutes a relatively large/relatively small value in this context? Before we answer this question however, it is important to recognise that the value of this sum depends on the units by which the variables of the model are measured. The first task is therefore to eliminate this problem. Durbin and Watson dealt with this problem by dividing the sum by the sum of the squares of the OLS residuals, hence:

$$DW = d = \frac{\sum_{t=2}^{n}(e_t - e_{t-1})^2}{\sum_{t=1}^{n}e_t^2} \qquad (2.42)$$

Both the numerator and the denominator have the same unit of measurement and so the ratio is unit free. We can now get back to the question of the magnitude of this ratio. To deal with this problem, Durbin and Watson expanded the ratio as follows:

$$DW = d = \frac{\sum\limits_{t=2}^{n} e_t^2}{\sum\limits_{t=1}^{n} e_t^2} + \frac{\sum\limits_{t=2}^{n} e_{t-1}^2}{\sum\limits_{t=1}^{n} e_t^2} + \frac{\sum\limits_{t=2}^{n} e_t e_{t-1}}{\sum\limits_{t=1}^{n} e_t^2}$$ (2.43)

They then argued that the value of the each of the first two ratios is approximately equal to one and the sum of the first two is therefore approximately 2. The third ratio is in fact the OLS estimator of the correlation coefficient. This coefficient lies between -1 and $+1$. We have therefore:

$$DW = d \cong 2(1 - \hat{\rho})$$

It is now possible to determine the range of the values of the DW ratio, as follows:

1   In the case of a perfect positive correlation the value of $\hat{\rho}$ would be approximately $+1$, and the value of the DW ratio is approximately zero.
2   In the case of a perfect negative correlation the value of $\hat{\rho}$ would be approximately -1 and the DW ratio would be approximately 4.
3   Finally, in the case of no autocorrelation the value of $\hat{\rho}$ would be approximately zero and the Durbin Watson ratio would be approximately 2.

The range of the values for the Durbin Watson ratio is therefore zero and 4. As a quick guide, zero could indicate to a strong positive autocorrelation, 4 a strong negative autocorrelation and 2 to lack of autocorrelation. This is a quick guide and what is needed is a test statistic, which can be used in all situations. For this purpose we need the sampling distribution of the Durbin Watson ratio, and some critical values. Durbin and Watson showed that their test statistic has not got a unique critical value, rather it lies between two limiting values under repeated sampling procedures. These limiting values are called the upper and the lower bound values and are denoted by du and d_L. Durbin and Watson then calculated these values for different sample sizes and alternative number of independent variables. These values were then tabulated in the table of DW, for different levels of significance of test. We can illustrate the test step-by-step as follows:

(1) We start the test by specifying the null and the alternative hypotheses:

H$_0$ = There is no autocorrelation in the model/data

H$_1$ = Autocorrelation does exist

or, more specifically

H$_{0:}$ $\rho = 0$

H$_{1:}$ $\rho \neq 0$

Although the test is essentially a two sided, in most applications H$_0$ is tested against a positive alternative, to reflect the kind of autocorrelation that expected.

(2) The DW test statistic is calculated as:

$$DW = d = \frac{\sum\limits_{t=2}^{n}(e_t - e_{t-1})^2}{\sum\limits_{t=1}^{n}e_t^2} \tag{2.44}$$

(3) We choose a particular level for test, say 5% and find the corresponding upper and lower bound values from the table of DW. The decision rule of the test is as follows:

Decision criterion

1   If DW is less than d_L reject H_0 in favour of positive autocorrelation.
2   If DW is between d_L and du the test is inconclusive. In this case use the du as the critical value of the test, this assumes autocorrelation exists. It is, however, recommended in this case, to use an additional test for autocorrelation to confirm this result.
3   If DW is between du and 4-du do not reject H_0. In other words autocorrelation does not exist.
4   If DW is between 4-du and 4-d_L the test is inconclusive. In this case, use 4-du as the critical value, in other words assume negative autocorrelation exits. It is advisable to carry out an additional test to confirm this result.
5   If DW value is between 4-d_L and 4 do not reject H_0, negative autocorrelation is present. In practice, the value of the DW test statistic is calculated by the computer package and is usually reported widely in regression software. All we need to do is to look at this value, if the value is close to 2 it is a good indication that autocorrelation does not exist, any other value signifies autocorrelation. We then need to confirm these results, using the tables for the Durbin Watson test statistic.

There are some shortcomings in this simple misspecification test. In particular, although the test is a first order test, a significant value of the DW test statistic is also consistent with higher order autocorrelation. In other words, if H_0 is rejected, practitioners should take this to mean that autocorrelation exists but the order of the autocorrelation is not known. Moreover the test is highly unreliable when a lagged value of the dependent variable appears as a regressor in the model. In this case the value of the DW test statistic would be biased towards 2, indicating no autocorrelation, despite the fact that autocorrelation is present. There are a number of alternative tests available appropriate for use in this situation.

4 THE DURBIN WATSON TEST IN PRACTICE

We start by assuming a first order autoregressive process as follows:

$$u_t = \rho u_{t-1} + e_t \qquad e_t \sim nid(0, \sigma_e^2)$$

Given this process, the H_0 and H_1 of the test may be written as follows:

$H_{0:} \rho = 0$

$H_{1:} \rho \neq 0$

The test statistic is:

$$DW = d = \frac{\sum_{t=2}^{n}(e_t - e_{t-1})^2}{\sum_{t=1}^{n}e_t^2} \tag{2.45}$$

The value of this statistic is reported by the Microfit software to be 0.55343. Using a 5% significance level the lower and the upper bound values for $n = 68$ and $k = 3$ and $K' = 3 - 1 = 2$ (where K is the number of parameters including the intercept term), are respectively: $d_L = 1.54$ and $du = 1.66$.

Decision criterion

Because $DW = d = 0.55343 < d_L = 1.54$, we reject H_0 at $d = 5\%$. H_0 is not consistent with the data. Positive autocorrection appears to be present.

It appears that there is autocorrelation in the model, the OLS regression results should therefore be treated as unreliable. This is what usually happens in practice. It is seldom the case that, within the framework of the traditional approach to econometric modelling, we succeed in the first attempt. The question now is how to respond to a significant value of the DW test statistic. We consider this now.

How to respond to a significant value of the DW test statistic

A significant value of the Durbin Watson test statistic implies that autocorrelation exists. However, neither the cause nor the order of the autocorrelation are known. In applied work autocorrelation is usually generated due to models' misspecification caused by one or more of the following factors:

1 Omissions of one or, more relevant regressors.
2 Dynamic misspecifications, for example, exclusion of lagged values of dependent and independent variables.
3 An incorrect functional form of the model, for example the correct model could be non-linear in parameters.

There are other causes of autocorrelation. For example, the prolonged influence of random shocks. This could be responsible for autocorrelation. In these situations we need to test for the specification errors first, correct the model using theory then deal with any residual autocorrelation left in the model.

Given that the nature and causes of autocorrelation are not clear, in practice, we need to follow a particular strategy to deal with the problem of autocorrelation. There is no specific strategy recommended and used by all practitioners. The strategy that we use follows the diagnostic tests. This strategy may be summarised as:

1   Attempt to find the order of autocorrelation first. For this purpose we use a general diagnostic test for autocorrelation, such as the Lagrange Multiplier test.
2   Use a general misspecification test for incorrect functional form such as Ramsey's regression specification error test or simply RESET.
3   Use another misspecification test this time for heteroscedasticity, which may be the cause of misspecification, particularly in using cross-section data.
4   If the diagnostic tests indicate the model is misspecified, try to correct the model from theory. This is particularly useful when the cause of the problem is omitted variables. In time series analysis, the cause of specification errors is often dynamic misspecification. In this type of study, researchers concentrate on the dynamics of the model and try to correct for it using some adjustment mechanism such as partial adjustment model.

We now discuss these diagnostic tests in detail.

The Lagrange Multiplier test (the LM test)

The LM test is a general test for the autocorrelation, used by the practitioners to detect the order of autocorrelation. This test is a large sample test and is reliable when a sufficiently large number of observations are available. Moreover, the larger the number of observations the more reliable the results of the test. As a crude rule of thumb, one needs at least 40 observations to use this test. To demonstrate the test, suppose that we wish to test for autocorrelation up to fourth order. Under this specification the disturbance term in any particular period depends on the disturbance terms in each of the previous four time periods. More specifically:

$$u_t = \rho_1 u_{t-1} + \rho_2 u_{t-2} + \rho_3 u_{t-3} + \rho_4 u_{t-4} + \epsilon_t \qquad (2.46)$$

$$\epsilon_t \sim \text{nid}(0, \sigma_t^2)$$

This pattern of autocorrelation could emerge, for example, when one using quarterly data. The disturbance term in a particular quarter would depend on the disturbance terms in the previous 4 quarters.

We start the test by specifying the null and the alternative hypotheses as:

$$H_0: \rho_1 = \rho_2 = \rho_3 = \rho_4 = 0$$

$$H_1: H_0 \text{ is not true}$$

If autocorrelation of up to fourth order exists, the regression model takes the following form:

$$\log M_t = \beta_1 + \beta_2 \log \text{GDP}_t + \beta_3 \log (P_m/P_D)_t + \rho_1 u_{t-1} + \rho_2 u_{t-2} + \rho_3 u_{t-3} + \rho_4 u_{t-4} + \epsilon_t \qquad (2.47)$$

We may call this equation the unrestricted form of the model, in a sense that there are no restrictions imposed on the parameters of the model.

If the null hypothesis could not be rejected, the model would be the regression equation we started with, hence:

$$\log M_t = \beta_1 + \beta_2 \log GDP_t + \beta_3 \log (P_m/P_D)_t + \epsilon_t \qquad (2.48)$$

We call this equation the restricted form of the model, in the sense that there are four parameter restrictions on parameters of the model (four zero restrictions), compared to the unrestricted form of the model. We are therefore essentially testing for parameter restrictions. We have already considered the F test for linear restrictions in the previous chapter. The test statistic is as follows:

$$F = \frac{[RSS_R - RSS_U]/d}{RSS_U/n - k} \sim F_{d, n-k} \qquad (2.49)$$

Decision rule

Using a 5% level of significance if, F(d, n–k) > F(critical value at 5%), reject the null hypothesis at the 5% level of significance, concluding that autocorrelation up to fourth order exists. Notice that we still do not know the exact order of autocorrelation. All we can say at this stage is that autocorrelation could be up to fourth order. To find the order of autocorrelation we need to test down. That is we start from a sufficiently high order, say order five and then test down using this procedure. If there is no autocorrelation of fifth order we test for fourth order and so on, until the order of autocorrelation is identified. We are now in the position to apply this test to the model. We do this in step-by-step fashion:

(1) Set up the null and the alternative hypotheses. Since we are dealing with quarterly data, we test for up to fourth order autocorrelation in the model/data. In fact Microfit 4 automatically tests for up to fourth order autocorrelation when inputted data is quarterly. In the case of annual and monthly data, Microfit automatically tests for first and up to twelfth order autocorrelation, respectively. A fourth order autoregressive model may be written as:

$$u_t = \rho_1 u_{t-1} + \rho_2 u_{t-2} + \rho_3 u_{t-3} + \rho_4 u_{t-4} + \epsilon_t$$
$$\epsilon_t \sim nid(0, \sigma_t^2) \qquad (2.50)$$

The null and the alternative hypotheses can now be written as:

H_0: $\rho_1 = \rho_2 = \rho_3 = \rho_4 = 0$

H_1: H_0 is not true

(2) The test statistic is:

$$F = \frac{[RSS_R - RSS_U]/d}{RSS_U/n - k} \sim F_{d, n-k} \qquad (2.51)$$

The value of this statistic is computed as F(4, 61) = 17.4855, where d = 4; n = 68; K = 7.

(3) We now choose a particular level of significance, say 5%, and find the critical value of the test from the table of the F distribution. The critical value is $F(4, 61) = 2.53$. Now because $F(4, 61) = 17.4855 > F(4, 66) = 2.53$, we reject the null hypothesis at the 5% level of significance, concluding that there is autocorrelation up to fourth order in the data/model.

The LM test therefore indicates a significant amount of autocorrelation in the model/data. Note that we have not yet determined the correct order of autocorrelation, all we have done is to confirm the result of the DW test that autocorrelation exits and it could be up to fourth order. We now go to the next step in the analysis and carry out a general misspecification test.

The Ramsey regression specification error test (RESET)

The RESET is a popular misspecification test used to test whether some unknown variables have been omitted from a regression model. The test can also be used to detect incorrect functional forms. This test is routinely carried out in practice by the econometric packages. For example, Microfit 4 reports the results of this test after the LM test as a main diagnostic test. We demonstrate this test with reference to the model of demand for imports. The regression model is:

$$\log M_t = \beta_1 + \beta_2 \log GDP_t + \beta_3 \log (P_m/P_D)_t + u_t \tag{2.52}$$

If the model is correctly specified, the expected value of the dependent variable is determined by the variables suggested by the theory, in this situation the disturbance term is a random variable capturing only the influence of the random terms on the dependent variable. In other words, if the model is correctly specified, there would not be a relationship between the expected value of the dependent variable and the disturbance term.

However, if the model is not correctly specified either due to omission of relevant regressors or incorrect functional form, the disturbance term would not only capture the influence of random terms on the dependent variable but also captures the influence of omitted variables. In other words, there are some variables included in the disturbance term which would have a systematic influence on the dependent variable. These unknown omitted variables, captured by the disturbance term, are the cause of model misspecification.

Following this line of reasoning, in developing a misspecification test, Ramsey recommends adding a number of additional terms to the regression model and then testing the significance of these. More specifically, he suggested that one needs to include in the regression model some functions of the regressors, on the basis that, if the model is misspecified, the disturbance term would capture these variables either directly or indirectly through other variables omitted from the regression. All we need to do now is test for the significance of these additional variables. If these additional variables are found to be significant, then the model is misspecified. Notice that the test does not tell us how to correct the model, it is designed to show that the model is misspecified. To correct the model, we need to use theory. In time series investigation, it is advisable to consider the dynamic structure of the model. The question which now arises is, how to perform a RESET in practice? In particular, what are these additional variables to include in the model to start the process? The additional variables are some function of the regressors. Therefore, we need a variable which depends on the regressors used in the model. An obvious candidate variable is the estimated expected value of the dependent variable obtained from the OLS estimation of the regression model under consideration.

With reference to our model of demand for imports, we have:

$$\log \hat{M}_t = \hat{\beta}_1 + \hat{\beta}_2 \log \text{GDP}_t + \hat{\beta}_3 \log (P_m/P_D)_t \tag{2.53}$$

The estimated expected value of the dependent variable ($\log \hat{M}_t$), depends on the regressors. Following Ramsey's recommendation we need a function of this variable to include as an additional variable in the model. In practice, normally we include a number of the powers of this variable as additional variables in the regression model to obtain the so called expanded equation; we then test for the significance of these additional variables using the F test for linear restrictions. If, the additional variables are found to be significant, one concludes that the regression model is misspecified. Regression packages, and in particular the Microfit 4 carries out this test automatically using the second power of the fitted value of the dependent variable as an additional regressor. In practice, however, it is useful to look at the graph of the OLS residuals against the fitted values to see whether we can detect a relationship. We can also count the number of turning points on the graph as a guide to the power of the fitted values to be included as additional variables. For example, normally, one turning point in the graph suggests a quadratic relationship, so we include the second power of the fitted value as an additional variable. Two turning points could be indicative of a third degree polynomial relationship so we include both the second and the third powers of the fitted values as additional variables in the regression, and so on.

Key steps needed for RESET tests

(1) Estimate the regression model by the OLS and make a note of the Residual Sum of Squares, call this RSS_R.

(2) Include a number of the powers of the fitted values (the squared values) as additional variables in the regression to obtain the expanded equation. With reference to the regression model the expanded equation is:

$$\log M_t = \beta_1 + \beta_2 \log \text{GDP}_t + \beta_3 \log (P_m/P_D)_t + \beta_4 [\log \hat{M}_t]^2 + u_t \tag{2.54}$$

(3) Estimate the expanded equation by the OLS and record the Residual Sum of Squares and call this RSS_U. Notice that the original regression model is in fact a restricted form of the expanded equation, where the coefficient of the additional variable is set to zero.

(4) Using the expanded equation as the unrestricted form of the model, and the original regression equation as the restricted form of the model, we test for the significance of additional variables. With reference to our model, we can specify the null and alternative hypotheses:

$H_0: \beta_4 = 0$

$H_1: \beta_4 \neq 0$

(5) This is the F test of linear restrictions. The test statistic is:

$$F = \frac{[RSS_R - RSS_U]/d}{RSS_U/n - k} \sim F_{d,\,n-k} \tag{2.55}$$

where d = number of parameter restrictions (the difference between the number of the parameters of the unrestricted model and the number of the parameters of the restricted model).

n = number of the observations,

K = number of the parameters of the unrestricted model.

(6) We need to compute the value of this test statistic. This is F(1, 64) = 2.6145, where d = 1 and n–k = 64.

Decision rule

Choosing a 5% level of significance: the critical value of the F test is approximately 4. Because F(1, 64) = 2.6145 < $F_{1,\,64}^{0.05}$ = 4, we do not reject the null hypothesis at the 5% level of significance. We now carry out yet another misspecification test, this time for heteroscedasticity before attempting to 'correct' the model. The RESET test appears to indicate that the model is not misspecified. In the light of the results of other misspecification tests (notably the DW test), this is surprising. However, this result cannot be relied on, because of the significant amount of autocorrelation in the data/model, detected by the DW and the LM tests. This situation clearly demonstrates one of the major shortcomings of the traditional (SG) approach to econometric modelling. In this approach, the reliability of each diagnostic test depends crucially on the results of all other tests carried out, and if any one of these tests breaks down, they all fail. Only when all the diagnostic test results are consistent and all assumptions are confirmed, we can consider that the regression results are reliable.

Testing for heteroscedasticity

One of the main assumptions of the linear regression model is the assumption of homoscedasticity or, constant variance. This assumption implies that the variance of the distribution of the dependent variable of the model, corresponding to each set of values of the independent variables, is constant. In other words, variations in the values of the independent variables are assumed not to affect the variance of the dependent variable. Symbolically, the assumption of the homoscedasticity may be written as:

Var(dependent variable) = σ^2 for all t = 1,..., n.

Alternatively, in terms of the variance of the distribution of the disturbance term, the assumption of homoscedasticity implies that the variances of the distributions of all the disturbance terms are the same. Symbolically,

Var(u_t) = σ^2 for all t = 1,..., n.

Notice that variance (dependent variable) = var(u_t) = σ^2

The failure of this assumption is called heteroscedasticity. In practice, heteroscedasticity could occur due to one, or combinations of, the following factors:

(a) In cross-sectional work, it is often the case that the values of the independent variables change significantly from one observation to the next. Significant variations in the independent variables cause the variance of the dependent variable/disturbance term to change resulting in heteroscedasticity. This particular form of heteroscedasticity is said to be caused by scale effects, spilling over from independent variables. For example, our model of consumption/income relationship, as income changes significantly from one observation to the next the variance of the expenditure (the dependent variable) tends to change. In particular, the higher the level of income the higher is the variance of the distribution of the consumption expenditure. This is because at higher levels of income consumers have greater scope/freedom to spend. The variance of consumption expenditure rises with income as the pattern of expenditure becomes more unpredictable.

(b) Although heteroscedasticity usually occurs in cross-sectional analysis due to scale effects, it can also be present in time series analysis. In particular the variance of the disturbance term could become smaller with the passage of time to reflect learning from the past errors or, better data collection procedures. Both these tend to reduce the variance of the disturbance term.

(c) Specification errors, due to omission of relevant regressors or incorrect functional form can also cause heteroscedasticity. For example, if a regressor has been left out of the model its influence is captured by the disturbance term, which in turn, could become heteroscedastic if the omitted variable is itself heteroscedastic. In practice it is therefore advisable to follow up a general misspecification test (e.g. RESET), with a test for heteroscedasticity, such as the Koenker-Bassett test, routinely carried out by Microfit 4.

The consequences of heteroscedasticity

Before we consider a number of tests carried out in practice to detect heteroscedasticity, it will be useful to discuss briefly the impact of heteroscedasticity on the OLS estimators. Generally speaking, these estimators are no longer best in the class of unbiased estimators and are not therefore reliable. In fact, it can be shown that another estimator, the Generalised Least Squares, is a superior estimator. More specifically, when heteroscedasticity is present it can be shown that:

1   The OLS estimators are unbiased but are no longer efficient, that is they tend to have relatively large variances.
2   The OLS estimator of the error variance is biased. The direction of the biased depends on the relationship between the variance of the disturbance term and the values taken by the independent variable causing heteroscedasticity.
3   The test statistics which make use of the OLS estimate of the error variance, including t and F tests are unreliable generating misleading results.
4   The coefficient of determination, R^2, will also be unreliable, usually over estimating the extent of a linear relationship between the dependent and the independent variables.

Given the serious nature of these consequences, it is clearly necessary to test for the heteroscedasticity in applied research, particularly, when cross-section data is used. We now consider a number of statistical tests designed to detect heteroscedasticity.

The Koenker Bassett test (KB)

This is a simple diagnostic test done automatically by regression packages. We demonstrate this test by using the regression results. Heteroscedasticity implies that the variance of the disturbance term is not constant over the data range. The Koenker-Bassett test assumes, in particular, that the variance of the disturbance term is a function of the regressors. The actual functional form needs not be specified. To get the test operational, we need a proxy for the non-observable variance and a variable which is influenced only by the regressors. The squares of the OLS residuals are usually used as a proxy for the variance. As far as the other variable is concerned, KB test uses the fitted values of the dependent variable (in our example $\log M$). The test then assumes a certain relationship between these two variables. Specifically, it is assumed that the squared residuals (the proxy for variance) is a linear function of the squared fitted values of the dependent variable used to capture the variations in all the regressors. We perform this test as follows:

$$e^2_t = a_0 + a_1[\log \hat{M}_t]^2 + \epsilon_t$$
$$\epsilon_t \sim nid(0, \sigma_t^2)$$

(2.56)

We need to test the significance of the parameter a, that is:

$H_0: a_1 = 0$

$H_1: a_1 \neq 0$

The test is a simple 't' test which is:

$$t = \frac{\hat{a} - a}{S\hat{E}(\hat{a})} \sim t_{n-k}$$

(2.57)

Given that the t test has only one degrees of freedom, the squared value of the t value is distributed as an F distribution with one degrees of freedom. In practice, one usually uses the F version of the test. The value of this statistic for the model of competitive imports is $F(1, 66) = 4.9546$. The critical value of the test at the 5% level of significance is approximately 4. We therefore reject H_0 concluding that there is evidence of heteroscedasticity present. Given the nature of the time series model it is likely that heteroscedasticity is due to the model misspecification.

The white test

This is yet another popular test for heteroscedasticity. In principle, it is similar to the Koenker-Bassett test, in that it examines if the variance of the disturbance term is a function of the regressors. More specifically, it tests to see if error variance is affected by the regressors. To use the test, the square of the OLS residuals in the regression model is regressed on the regressors of the model, squared values and cross products. The R^2 of this equation times the number of observations, n, can be shown to be distributed as a Chi-square distribution for large samples. For a large value of the test statistic we reject the null hypothesis and conclude that heteroscedasticity is a problem.

A test for normality assumption: The Bera-Jarque (BJ) test

An important assumption of the classical linear regression model is the assumption that the dependent variable/disturbance terms are normally distributed. Given this assumption the OLS estimators are also normally distributed and the specification/misspecification tests outlined have associated 't', 'F' and 'χ^2' distributions. Given the crucial role of the diagnostic tests in the analysis of the results, it is important to carry out tests to confirm the assumption of normality. A popular test for this purpose is the Bera-Jarque (1982) test. The BJ test is designed to track departures from the normal distribution by checking the skewedness and the kurtosis of the distribution of the OLS residuals. The BJ test statistic is given as follows:

$$BJ = [n/6\,SK^2 + n/24\,(EK - 3)^2] \tag{2.58}$$

where n equals to number of observations, SK = a measure of skewedness of the distribution and EK = a measure of kurtosis of the distribution.

It can be shown that the BJ test statistic, under repeated sampling and the null hypothesis that the OLS residuals are normally distributed, have a chi-square distribution with 2 degrees of freedom. Microfit 4 reports the value of the BJ test statistic with other diagnostic tests. In practice, all we need to do is to compare the BJ test value with the theoretical value of χ^2 with 2 degrees of freedom, at a prespecified level of significance. If the BJ test statistic is greater than the corresponding χ^2 at a prespecified level of significance, for example 1% or 5%, we reject the normality assumption.

The BJ test is highly sensitive to the presence of the outliers observations which generate large OLS residuals. The rejection of the normality assumption could therefore signal the existence of the outliers observations or, breaks in the data. These effects may be further investigated by employing dummy variables in the regression models (see Chapter 3). The BJ test may be performed as follows:

H_0: the disturbance terms are normally distributed (the OLS residuals are normally distributed).

H_1: H_0 is not true

The test statistic is: χ^2

$$BJ = [n/6\,SK^2 + n/24\,(EK - 3)^2] \sim \chi_2^2 \tag{2.59}$$

The value of the BJ test statistic reported by the Microfit 4 is 0.81353. Because $BJ = 0.81353 < \chi_2^{2(0.05)} = 5.99$, we do not reject H_0 at $\alpha = 5\%$, concluding that the normality assumption is maintained.

5 THE (S-G) APPROACH TO 'CORRECTING'/GENERALISING THE MODEL

The regression models derived from theory are long-run models. Based on theory they show functional relationships among a number of variables in the long run. The time series data that are used in regression models are, however, dynamic showing the magnitude of economic variables over specified intervals of time. There is no guarantee that these observed magnitudes are long-run values for the variables under consideration, or that interactions among economic variables are completed. There could be some lagged responses between the dependent variable and the regressors specified in the model, due to slow adjustment processes. The data captures the lagged responses, but the regression model excludes them. There is, therefore, an inconsistency between long-run static econometric models and the data used for estimation. This inconsistency between econometric model and the data is shown in the results of diagnostic tests which usually, in time series analysis, indicate inadequacy of the static econometric model in capturing the main features of data generation process.

To modify the model to replicate the data generation process, the first step is to look for dynamic misspecification as the cause of inadequacy. In particular, attempts are made to introduce lags among economic variables to capture the short-run dynamic nature of the data generation process. A particular procedure, used frequently by practitioners of the traditional type, is to assume partial adjustment mechanisms which are then linked to the long-run regression model. The practitioner then estimates the short run model and uses the results to derive the implied long-run parameters. In what follows we demonstrate this method by using the log linear model of imports.

The static econometric model under consideration can be seen as portraying the long run paths of desired levels of imports. In other words, theory suggests that in the long-run/desired levels of imports depend on the level of expenditure/income and relative prices term. The desired or, the long-run level of imports is not observable and we need a mechanism which can be used to convert the model into data values in the data set. One such a mechanism is a partial adjustment scheme. The basic ideas are as follows. The actual change in the dependent variable suggested by the data values is assumed to be a function of the desired change, due to sluggish conditions in the economy and slow adjustment processes among economic variables. Symbolically, in terms of model of imports, we can write:

$$\underset{\text{actual change}}{(m_t - m_{t-1})} = \underset{\text{desired change}}{\alpha(m^D t - m_{t-1})} \tag{2.60}$$

where $0 < \alpha \le 1$ is the speed of the adjustment parameter. The closer α is to unity the quicker the response of economic variables. Moreover, the closer α is to unity the more efficient the economy. Economies with institutional rigidities, labour market rigidities and inefficient market mechanisms tend to be sluggish, reacting slowly to required/desired changes. For these economies the parameter α tends to be close to zero, indicating a slow and prolonged adjustment process. We can now combine the partial adjustment process with the regression model to generate a short-run model as follows:

$$m_t = \alpha m^D t + (1 - \alpha)m_{t-1} \tag{2.61}$$

substitution for $m^D{}_t$ from the regression model into above results in:

$$m_t = \alpha[\beta_1 + \beta_2 GDP_t + \beta_3 P_t + u_t] + (1 - \alpha)m_{t-1} \tag{2.62}$$

$$m_t = \alpha\beta_1 + \alpha\beta_2 GDP_t + \alpha\beta_3 P_t + (1 - \alpha)m_{t-1} + \alpha u_t \tag{2.63}$$

where all variables are in logs and $P = P_M/P_D$.

The appearance of the m_{t-1}, the lagged dependent variable, as a regressor implies that this is a short-run dynamic model. This model is called an autoregressive dynamic short-run model. Notice that if u_t is $(0, \sigma^2)$, αu_t would also be a white noise process. We can therefore estimate the model by the OLS, find the estimates of the short-run parameters and then use the values to estimate the implied long-run parameters. This two-step method is what is done in practice. We illustrate the procedure now.

First let:

SR		LR
α_1	=	$\alpha\beta_1$
α_2	=	$\alpha\beta_2$
α_3	=	$\alpha\beta_3$
α_4	=	$(1 - \alpha)$
v_t	=	αu_t

The SR model may now be written as:

$$m_t = \alpha_1 + \alpha_2 GDP_t + \alpha_3 P_t + \alpha_4 m_{t-1} + v_t \tag{2.64}$$

The new results from Microfit 4 are illustrated in Table 2.1.

Dependent variable is $m_t \log M_t$, 67 observations used for estimation from 1980Q2 to 1996Q4.

Table 2.1 OLS estimation from Microfit 4

Regressor	Coefficient	Standard error	T-ratio
C	−5.5571	1.2243	−4.5391
log GDP	0.81046	0.17343	4.6731
log p	−0.011308	0.056721	−0.1993
log m_{t-1}	0.6178	0.084666	7.2969
R-squared	0.98973	F-statistics	$F(3, 63) = 2023.1$
R-Bar squared	0.98924	SE of Regression	0.026132
Residual Sum of Square	0.04302	Mean of Dependent Variable	10.3558
SD of dependent variable	0.25189	Maximum of Log Likelihood	151.1823
DW – statistic	1.8125		

Test statistics	F version
A: Serial Correlation	$F(4, 59) = 2.2653$
B: Functional Form	$F(1, 62) = 0.3023$
C: Normality	$\chi^2(2) = 4.95$
D: Heteroscedasticity	$F(1, 65) = 3.65$

Where:
A: Lagrange multiplier test for serial correlation.
B: Ramsey's reset test using the square fitted values.
C and D : Based on the regression of squared residuals or squared filled values.

(a) The regression results

First, we note that the estimated slope coefficients' signs are in accordance with economic theory. In particular the estimated (partial) income elasticity of demand for imports is positive (0.81046) and price elasticity of demand is negative (-0.011308). The magnitude of these coefficient are, however quite small reflecting the short-run nature of the model. Nevertheless the SR price elasticity of demand is estimated at only -0.011308 which indicates a very small response in demand for imports due to changes in relative prices. Moreover, the T-ratio corresponding to the coefficient of $\log p$ appears to be very small indeed (-0.19937) and statistically insignificant. This result suggests that UK demand for imports appears to be non-price sensitive and is a surprising result and at odds with economic theory. It is worth reviewing literature concerning the estimate of the price elasticity of demand for imports, before accepting/reporting this result.

(b) Diagnostic tests

The DW test is inappropriate in this case, as the lagged dependent variable is a regressor. In this situation we can either consider the result of the Lagrange Multiplier test or carry out the Durbin 'h' test, which is a test for the 1st order autocorrelation in the autoregressive models. We first explain the Durbin 'h' test and then consider the LM test statistic.

The Durbin 'h' test

This is a test for the first order autocorrelation and is carried out when the lagged dependent variable is one of the regressors of the model. The test is essentially a large sample test and the more observations are used the more accurate the test results. Given a first order autoregressive process the null and the alternative hypotheses of the test are written as follows:

$$\text{Given AR(1)} u_t = \rho u_{t-1} + \epsilon_t \qquad (2.65)$$

$$\epsilon_t = u \sim \text{nid}(0, \sigma^2)$$

$$H_0: \rho = 0$$

$$H_1: \rho \neq 0$$

Durbin has shown that under H_0, the following test statistic, when a large data set is used, has a standard normal distribution:

$$h = \hat{\rho} \sqrt{\frac{n}{1 - n\,\text{vâr}(\alpha_4)}} \sim N(0, 1) \qquad (2.66)$$

where $\hat{\rho}$ = estimate of first order correlation coefficient given by $\hat{\rho} = 1 - \text{DW}/2$. n is the sample size, vâr(α_4) is the estimated variance of the lagged dependent variable. For example, in the model of imports vâr(α_4) = $(0.084666)^2$. To apply the test we need to calculate the value of the h statistic and then compare it with the critical values, at a prespecified level of significance, of the normal distribution. For example, at $\alpha = 5\%$, the critical values are ± 1.96. Now if $-1.96 \leq h \leq 1.96$ we do not reject H_0, concluding that H_0 is consistent with the data and first order autocorrelation does not appear to exist.

As pointed out h test is essentially a large sample test and should only be used when a large number of observations are available. We now apply this test to the model under consideration:

$$h = 1 - \frac{1.8125}{2} \sqrt{\frac{67}{1 - 67(0.084666)^2}} = 1.0644 \tag{2.67}$$

A decision rule: H_0 is consistent with the data at $\alpha = 5\%$ and first order autocorrelation does not appear to exist.

The LM test is frequently used to test for higher order autocorrelation. Again it is a large sample test and its use should be restricted to when a large data set is available. Given that we have used quarterly observations (Microfit 4), automatically test for up to fourth order autocorrelation Chi-sq (4) or F(4, 57). The first number (4), is the number of linear restrictions which have been imposed on the parameters of the model. The underlying assumptions and procedures are as follows:

Assume AR (4):

$v_t = \rho_1 u_{t-1} + \rho_2 u_{t-2} + \rho_3 u_{t-3} + \rho_4 u_{t-4} + \epsilon_t$

$\epsilon_t \sim \text{nid}(0, \sigma^2_t)$

$t = 1, 2, 3 \ldots$

under this specification the unrestricted form of the model is:

$\log m_t = \alpha_1 + \alpha_2 \log \text{GDP}_t + \alpha_3 \log P_t + \alpha_4 \log m_{t-1} + \rho_1 u_{t-1} + \rho_2 u_{t-2} + \rho_3 u_{t-3} + \rho_4 u_{t-4} + \epsilon_t$ (2.68)

H_0: $\rho_1 = \rho_2 = \rho_3 = \rho_4 = 0$ [4 linear restrictions]

H_1: H_0 is not true

The restricted form of the model is thus, the original model:

$\log m_t = \alpha_1 + \alpha_2 \log \text{GDP}_t + \alpha_3 \log P_t + \alpha_4 \log m_{t-1} + \epsilon_t$

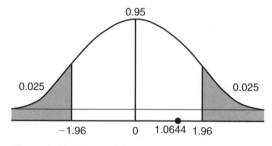

Figure 2.11 The Durbin 'h' test

Test statistic:

$$F = \frac{RSS_R - RSS_U/d}{RSS_U/n - k} \sim F(d, n - k) \tag{2.69}$$

This value is calculated by Microfit 4 as $F(4, 59) = 2.2653$. Note: we have 67 observations and $K = 8$ (parameters in the unrestricted model).

Decision rule

Because $F(4, 59) = 2.2653 < F(4, 59)^{0.05} = 2.53$, we do not reject H_0 at $\alpha = 5\%$. H_0 is consistent with the data. Therefore, according to this result autocorrelation up to fourth order does not exist in this model. Note $F(4, 59)^{0.05} = 2.53$ is the theoretical value of $F(4, 59)$ at 5%, obtained from the F distribution table.

Test for specification error: the Ramsey's RESET

This is a misspecification test. As pointed out previously, the square of the fitted values are added to the original equation to generate an expanded equation. One then tests for the significance of the addition terms. These procedures are demonstrated below:

$$\log m_t = \alpha_1 + \alpha_2 \log GDP_t + \alpha_3 \log P_t + \alpha_4 \log m_{t-1} + \alpha_5 [\log \hat{m}_t]^2 + u_t \text{ (expanded equation)}$$

$H_0: \alpha_5 = 0$

$H_1: \alpha_5 \neq 0$

Test statistics:

$$F = \frac{RSS_R - RSS_U/d}{RSS_U/n - k} \sim F(d, n - k) \tag{2.70}$$

The value of this test statistic is computed as chi-sq $(1) = 0.32608$ or $F(1, 62) = 0.30323$.

Notice that the first degrees of freedom (one) is the number of linear restrictions on parameters in the model $(d = 1)$. The theoretical value of $F(1, 62)$ at $\alpha = 5\%$ is approximately 4.

Decision rule

Since $F(1, 62) = 0.3023 <^{0.05} F(1, 62)$, we do not reject H_0 at $\alpha = 5\%$. H_0 is therefore consistent with the data and the model appears to be well specified according to the RESET 'test'.

The Koenker–Bassett test for heteroscedasticity

This is another diagnostic/misspecification test. It is specifically designed to detect heteroscedasticity in the data. It is based on the regression of the squared residuals on squared fitted values. It can be demonstrated as follows:

$$\mathrm{Var}(V_t) = F(\log \mathrm{GDP}, \log P, \log M_{t-1})$$

$$\left|\mathrm{Proxy} \qquad\qquad \right| \mathrm{Roxy}$$

$$e^2_{\mathrm{OLS}} \qquad\qquad \log \hat{m}_t$$

or

$$e^2_{\mathrm{OLS}} = \beta_1 + \beta_2[\log \hat{m}_t]^2 + u_t$$

$$H_0: \beta_2 = 0$$

$$H_1: \beta_2 \neq 0$$

Test statistics:

$$F = \frac{RSS_R - RSS_U/d}{RSS_U/n - k} \sim F(d, n - k) \tag{2.71}$$

The value of the above test statistics is computed as $F(1, 65) = 3.65$. The critical value of the $F(1, 65)$ at $\alpha = 5\%$ is approximately 4.

Decision rule

Since $F(1, 65) = 3.65 < 0.05\ F(1, 65) = 4$

We do not reject H_0. H_0 is therefore consistent with the data. It appears that there is no evidence of heteroscedasticity in the model/data and the model is well specified.

Test for normality

The BJ test for normality is conducted as follows:

H_0: Disturbance terms are normally distributed

H_1: H_0 is not true

The BJ test statistics:

$$BJ = [n/6\,SK^2 + n/24\,(EK - 3)^2] \sim \chi_2^2 \tag{2.72}$$

The BJ statistic is computed by the Microfit 4 as 4.96. The chi-square value with 2 degrees of freedom at $\alpha = 5\%$ is 5.99.

Decision rule

Because BJ $= 4.95 < \chi_2^{2(0.05)} = 5.99$. We do not reject H_0, concluding that H_0 is consistent with the data. The normality assumption is therefore appear to be consistent with the data.

Based on the results of the diagnostic tests, the short-run partial adjustment model appears to be well specified and explains the short-run data generation process. The only point of concern is the statistical insignificance of the relative price term, which needs further investigation.

We can now use the SR estimated coefficients to derive the long-run parameters as follows:

$\hat{\alpha}_1 = -5.5571$

$\hat{\alpha}_2 = 0.81046$

$\hat{\alpha}_3 = -0.011308$

$\hat{\alpha}_4 = 0.61780$

Since $\alpha_4 = (1 - \hat{\gamma})$, $\hat{\gamma} = 1 - 0.6178 = 0.3822$ (2.73)

The coefficient of adjustment/speed is indicating to a fairly sluggish adjustment process. The estimate of other long-run parameters are as follows:

$\hat{\beta}_1 = -5.5571/0.3822 = -14.54$

$\hat{\beta}_2 = 0.81046/0.3822 = 2.2$

$\hat{\beta}_3 = -0.011308/0.3822 = -0.03$

where $\hat{\beta}_2$ and $\hat{\beta}_3$ are the estimates of LR income elasticity and the price elasticity of demand for imports, respectively. While, the LR income elasticity appear to be consistent with previous studies, the price elasticity of demand, however, is only estimated at -0.03, reflecting very little response in demand for import to changes in relative prices. While the sign of this coefficient is consistent with the theoretical consideration, the magnitude appears to be too small. In conclusion, income seems to be the main variable influencing the level of imports both in short and long run. In particular, according to the empirical results, for every one percentage change in the level of national income/GDP, we expect that, on average, imports to change by 2.2%. Imports are, therefore, considered to be highly elastic. Given this result, unless exports also increase at the same rate per period, the economy could be faced with the balance of trade problems. Moreover, the devaluation, the traditional remedy to deal with balance of payment problems would not be an effective policy option, given a high-price inelastic import demand function.

(a) Tests for structural breaks and stability

Here we consider a number of diagnostic tests routinely carried out in applied research identifying specific problems including structural breaks and measurement errors.

In time series applications it is advisable to test for the parameter stability over the period of investigation. This is very important when long-run time series are being used for estimation. The fundamental motive for this is to see whether the underlying regression model has remained unchanged. For this to be the case, the parameters of the model must have remained constant.

Econometric models do fail and as a result, the parameters, linking the variables, do change drastically. A breakdown in economic relationships occurs due to structural breaks in the economy. For example, significant currency crises, external economic shocks such as oil embargoes and sanctions, or due to political changes which usually bring about structural changes in the economy. Whatever the cause, a structural break results in the failure of economic relationships, misspecified econometric models and faulty inference. In practice, it is therefore advisable to carry out diagnostic tests for parameter stability.

There are two types of diagnostic tests available:

1   Chow 'type' test: designed to test for parameter stability when a possible break point in the data/model can be identified 'a priori'.
2   Tests based on recursive estimation methods. These tests include cusum and cusum Q test and are carried out when the break point in the data/model is not known 'a priori'.

We provide a brief outline of each 'type' of test.

Chow test for parameter stability

The Chow test is designed to test for parameter stability. In this situation, the aim of the analysis is to see if the parameters of the model have been constant over the period of investigation. Suppose that there are n observations available and it is thought a structural break in the economy has occurred at the end of period n_1. The structural break divides the data set into n_1 observation before the possible break in the data and n_2 observations after the possible break in the data, where $n_1 + n_2 = n$, the total number of observations. Provided that we have sufficient degrees of freedom (i.e. n_1 and n_2 are long periods of time), the Chow test for parameter stability is done as follows:

a   Estimate the model by OLS, using n observations to generate the residual sum of squares RSS.
b   Estimate the model by OLS, using n_1 observations, to generate the residual sum of squares for the period before the 'break', RSS_1.
c   Estimate the model by OLS, using the n_2 observation to generate the residual sum of squares, for the period after the 'break', RSS_2.

Chow argued that, in the absence of a structural break, the residual sum of squares of the entire period RSS, must be approximately the sum of the residual sum of squares of two sub-sample periods; so:

$$RSS \cong RSS_1 + RSS_2 \qquad (2.74)$$

Or, simply the difference between RSS and $(RSS_1 + RSS_2)$ must not be statistically significant. A test statistic is therefore needed based on $RSS - (RSS_1 + RSS_2)$. Chow showed that the following test statistic has an 'F' distribution:

$$F = \frac{RSS - (RSS_1 + RSS_2)/K}{(RSS_1 + RSS_2)/n - 2K} \sim F(K, n - 2k)$$

where the numerator has a chi-distribution with K degrees of freedom and the denominator has a chi-square distribution with $n - 2k$ degrees of freedom. K is the number of parameters of the model (including the intercept term) and n is the number of observations. The 'F' test statistic is used to test for the stability of the regression parameters. The test however, is only reliable when the variance of the disturbance term has also remained unchanged. It is therefore customary to carry out a test to check for changes in the variance of the disturbance term, and then use Chow's test. To illustrate the Chow test, we consider the regression model:

$$Y_t = \beta_1 + \beta_2 X_{2t} + \beta_3 X_{3t} + u_t \tag{2.75}$$

$$u_t \sim nid(0, \sigma_2)$$

$$t = 1, 2, \ldots, n.$$

Suppose that a possible break point in the data is at the end of period n_1. The regression model for the period $n_{1+1}, \ldots, n$, may be expressed with 'new' parameters as follows:

$$Y_t = \alpha_1 + \alpha_2 X_{2t} + \alpha_3 X_{3t} + u_t \tag{2.76}$$

$$u_t \sim nid\ (0, \sigma_2^2)$$

$$t = n_1 + 1, \ldots, n.$$

The Chow test is as follows:

$$H_0: \sigma_1^2 = \sigma_2^2$$

$$H_0: \sigma_1^2 \neq \sigma_2^2$$

We first test for H_0, if H_0 is not rejected, we then test for parameters' stability. The test statistic for this is based on the estimation of the error variance, using n_1 and n_2 observations in turn. Hence:

$$F = \frac{RSS_2/n_2 - k}{RSS_1/n_1 - k} \sim F(n_2 - k, n_1 - k) \tag{2.77}$$

where:

RSS_2 = residual sum of squares obtained from the OLS regression of the model over the period $n_{1+1}, \ldots, n$, using n_2 observations.

$\quad RSS_1$ = residual sum of squares obtained from the OLS regression of the model over the period $t = 1, \ldots, n_1$, using n_1 observations.

If $F > F^{\alpha}_{n_2-k, n_1-k}$ reject H_0 at $\alpha\%$ level of significance. If H_0 is not rejected, use the Chow test for the parameter stability as follows:

H_0: $\beta_1 = \alpha_1$; $\beta_2 = \alpha_2$; $\beta_3 = \alpha_3$;

H_1: H_0 is not true

Test statistic:

$$F = \frac{RSS - (RSS_1 + RSS_2)/K}{(RSS_1 + RSS_2)/n - 2K} \sim F(K, n - 2k) \tag{2.78}$$

where in the example K = 3,

if $F > F^{\alpha}_{k, n-2k}$ reject H_0 at $\alpha\%$ level of significance concluding that H_0 is not consistent with the data and parameters have changed over the simple period. Notice that the Chow test is essentially a joint test requiring a constant variance for the disturbance term over the sample period. There are occasions when the break point in the data is such that it does not provide sufficient degrees of freedom for two separate OLS regressions. This situation occurs when either $n_1 < K$ or $n_2 < K$. Supposing $n_2 < K$, the Chow test procedure may be summarised as follows: ($n_1 < K$) may be dealt with in the same way):

(1) First, test for constant error variance:

H_0: $\sigma_1^2 = \sigma_2^2$

H_1: $\sigma_1^2 \neq \sigma_2^2$

The test statistic for this test is:

$$HF(n_2) = 1/\frac{RSS_1}{n_1 - K} \cdot \sum_{t=1}^{n2} e_{2t}^2 \sim \chi_{n_2}^2 \tag{2.79}$$

where $\sum_{t=1}^{n2} e_{2t}^2$ is the sum of squares of the one-step-ahead forecast errors. To obtain e_{2t}, we first estimate the regression by OLS over the n_1 sub-sample. We then use these parameter estimates together with the value of the regressors in each period, over $n + 1, \ldots, n$, to obtain one-step-ahead forecast errors.

Decision rule

If $HF(n_2) > \chi_{n_2}^{2(\alpha)}$, reject H_0 at $\alpha\%$ level of significance. This test is known as Hendry Forecast (HF) test. If H_0 cannot be rejected go to the next step as follows:

(2) Second:

H_0: $\beta_1 = \alpha_1$; $\beta_2 = \alpha_2$; $\beta_3 = \alpha_3$;

H_1: H_0 is not true

Test statistic:

$$F = \left(\frac{RSS - RSS_1}{RSS_1} \right) \left(\frac{n_1 - K}{n_2} \right) \tag{2.80}$$

If $F > F^{\alpha}_{n_2 - k, n_1 - k}$ reject H_0 at $\alpha\%$ level of significance concluding that H_0 is not consistent with the data.

Recursive estimation

The key shortcoming of the Chow test is that in practice one possible break point in the data/model must be identified from the outset. In many applications such a priori information is not available. In these situations, it is advisable to complement the Chow test with one or more diagnostic tests based on the recursive estimation methods. Recursive estimation involves a series of OLS estimations. In practice we start with a minimum number of observations, usually as low as the number of the parameters of the model, K_k. We then estimate the model on the basis of K, K + 1, K + 2,..., up to n. This procedure generates a time series of OLS estimates. If the model is structurally stable, the variations in all parameter estimates must be small and random. On the other hand if parameter estimates tend to change significantly and systematically, it could indicate a structural break and some underlying misspecification. The recursive estimation methods generate recursive residuals upon which a number of diagnostic tests have been based.

Cusum and Cusum Q tests for parameter stability

Using the hypothetical regression model used previously, let:

$$V_t = Y_t - (\hat{\beta}_1 + \hat{\beta}_2 X_{2t} + \hat{\beta}_3 X_{3t}) \tag{2.81}$$

for $t = K + 1, \ldots, n$.

Where:

$\hat{\beta}_1, \hat{\beta}_2, \hat{\beta}_3$ are the OLS estimates obtained from K observations.

It can be shown that a normalised form of the recursive residuals has the following distribution:

$$W_t \sim N(0, \sigma^2)$$

where W_t is a standardised recursive residual.

Cusum test is based on:

$$\text{Cusum}\, t = (1/s) \sum_{i=k+1}^{t} W_i \tag{2.82}$$

Where s is the full sample estimation of the standard error of regression

$$\left(\text{i.e. } s = \sqrt{\frac{\sum_{i=1}^{n} e_i}{n-k}} \right).$$

If the residuals are random and small in magnitude we would expect Cusum statistic to remain close to zero; any systematic departure from zero is taken to indicate parameter instability/misspecification. Microfit 4 produces a graph of Cusum statistic against time. In practice, a systematic trend in the graph of Cusum statistic against time is indicative of failure in the regression.

The Cusum Q statistic is defined as:

$$\text{Cusum Q} = \left(\sum_{i=k+1}^{t} W_i^2 \bigg/ \sum_{t=k+1}^{n} W_i^2 \right) = RSS_t / RSS_n \tag{2.83}$$

where:

RSS$_t$ = residual sum of squares of the recursive residuals.

RSS$_n$ = residual sum of squares of the OLS residuals for the full sample period.

The Cusum Q statistic lies between 0 and 1. A random dispersion of the Cusum Q statistic close to zero, within the band of zero and one, is indicative of parameter stability. However, both Cusum and Cusum Q statistics are essentially used in practice as diagnostics indicative of parameter stability. Tests based on these statistics have low power (are not reliable for small samples). It is advisable to complement these parameter stability indicators with a formal Chow test.

(b) Measurement errors: errors in the variables

Consider the following regression model of the UK food exports:

$$Y_t = \alpha + \beta q_t + \gamma p_t + u_t \tag{2.84}$$
$$u_t \sim \text{nid}\,(0, \sigma^2)$$

where:

Y_t = level of UK food exports

q_t = level of world food output

p_t = an index of world food prices relative to the UK food prices.

On the theoretical grounds we expect that slope parameters β and γ to be positive, implying that Y_t will change in the same direction with both, q_t and p_t.

Examining the data requirements of the model, it is likely that all three variables are measured with errors. Does this matter? As far as the dependent variable of the model is concerned, the measurement errors of calculating/gathering data on this variable are captured by the disturbance term, u_t. As long as these errors are normally and independently distributed, the OLS method is still BLUE and so we should not encounter specific problems. The story, however, is different as far for the regressors are concerned. Here, the OLS estimators are biased, inconsistent and unreliable.

To fix these ideas, let:

$q*_t$ = true value of q_t

$p*_t$ = true value of p_t

We assume that both q_t and p_t are measured with random measurement errors, under this framework, the observed values of the regressors may be expressed as follows:

$$q_t = q_t^* + u_{1t} \tag{2.85}$$

and $p_t = p*_t + u_{2t}$ where u_{1t} and u_{2t} are normally and independently distributed random measurement errors with the following characteristics:

$$u_{1t} \sim nid(0, \sigma^2_{u_{1t}}) \text{ and } u_{2t} \sim nid(0, \sigma^2_{u_{2t}}) \text{ also } cov(u_{1t}u_{2t}) = 0$$

The 'true' regression model is as follows:

$$Y_t = \alpha + \beta q_t^* + \gamma p_t^* + \mu_t$$

Substitution for the observed values of the regressors into the above equation yields:

$$Y_t = \alpha + \beta(q_t - u_{1t}) + \gamma(p_t - u_{2t}) + u_t \qquad Simoue \tag{2.86}$$

or

$$Y_t = \alpha + \beta q_t + \gamma p_t + u_t - \beta u_{1t} - \gamma u_{2t} \tag{2.87}$$

let $w_t = u_t - \beta u_{1t} - \gamma u_{2t}$, a composite error term.

The regression model may therefore be expressed as follows:

$$Y_t = \alpha + \beta q_t + \gamma p_t + w_t \tag{2.88}$$

Is it possible to estimate this model by the OLS method? The OLS method is appropriate only if all standard assumptions are valid. If there are indeed random measurement errors in the regressors, the regressors cannot be assumed to be fixed under repeated sampling procedures. In fact, they are stochastic/random variables. Moreover, the regressors are correlated with the composite error term w_t, violating a standard assumption of the classical linear regression model. In this case, it can be shown that OLS estimators are biased and inconsistent. Applying the OLS method, therefore, results in unreliable estimates and faulty inference. In research, when we are dealing with regressors which are likely to have been measured with errors, it is crucial that we undertake a test for measurement errors. If measurement errors are present, we use an alternative estimation method.

Instrumental variable (IV) estimation methods

The problem with the regression model under consideration is that it includes regressors, q_t and p_t each containing random measurement errors resulting in inconsistent OLS estimators. The inconsistency of the OLS estimation is due to existence of correlation between each regressor and the composite error term w_t. Within the framework of the IV method we replace each one of the stochastic regressors with an 'instrument'. An instrument for a stochastic/random regressors is a variable which is highly correlated with the regressor but not correlated with the disturbance term. With regard to the regression model under consideration, we need two instruments, one for q_t and one for p_t with the following characteristics:

(Instrumental for q_t)
A variable highly correlated with q_t but not correlated with w_t

(Instrumental for p_t)
A variable highly correlated with p_t but not correlated with w_t

Finding an instrument for stochastic variables appearing in the model is a difficult task in practice. However, with respect to the model under consideration, we use the lagged values of q_t and p_t as the respective instruments. Since $Cov(q_t q_{t-1})$ is likely to be high,

And

$$Cov(q_{t-1} w_t) = 0, \text{ because } Cov(u_{1t} u_{1t-1}) = 0 \tag{2.89}$$

Also

$$Cov(p_t p_{t-1}) \text{ is likely to be high}$$

And

$$Cov(p_{t-1} w_t) = 0, \text{ because } Cov(u_{2t} u_{2t-1}) = 0.$$

Notice that q_{t-1} and p_{t-1} are chosen as instruments, because measurement errors are assumed to be random. This assumption implies zero covariance between each of the instruments and the composite error term.

Given the instruments the regression model is then estimated by the IV method. This method is similar to the OLS, in that the estimators are obtained such that the residual sum of squares is minimised. This procedure generates so-called 'normal' equations, as follows:

$$\Sigma y_t = n\alpha + \beta\Sigma q_t + \gamma\Sigma p_t \qquad \ldots[1] \qquad (2.90)$$

$$\Sigma y_t q_{t-1} = \alpha\Sigma q_{t-1} + \beta\Sigma q_t q_{t-1} + \gamma\Sigma p_t q_{t-1} \qquad \ldots[2] \qquad (2.91)$$

$$\Sigma y_t p_{t-1} = \alpha\Sigma p_{t-1} + \beta\Sigma q_t p_{t-1} + \gamma\Sigma p_t p_{t-1} \qquad \ldots[3] \qquad (2.92)$$

The solution to these three simultaneous equations generates the instrumental variable estimators. Most econometric software have a routine for the IV estimation. In practice once the instruments are identified, the estimation process itself is done by the appropriate packages.

Testing for measurement errors

To develop a diagnostic test for measurement errors, we consider the model for the food exports:

$$Y_t = \alpha + \beta q_t + \gamma p_t + w_t \qquad (2.93)$$

The OLS estimators of the parameters are denoted as follows:

$$\hat{\alpha}_{OLS}, \hat{\beta}_{OLS}, \hat{\gamma}_{OLS}$$

If random measurement errors are present these estimators are inconsistent.

Using q_{t-1} and p_{t-1} as instrument of q_t and p_t, respectively, we can generate the IV estimators of parameters. These are denoted as follows:

$$\hat{\alpha}_{IV}, \hat{\beta}_{IV}, \hat{\gamma}_{IV}$$

If measurement errors are not present, both methods of estimation generate consistent estimators. This observation enables us to formulate the null and alternative hypotheses of the test as follows:

H_0: no measurement errors in the regressor

H_1: H_0 is not true

Or

H_0: $u_{1t} = u_{2t} = 0$

H_1: H_0 is not true

Or

H_0: $p\lim(\hat{\alpha}_{IV} - \hat{\alpha}_{OLS}) = 0$;

$p\lim(\hat{\beta}_{IV} - \hat{\alpha}_{OLS}) = 0$;

$p\lim(\hat{\gamma}_{IV} - \hat{\gamma}_{OLS}) = 0$

H_1: H_0 is not true

The final format of H_0 makes the point that in the absence of measurement errors, the OLS and the IV estimators are both consistent estimators. In the absence of measurement errors the two estimators are asymptotically identical. It can be shown that, the difference between the OLS and IV estimators is zero, if the instruments and the OLS residuals are uncorrelated. To carry out a diagnostic test for the measurement errors the instruments are added to the regression model to generate an expanded equation. The joint significance of instruments are then tested by the familiar F test for parameter restrictions. If the instruments are found to be not significant it is taken to mean that they are uncorrelated with the OLS residuals. Therefore the IV and the OLS estimators are essentially the same implying that there is no measurement errors in the regressors. In this situation IV and OLS estimators are consistent, however, the OLS estimators have smaller variances. We now illustrate the diagnostic test for the measurement error:

(1) State the null and alternative hypotheses.

H_0: there are no measurement errors in the regressors

H_1: H_0 is not true

Or

H_0: $u_{1t} = u_{2t} = 0$

H_1: H_0 is not true

(2) Estimate the expanded equation by the OLS

$$Y_t = \alpha + \beta q_t + \gamma p_t + \beta_1 q_{t-1} + \beta_2 p_{t-1} + w_t \tag{2.94}$$

Using $n = 45$ observations the OLS estimate of the expanded equation is:

$$Y_t = 29.6 + 0.48q_t + 0.30p_t + 0.20q_{t-1} + 0.30p_{t-1}; R^2 = 0.90 \tag{2.95}$$
$$\quad\;\; (5.8) \quad (0.14) \quad\; (0.51) \quad\;\; (1.01)$$

(3) Test for the joint significance of the instruments that is:

H_0: $\beta_1 = \beta_2 = 0$

No measurement errors [i.e. instruments and the OLS residuals (w_t) are not correlated]

H_1: H_0 is not true

Measurement errors are present.

(4) Estimate the restricted model by OLS:

$$Y_t = \alpha + \beta q_t + \gamma p_t + w_t \tag{2.96}$$

$$\hat{y}_t = 38.0 + 0.71q_t + \;\; 0.95p_t \tag{2.97}$$
$$\quad\;\; (5.0) \quad (0.12) \quad\quad (0.42)$$

$R^2 = 0.86$

$n = 45$

(5) Test statistic:

$$F = \frac{[R^2_u - R^2_R]/d}{[1 - R^2_u]/n - k} \sim F(d, n - k) \tag{2.98}$$

Using the numerical values, we have:

$$F = \frac{0.9 - 0.86}{1 - 0.9} \times \frac{45 - 5}{2} = 8 \tag{2.99}$$

Notice that n = 45; k = 5 and d = 2

Decision criterion

If $F > F^{\alpha}_{d, n-k}$ reject H_0 at α level of significance, concluding that H_0 is not consistent with the data. Measurement errors are present in the regressors. Rejection of the null hypothesis implies that OLS estimators are biased and inconsistent and IV method should be employed.

In our example:

$$F^{0.05}_{2,40} = 3.23$$

Because $F = 8 > F^{0.05}_{2,40} = 3.23$ we reject H_0 concluding that measurement errors are present in the regressors.

DISCUSSION QUESTIONS

1   Explain steps taken in an econometric analysis based on the traditional approach. Illustrate your analysis with reference to consumption function.
2   Explain the type of criteria you would employ to evaluate the regression results.
3   Explain what you understand by a statistical test of an hypothesis. Discuss how you would carry out a test of parameter significance.
4   Explain what you understand by a specification test. What is the basic difference between a specification and a misspecification test?
5   Explain the DW test, the LM test and the 'h' test for serial correlation. Under what situation do you employ them? What are the main shortcomings of these tests?
6   Explain the RESET test and how it might be applied.
7   Explain how you would respond to a significant value of the Durbin Watson (DW) test statistic in an applied econometric research.

ADDITIONAL READING

See, W. H. Greene, Prentice Hall, 2000. For some of technical material used in this chapter. Also see P. Kennedy, A guide to Econometrics, Blackwell. 1999. Also see (H. Seddighi and A. Abbott) 'Aggregate Imports and Expenditure Components in the UK. Journal of Applied Economics, September 1996.

3 Qualitative variables in econometric models

INTRODUCTION

The econometric models are derived on the basis of deductive reasoning, which identifies interrelationships among a number of variables. These variables are measured by conventional time series or cross-section data. However, qualitative variables, those variables which cannot be measured by quantitative data, can also have a significant impact on the dependent variable under consideration. For example, in cross-sectional analysis of household consumption expenditures, qualitative variables, such as location, the level of education of household members, and the gender mix of the household can influence the household consumption. Here, we provide a discussion on the treatment of qualitative variables as regressors in regression models.

Qualitative variables can also appear as dependent variables in regression models. For example, in a study of incidence of private health insurance based on, cross-section data, the dependent variable in the model is expressed as either an individual possessing private insurance or not, given his/her characteristics including income, and occupation. In these this type of models, the dependent variable essentially represents the response of an individual to a particular question being asked. The answer to the question is either positive or negative, representing qualitative variables. In recent years, due to availability of large scale microeconometric data, these models have become popular particularly in the area of market research.

Key points

- Dummy variables
- Seasonal adjustment of data
- Logit/probit model
- Panel data

1 DUMMY VARIABLES AND REGRESSION ANALYSIS

Dummy variables are used in regression analysis to take account of the impact of qualitative variables on the dependent variable. They are binary variables taking only the values of unity and zero. Unity is used to denote the occurrence of an event/characteristic while zero signifies the absence of a qualitative characteristic.

Dummy variables are used in cross-section and time-series regression models. In a time-series model they are employed for seasonal adjustments of data. We now demonstrate the application of the dummy variables in regression/econometric models.

(a) Impact of location on consumption

The location of households can have a significant impact on household consumption. Location is a qualitative variable and to capture its impacts, we need to employ dummy variables. In the example of family consumption expenditure, we distinguish between two locations; south and north to bring the location variable into the analysis. We define the following location dummy variable:

Di = 1, if the household is located in the south

    = 0, otherwise if the household is located in the north

Suppose the regression model is specified as:

$C_i = \beta_1 + \beta_2 Y_i + u_i \qquad u_i \sim nid(0, \sigma^2)$

C_i = the ith household consumption expenditure

Y_i = the ith household income

How does location of the household influence consumption expenditure? We can distinguish between three types of impact:

1   location only influences the level of autonomous expenditure, the intercept term β_1;
2   location only influences the marginal propensity to consume, the slope parameter β_2;
3   location influences both the intercept term and the slope parameters of the regression model.

We consider each in turn:

a. location influences the level of autonomous consumption expenditure: Figure 3.1 depicts the population regression of the two different locations.

Note that β_2 is the marginal propensity to consume in both locations. The slope of the lines, are the same, while the level of autonomous consumption expenditure is assumed to be higher in the south. Thus, households with the same level of income, on average, have higher autonomous expenditure in the south, because the cost of living in the south is higher. The location variable is, therefore, assumed to shift the southern household's consumption function upwards as depicted in Figure 3.1. Note that, the basic idea is that location of the household is a relevant regressor in cross-section studies of the type we are studying. This is an hypothesis and should be tested later. We have two regression models:

$C_i = \alpha_1 + \beta_2 Y_i + u_i$ Southern household   (3.1)

$C_i = \beta_1 + \beta_2 Y_i + u_i$ Northern household   (3.2)

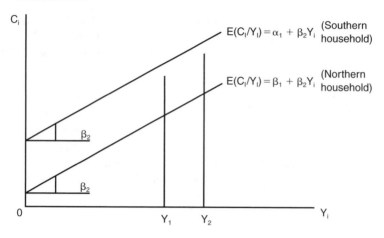

Figure 3.1 Intercept dummies: southern and northern locations

A way of proceeding would be to run two regressions, one for the south and one for the north, then test to see if the difference between α_1 and β_1 is statistically significant. This procedure, however, is seldom done in practice, since it involves two regressions. The 'correct' way is to use the location dummy variable to combine the two models and then run one regression to test the significance of location. This can be done in two ways.

Base groups: Regression with an intercept term

The basic idea here is to combine the two models, using one as a base group. For example, we can use 'northern household' group as a base group and compare the southern households with this group.

A combined model

Using northern group as a 'base' group, the location dummy may be used to derive the combined model as follows:

$$C_i = \beta_1 + \beta_2 Y_i + (\alpha_1 - \beta_1)D_i + u_i; \; i = 1, \ldots, n \tag{3.3}$$

Base Group   Deviation
(Northern   from the
Household)   Base group

If the household is located in the north, $D_i = 0$, and the regression model will be $C_i = \beta_1 + \beta_2 Y_i + u_i$, which is the northern household group regression equation. If the household is located in the south, then $D_i = 1$ and the regression model will be $C_i = \alpha_1 + \beta_2 Y_i + u_i$. In this example, the location variable only effects the intercept terms of the model, i.e. the level of autonomous expenditure. The location variable, can, however, also influence the slope parameter.

(b) Where the location variable affects the intercept term and the slope parameters

It can be argued that not only is the level of autonomous expenditure influenced by the location but also the marginal propensity to consume. In this case, the regression model for each region may be written as follows:

$$C_i = \alpha_1 + \alpha_2 Y_i + u_i \qquad \text{Southern} \qquad (3.4)$$

$$C_i = \beta_1 + \beta_2 Y_i + u_i \qquad \text{Northern} \qquad (3.5)$$

A combined model

Using northern group as a 'base' group, the location dummy may be used to derive the combined model as follows:

$$C_i = \beta_1 + \beta_2 Y_i + (\alpha_1 - \beta_1) D_i + (\alpha_2 - \beta_2) D_i Y_i + u_i \qquad (3.6)$$

(Base Group)

How does this combined model work? We start with Northern household in this case $D_i = 0$ and the model is:

$$C_i = \beta_1 + \beta_2 Y_i + u_i \qquad \text{Northern household/Base group} \qquad (3.7)$$

If, the household is located in the South, $D_i = 1$ and

$$C_i = \beta_1 + \beta_2 Y_i + (\alpha_1 - \beta_1) + (\alpha_2 - \beta_2) Y_i + u_i \qquad (3.8)$$

Or,

$$C_i = \alpha_1 + \alpha_2 Y_i + u_i \qquad \text{Southern} \qquad (3.9)$$

The combined model may then be estimated by the OLS method using a cross-section data set. The model can be used to test the hypothesis that location makes a difference to household consumption. We can conduct the following test:

(c) Significance tests for location on autonomous expenditure

We set the null hypothesis as:

$$H_0: \alpha_1 - \beta_1 = 0$$
$$H_1: \alpha_1 - \beta_1 \neq 0$$

the level of autonomous expenditure is the same for both Northern and Southern households.

Let $\gamma = \alpha_1 - \beta_1$, where γ is the parameter of the dummy variable, D_i, in the combined equation. The test statistic may be written as:

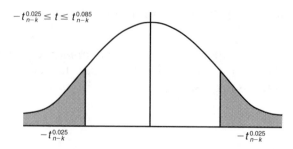

$-t_{n-k}^{0.025} \leq t \leq t_{n-k}^{0.085}$

$-t_{n-k}^{0.025}$

$-t_{n-k}^{0.025}$

Figure 3.2 Critical region for t-distribution

$$t = \frac{\gamma - \hat{\gamma}}{\hat{SE}(\hat{\gamma})} \sim t_{n-k} \tag{3.10}$$

To proceed with the test, we estimate the equation by OLS and calculate the value of the 't' statistic. If the calculated value of t falls in the rejection region, we reject H_0 at $\alpha = 5\%$, concluding that the difference in the level of autonomous expenditure in the two regions is statistically significant. We conduct a similar test of significance on the slope parameter, $(\alpha_2 - \beta_2)$, to see if there are differences between the marginal propensity to consume in the two regions. Finally, we can conduct a joint test of significance.

H_0: $(\alpha_1 - \beta_1) = 0$; $(\alpha_2 - \beta_2) = 0$

H_1: H_0 is not true

Test statistic:

$$F = \frac{[RSS_R - RSS_U]/d}{RSS_U/n - k} \sim F_{d,n-k} \tag{3.11}$$

Where; RSS_R = residual sum of square of the restricted model, RSS_U = residual sum of square of the unrestricted model, d is the number of restriction (i.e. $d = 2$) and n–k is the degrees of freedom of the unrestricted regression equation. To proceed with the test, we need to estimate the unrestricted and the restricted equations, calculate the value of the test statistic and compare the value with the critical value of the test, say at the 5% level of significance; if:

$$F \geq F_{n-k}^{0.05}$$

We reject the null hypothesis, at $\alpha = 5\%$, concluding that the location of household has a significant impact on household consumption.

(d) No base group: Regression with no intercept term

We can combine various groups, without using a particular group as a base. Assuming that location only influences the level of autonomous expenditure, we may write:

$$C_i = \alpha_1 + \alpha_2 Y_i + u_i \qquad \text{Southern}$$

$$C_i = \beta_1 + \beta_2 Y_i + u_i \qquad \text{Northern}$$

There is no base group, so we need to define two dummy variables one for each group:

Let $D_{1i} = 1$ if the *i*th household is located in the North
      $= 0$ otherwise

and

Let $D_{2i} = 1$ if the *i*th household is located in the South
      $= 0$ otherwise

The combined regression model may now be written as:

$$C_i = \alpha_1 D_{1i} + \beta_2 Y_i + \beta_1 D_{2i} + u_i; \qquad i = 1, \ldots, n \qquad (3.12)$$

Let us now see how it works. If a household is located in the North, $D_{1i} = 1$ and $D_{2i} = 0$, we get:

$$C_i = \alpha_1 D_{1i} + \beta_2 Y_i + u_i \qquad \text{Northern household} \qquad (3.13)$$

If a household is located in the south, $D_{1i} = 0$ and $D_{2i} = 1$, and we get the southern household model. Note that, within this framework, the regression model is estimated with no intercept. The lack of an intercept term can cause problems for some regression packages. It is therefore advisable, to select a base group and run a model with an intercept term.

(e) A number of qualitative variables influencing consumption

In addition to the location of households suppose we have information on more qualitative variables, for example:

a   the gender of the head of household: female; male
b   the age of the head of household: below 25 years; between 25 and 50; above 50
c   the level of education of head of household: No university degrees; university degrees.

There are now, therefore, four qualitative variables including the location of the household. In practice, we are interested to see whether any one or, all of variables, influence consumption. To this end we need to introduce a number of dummy variables as follows:

Location dummy

    $D_{1i} = 1$ if the *i*th household is located in the south
        $= 0$ otherwise

$D_{2i} = 1$ if the ith head of household is female
= 0 otherwise

$D_{3i} = 1$ if the age of head of the household is 25 or less
= 0 otherwise

$D_{4i} = 1$ if the age of head of household is more than 25 but less than 50
= 0 otherwise

$D_{2i} = 1$ if the ith head of household has a degree
= 0 otherwise

To proceed we choose a 'base group'. We define the 'base group' to be the one for which the values of each dummy variable is zero. Within this framework the 'base group' household is defined to be located in the north, the head of the household is male, the age of the head of household is 50 years or over, and the head of household has not got a university degree. The regression model of the base group may therefore be written as follows:

$$C_i = \beta_1 + \beta_2 Y_i + u_i \qquad i = 1, \ldots, n \qquad (3.14)$$

The above regression model is then used to estimate the difference between autonomous expenditure for each group of households and that of the base group as follows:

$$C_i = \beta_1 + \beta_2 Y_i \qquad\qquad + (\alpha_1 - \beta_1)D_{1i} + (\alpha_2 - \beta_1)D_{2i}$$

Base Group   The difference in   Difference due
                          autonomous   to gender
                          expenditure due
                          to location

$$+ (\alpha_3 - \beta_1)D_{3i} + (\alpha_4 - \beta_1)D_{4i} + (\alpha_5 - \beta_1)D_{5i} + u_i \qquad (3.15)$$

Difference due   Difference due to   Difference due
to age (less   age (25 years   to education
than 25 years) $\leq$ age $\leq$ 50 years)   level

Having estimated the model, we undertake 't' and 'F' tests to measure the significance of each qualitative variable (t-test) as well as to test their collective significance (F-test).

Notice that, when we choose a base group, the regression model has an intercept term. Moreover, the number of dummy variables needed in each category is one minus the number of categorical variables. For example, taking into account the three categories of age, as defined above, we need to define only two dummy variables, when a base group is used.

(f) Seasonal adjustments of data using seasonal dummy variable

Many economic variables exhibit seasonal variations. So values change with the seasons. For example, consumer expenditure tends to increase in the last quarter of the year compared with others. Dummy variables are frequently used to capture seasonal components.

Consider the demand for textiles. The demand equation for textiles may be specified as:

$$Y_t = \beta_1 + \beta_2 X_{2t} + \beta_3 X_{3t} + u_t \tag{3.16}$$

where:

Y_t = real per capita expenditure on textiles

X_{2t} = real per capita income

X_{3t} = relative prices of textiles

All variables are measured in natural logs. Based on 28 quarterly seasonally unadjusted observations, we obtain the following OLS results

$$Y_t = 1.370 + 1.140\ X_{2t} - 0.830\ X_{3t} \tag{3.17}$$
$$\quad\ (0.310)\quad (0.160)\qquad (0.140)$$

RSS = 0.75, n = 28

To see if the demand for textiles has seasonal components, we use the fourth quarter of the year as the 'base' period. We then use seasonal dummy variables to compare other quarters with the fourth quarter:

D_{1t} = 1 if, t is the first quarter
    = 0 otherwise

D_{2t} = 1 if, t is the second quarter
    = 0 otherwise

D_{3t} = 1 if, t is the third quarter
    = 0 otherwise

The regression model, including these dummy variables is specified as:

$$Y_t = \beta_1 + \beta_2 X_{2t} + \beta_3 X_{3t} \quad +(\alpha_1 - \beta_1)D_{1t}$$

Base Period   Deviation of the
first quarter from
'base period'

$$+ (\alpha_2 - \beta_2)D_{2t} \qquad\qquad + (\alpha_3 - \beta_3)D_{3t} + u_t \tag{3.18}$$

Second quarter   Third quarter
compared with   compared with
the based quarter   the based quarter

Now consider how the regression model works. When the observations relate to the first quarter of the year, $D_{1t} = 1$ and $D_{2t} = D_{3t} = 0$; we get the following model, corresponding to the first quarter:

$$Y_t = \alpha_1 + \beta_2 X_{2t} + \beta_3 X_{3t} + u_t \tag{3.19}$$

First quarter

$$D_{1t} = 1$$

$$D_{2t} = 0$$

$$D_{3t} = 0$$

The regressions for the second and the third quarter of the year are obtained similarly as:

$$Y_t = \alpha_2 + \beta_2 X_{2t} + \beta_3 X_{3t} + u_t \tag{3.20}$$

Second quarter

$$D_{1t} = 0$$

$$D_{2t} = 1$$

$$D_{3t} = 0$$

and

$$Y_t = \alpha_3 + \beta_2 X_{2t} + \beta_3 X_{3t} + u_t \tag{3.21}$$

Third quarter

$$D_{1t} = 0$$

$$D_{2t} = 0$$

$$D_{3t} = 1$$

The seasonal regression model was estimated by OLS generating the following output:

$$Y_t = 1.20 + 1.10X_{2t} - 0.75X_{3t} + 0.28D_{1t} - 0.38D_{2t} - 0.45D_{3t} + e_t \tag{3.22}$$
$$\quad (0.210) \quad (0.245) \quad (0.0620) \quad (0.028) \quad (0.031) \quad (0.042)$$

$$RSS = 0.65$$

$$n = 28$$

From these results, the impact of autonomous expenditure on textiles goods in the fourth quarter of the year is estimated at 1.20 per capita. The coefficient of D_{1t}, 0.28 shows how much this figure changes in the first quarter of the year, compared to the fourth quarter. It appears, therefore, that, autonomous per capita expenditure in the first quarter, compared to the fourth quarter, rises by 0.28. Then in the second and third quarters compared to the fourth, it falls by 0.38 and 0.45, respectively. We also notice each seasonal component is statistically significant. To check, if there is a statistically significant seasonal variation in expenditure on textiles, we perform a joint test of significance:

$H_0: (\alpha_1 - \beta_1) = (\alpha_2 - \beta_1) = (\alpha_3 - \beta_1) = 0$

(no seasonal variation)

$H_1: H_0$ is not true

Test statistic:

$$F = \frac{[RSS_R - RSS_U]/d}{RSS_U/n - k} \sim F_{d,n-k} \qquad (3.23)$$

Substitution of values in (3.23) yields:

$$F = \frac{[0.75 - 0.65]/3}{0.65/[28 - 6]} \qquad (3.24)$$

$$= 14/39$$

$$\cong 0.3$$

Comparing this figure with the theoretical value of:

$$F_{3,22}^{0.05} = 23$$

We cannot reject the null hypothesis. Therefore, there is no seasonal effects present in the data.

2 QUALITATIVE DEPENDENT VARIABLE MODELS

In recent years large-scale cross-section data sets containing several hundreds and even thousands of observations on the characteristics of individuals, firms and even towns/cities/regions have become available. The availability of this type of data and advances in computing have made it possible for practitioners to perform empirical investigations in the fields of social science, economics and marketing. A distinguishing feature of this type of investigation involving regression models, is that the dependent variable is specified as a qualitative variable, representing the response of an individual or firm to particular questions. For example, in an investigation on the incidence of R&D activities by firms, a firm is either undertaking R&D activities or not, given its characteristics (e.g. sales, exports, etc.). The dependent variable, the R&D status of the firm, is dichotomous, representing the response of the firm to the question: are you undertaking R&D activities? The response is either positive or negative. The dependent variables is, therefore, a dummy variable taking two values: if the answer is positive, it takes a value of one, otherwise it takes a value of zero. Alternatively, in a study of incidence of private health insurance using a large sample of cross-section data on the characteristics of individuals, an individual either has private health insurance or not, given his or her characteristics (e.g. income, occupation, age, etc.) The question being asked of each individual is: do you have a private health insurance? The response is either positive or negative. If it is found to be positive, the dependent variable, the insurance ownership status of an individual, takes a value of one, otherwise it takes a value of zero. In these examples, the dependent variable is defined on the characteristics of individuals/firms. It is a dichotomous qualitative variable eliciting a 'yes' or a 'no' response. Qualitative response models involve a number of interesting problems concerning estimation, interpretation and the analysis.

(a) The linear probability model (LPM)

We introduce the LPM, by considering the incidence of private health ownership.

> Let Y = an individual's status of private health insurance
> if, Y_i = 1 the ith individual has a private health insurance
> = 0 the ith individual does not have a private health insurance

It is assumed that Y depends on income, thus:

$$Y_i = \beta_1 + \beta_2 X_i$$

We introduce in addition a disturbance term u_i to take into account the influence of other factors, so:

$$Y_i = \beta_1 + \beta_2 X_i + u_i \tag{3.25}$$

where:

$$E(u_i) = 0, \; Var(u_i) = \delta^2, \; cov = (u_i \, u_j) = 0$$

A model, where a dichotomous dependent variable is expressed as a linear function of one or, a number of regressors is called a linear probability model. This is because the expected value of Y_i is the conditional probability that, the individual runs a private health insurance (i.e. $Y_i = 1$, given the individual's income (given x_i). To see this, let P_i = probability that $Y_i = 1$ (the response is positive). In this case, $1 - P_i$ = probability that $Y_i = 0$, the response is negative. The probability distribution of Y_i is:

Y_i	*Probability*
$Y_i = 1$	P_i
$Y_i = 0$	$1 - P_i$

The mathematical expectation of Y_i is therefore: $E(Y_i/X_i) = P_i$, hence the conditional expectation of Y_i, given X_i is the probability of a positive response, given income. Notice also that, under $E(u_i) = 0$, using LPM, we have:

$$E(Y_i/X_i) = \beta_1 + \beta_2 X_i \tag{3.26}$$

Therefore, within a LPM model, we have:

$$E(Y_i/X_i) = \beta_1 + \beta_2 X_i = P_i \tag{3.27}$$

In a LPM, a key aim is to estimate the conditional probability of a positive response of, an event occurring, given certain characteristics. This is done by estimating two unknown parameters β_1 and β_2.

Problems associated with the LPM

Given a LPM model, $Y_i = \beta_1 + \beta_2 X_i + u_i$, our aim is to estimate the conditional probability of a positive response. That is $P_i = E(Y_i = 1) = \beta_1 + \beta_2 X_i$; where $0 \le P_i \le 1$.

This is done by estimating β_1 and β_2 using data on X_i and Y_i. However, there is no mechanism in the model to ensure that the estimated probabilities obtained from estimate of β_1 and β_2 are within the permissible range of zero and one. This is an obvious problem in the use of LPMs. In addition, the assumption of linearity may be considered to be rather unrealistic. To see this, notice that, β_2 can be expressed as:

$$\beta_2 = \Delta P_i / \Delta X_i$$

That is β_2 measures the change in probability due to a unit change in income. Since, parameter β_2 is a constant, the change in the probability at all levels of income is in fact assumed to be constant. This feature is rather unrealistic. At low levels of income up to a certain level of income, it is reasonable to assume that, incremental changes in income has no effect on the probability. Once income reaches a certain level, however, there is a strong probability that it will change at an increasing rate with income. Once a certain high level of income is reached, there is no effect on the probability as a result of the change in income. This type of non-linear relationship between the probability P_i and income may be depicted as follows:

The S shaped curve depicted in the Figure 3.3 can be expressed by a logistical function as:

$$P_i = E(Y_i = 1) = 1/(1 + e^{-(\beta 1 + \beta 2 Xi)})$$

For all values of X_i, the logistical function falls between zero and one. Given this property and non-linearity, the logistical function is frequently used to estimate conditional probabilities in models involving qualitative dependent variables. In addition to these problems other issues arise with LPMs.

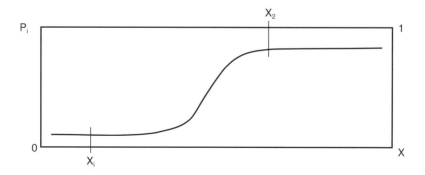

Figure 3.3 Non-linear probabilities between health insurance and income

---

Chapter 3: Key terms

Dummy variables   Used to capture the impact of qualitative variables on the dependent variable of a regression model.

A shift dummy   Used to capture the impact of a qualitative variable on the intercept term of a regression model.

Qualitative dependent variable models   Models in which the dependent variable is qualitative/dichotomous.

LPM   Linear Probability Model. A linear model designed to estimate the conditional probability of a positive response/presence of a characteristic.

The logit model   A non-linear probability model based on a logistic curve designed to estimate the conditional probability of a positive response/presence of a characteristic.

The probit/normit model   A non-linear probability model based on the normal cumulative distribution designed to estimate the conditional probability of a positive response/presence of a characteristic.

Panel data   A data set normally consisting of a large number of observations on cross-sectional units over a relatively short period of time.

A panel data study   A study based on a panel data set to capture individual cross-sectional effects.

Fixed effects model   Individual effects are captured by shifts in the intercept term of the model.

LSDV   Least squares dummy variable model, used with fixed effects model.

Random effect model (REM)   Individual effects are randomly distributed over cross-sectional units.

Heteroscedasticity

LPM is associated with heteroscedasticity in the disturbance term. In other words, under LPM specification, the distribution of disturbance terms of the model are inevitably heteroscedastic. The estimation of the model by OLS therefore results in inefficient estimators. To see why the disturbance term is heteroscedastic, we derive the variance.

The disturbance term takes only two values with the following probabilities:

when $Y_i = 1$

$$u_i = 1 - \beta_1 - \beta_2 X_i \qquad \text{Probability } (P_i) \qquad (3.28)$$

when $Y_i = 0$

$$u_i = -\beta_1 - \beta_2 X_i \qquad \text{Probability } (1 - P_i) \qquad (3.29)$$

$$\text{Var}(u_i) = (1 - \beta_1 - \beta_2 X_i)^2 P_i + (-\beta_1 - \beta_2 X_i)^2 (1 - P_i)$$

Also, we know that $P_i = E(Y_i/X_i) = \beta_1 + \beta_2 X_i$. Substitution for P_i into the above results in:

$$
\begin{aligned}
\text{Var}(u_i) &= (1 - \beta_1 - \beta_2 X_i)^2 (\beta_1 + \beta_2 X_i) + (-\beta_1 - \beta_2 X_i)^2 (1 - \beta_1 - \beta_2 X_i) \\
&= (1 - \beta_1 - \beta_2 X_i)(\beta_1 + \beta_2 X_i)[1 - \beta_1 - \beta_2 X_i + \beta_1 + \beta_2 X_i] \qquad (3.30) \\
&= (1 - P_i)P_i \qquad \textit{Heteroscedastic}
\end{aligned}
$$

In addition to the existence of heteroscedasticity, the disturbance term is not normally distributed and, unless, large observations are available, statistical tests based on normal distribution cannot be used in LPMs.

(b) The logit model

The logit model is a popular qualitative dependent variable model used in practice. The attractions of the model are:

1 Being based on the logistical curve, for all values of the regressors, the value of the dependent variable (the probability of positive response) falls between zero and one.
2 The probability function is a non-linear function following a logistical curve. This is a more realistic pattern of change in the probability compared to LPMs.
3 The estimation of the logit model is quick and easy and there are many packages available for the estimation of logit type models (SHAZAM, Microfit 4).

Estimation of a logit model

To illustrate the technique, we consider the example of private health insurance once again. We have:

$$P_i = P_r(Y_i = 1) = E(Y_i/X_i) = 1/(1 + e^{-(\beta_1 + \beta_2 X_i)}) \qquad (3.31)$$

Equation (3.31) gives the probability of a positive response. Under this specification the probability of a negative response is:

$$1 - P_i = 1 - 1/(1 + e^{-(\beta_1 + \beta_2 X_i)}) = e^{-(\beta_1 + \beta_2 X_i)}/(1 + e^{-(\beta_1 + \beta_2 X_i)}) \qquad (3.32)$$

Division of P_i by $1 - P_i$ gives the odd ratio in favour of an individual possessing a private health insurance, as follows:

odds ratio $= P_i/1 - P_i = e^{\beta_1 + \beta_2 X_i}$

To estimate the model, we take the natural log of both sides; hence:

$$\log (P_i/1 - P_i) = \log e^{\beta_1 + \beta_2 X_i}$$
$$L_i = \log (P_i/1 - P_i) = \beta_1 + \beta_2 X_i \qquad (3.33)$$

where, L_i is the logit function. In practice, the model is usually estimated by maximum likelihood methods. However, in principle, it may be estimated by OLS methods, provided that a large data set is available. To illustrate the OLS estimation from the following data set we have:

Income category	N	Number of people having PI	Frequency
X_1	N_1	n_1	$P_1 = n_1/N_1$
X_2	N_2	n_2	$P_2 = n_2/N_2$
–	–	–	–
–	–	–	–
X_u	–	–	–

These relative frequencies are then used as proxies for the probabilities in the logit model:

$$L_i = \ln(\hat{p}_i/1 - \hat{p}_i) = \beta_1 + \beta_2 X_i + u_i \qquad (3.34)$$

where u_i is a disturbance term. Given the nature of the model, u_i is heteroscedastic with the following properties:

$$u_i \sim N[0, 1/[N_i P_i(1 - P_i)]]$$

To generate efficient estimates the logit model is usually divided by $1/[N_i P_i(1 - P_i)]$ (multiply the dependent/independent variable by $N_i P_i(1 - P_i)$). This procedure usually succeeds in removing heteroscedasticity. The model can then be estimated by the OLS method.

The probit/normit model

The logit model provides a convenient way of estimating the dummy dependent variable models. This model is based on the cumulative logistical function which is 'S' shaped and falls between zero and one. Another cumulated density function (CDF) which has similar characteristics to those of a logistical function is the normal cumulative density function. This is another 'S' shaped curve lying between zero and one. The normal CDF is also frequently used by the practitioners for estimation of the dummy dependent variable models. The dummy dependent variable regression models based on the normal cumulative density function are called 'probit' or sometimes normit models. Thus the incidence of private

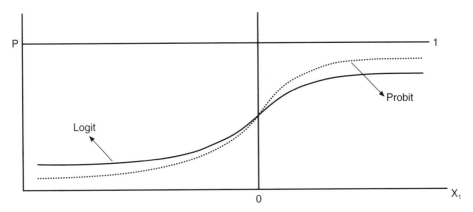

Figure 3.4 A normal CDF plotted against a logistical CDF

health insurance, the corresponding normal cumulative density function may be written as:

$$P_i = P_r(Y_i = 1) = \frac{1}{\sqrt{2\pi}} \int_{-\infty}^{\beta + \beta_2 x_i} e^{-t^2/2} dt \qquad (3.35)$$

where t is a standardised normal variable that is $t \sim n(0, 1)$, P_i = probability of a positive response, given X_i. The Figure 3.4 presents a normal CDF which is depicted against a logistical CDF.

As seen in Figure 3.4 there is little difference between the two CDFs. Normally, a logit model has flatter tails compared with probit models. That is, the conditional probability P_i approaches one or zero at a slower rate in logit models compared to probit models. Given slight differences between the two CDFs, the choice between probit and the logit models is essentially a matter of convenience.

Logit/probit models in practice

In regressions involving a dichotomous dependent variable, logit/probit models are used to estimate the conditional probability that a particular characteristic is present. For example, in a study of whether or not a firm is engaged in research and development (R&D) activities, we can identify the main characteristics, and based on these estimate the conditional probability that a firm is active in R&D. To illustrate the methodology we consider the following example:

The dependent variable

$Y_i = 1$ if the *i*th firm is engaged in R&D

$Y_i = 0$ otherwise

The independent variable/regressors

Here we need to use theory to identify the factors influencing a firm's R&D activities. According to the theory, some of these factors may be listed as:

- X_1 = firm size, we distinguish between small, medium and large firms so:
  large firms are where turnover/sales exceeds $1 million;
  medium firms are where turnover/sales exceeds $100,000;
  small firms are where turnover/sales are less than $100,000.
- X_2 = export intensity, we distinguish the following categories:
  high: where 50% or more, of sales are for exports;
  medium: where export intensity is between 10%–50% of sales;
  low: where exports intensity is less than 10% (less than 10% of sales are exported).
- X_3 = technological opportunity, here we consider the following two categories:
  High: where a firm operates in a high/modern technological sector (software firms);
  Low: where a firm operates in a low technological sector (metal work).

[3] A model – to illustrate the basic ideas a logit model can be used:

$Pr(Y_i = 1/\text{firms size, export intensity, technology opportunity})$

$$= pr(Y_1 = 1/X_1, X_2, X_3) = \frac{1}{1 + e^{-(\beta_0 + \beta_1 X_{1i} + \beta_2 X_{2i} + \beta_3 X_{3i})}}$$

Where: β_0, β_1, β_2, β_3 are the unknown parameters of the logit model. The model suggests that the probability of a firm undertaking R&D activities $Pr(Y_i = 1)$, depends on the size of the firm, export intensity and technological opportunity. The aim of the investigation is to estimate this conditional probability on the basis of a set of observations on Y_i, X_{1i}, X_{2i} and X_{3i}.

To obtain data, we design a questionnaire to obtain information on R&D activities, size of firm, export intensity and technological opportunities. By doing this we obtain a sample data of, say 130, firms, containing information on these variables. Given that, all regressors are categorical variables, we assign the following values to each variable as:

- X_{1i} = 1 if firm is small
  = 2 if firm is medium
  = 3 if firm is large

- X_{2i} = 1 if export intensity is low
  = 2 if export intensity is medium
  = 3 if export intensity is high

- X_{3i} = 1 if in a low technology sector
  = 2 if in a high technology sector

Finally;

- Y_i = 1 if firm is R&D active
  = 0 if firm is not R&D active.

There are $3 \times 3 \times 2 = 18$ different types of firm. For each type we calculate the conditional probability of a firm being engaged in R&D activities.

The estimation is done using an appropriate econometric package (for example Microfit 4 or Shazam). A typical output for our example may be listed as follows:

Variable	Estimated coefficient	Standard error	t-ratio
X_1: β_1	0.0458	0.34055	0.13477
X_2: β_2	1.1050	0.34939	3.1628
X_3: β_3	1.3548	0.48041	2.8201
constant: β_0	−4.7311	0.98294	−4.8132

We notice that X_2 and X_3 are statistically significant (relatively high t-ratios), whereas X_1 appears to be statistically insignificant. However, recall that a firm is identified by all three variables jointly. Therefore we are interested in estimates of all slope coefficients (β_1, β_2, β_3) rather than each individual case. The estimated coefficients are then used in the logit model to generate predicted probabilities. Table 3.1 below presents the predicted probabilities of a selection of firms.

According to these results, firms with the highest predicted probability of undertaking R&D are those with high-turnovers, high export intensities and high technological opportunity (firm 1). Firms with low turnovers, low export intensity and low technological opportunity have the lowest predicted probability of undertaking R&D (only 9%). Analysing the top six firms, that is, those firms with a predicted probability of approximately 50% and more, it can be seen that all such firms have high or medium export intensities. While five have high sales and three operate within a high technological environment. The analysis clearly identifies exporting firms as being likely to be involved in R&D.

In most applications, the logit and probit models generate very similar results. The probit results for this example are essentially the same. If a researcher considers that movement towards probability of one or zero, after certain values of the regressors have reached, occurs quickly then a probit model provides a better approximation to the data generation process. Otherwise, the logit model is preferred and more commonly used.

Panel/longitudinal data

Panel data sets provide rich environment for researchers to investigate issues which could not be studied in either cross-sectional or time series settings alone. In a

Table 3.1 Logit probability of R&D activity and firm characteristics

Firm	Probability	X_1			X_2			X_3		
		3	2	1	3	2	1	3	2	1
1	0.80710	*			*				*	
2	0.58085	*				*			*	
3	0.56964		*				*		*	
4	0.51912	*			*					*
5	0.50765		*		*					*
6	0.49618			*	*					*
7	0.31439			*		*			*	
8	0.10588			*		*				*
9	0.10161		*			*				*
10	0.09749		*			*				*

typical panel data study there are a large number of cross-sectional units, for example a large number of individuals, firms or even regions/countries and only a few periods of time. Researchers collect data on the characteristics of the cross-sectional units over a relatively short period of time to investigate change in behaviour or potential of a typical cross-section unit. The following are a number of examples based on panel data set.

1 Earning potential of the university graduates

In a panel study of earning potential of university graduates researchers collect data on the characteristics of a large number of graduates, say 1,000. These characteristics can include, gender, age, type of degree, employment etc. The graduates are then followed over time (usually over a short span of time, i.e. 1–3 years). The researchers then collect data on graduates' earnings, promotions etc, to investigate the impact of each one of the above characteristics on earning potential. Such a panel data set consist of 1,000 observations on the cross-sectional units, graduates, over a relatively short period of time, 1–3 years.

2 Effectiveness of negative income tax

A frequently cited panel data study is a study concerning the effectiveness of negative income tax in the USA in the early 1970s. In their investigation, researchers collected data on thousands of individuals/families over 8–13 quarters. The information included in the panel data consisted of individuals' characteristics including consumption expenditure, age, gender, number of children etc. The focus of the investigation was whether or not negative income tax has influenced the consumption expenditure of a typical individual/family over time.

3 The neoclassical growth theory and the 'convergence' hypothesis

A frequently debated topic in the neoclassical growth theory is the issue of convergence, put forward by Solow (1956). According to the convergence hypothesis the per capita growth rate of GDP of the poorer nations will converge to that of richer nations over time. In particular, the lower the initial level of GDP is the faster the economy will grow. This process will eventually result in an absolute convergence. To test this hypothesis one needs a panel data set consisting of a sample of a large number of countries over a specific period of time. The panel data set provides information on the characteristics of a large number of countries (the cross-sectional units) over a specific period of time. The convergence hypothesis cannot be investigated either in cross-sectional or time series settings alone.

4 Firms' R&D expenditure

A panel data set consisting of a large number of firms can be used to investigate factors influencing the rate of growth of the firms' R&D expenditure. Such panel data set could consist of observations on the characteristics of a large number of firms' over a relatively short period of time. In all above examples panel data sets provide information on a large number of cross-sectional units over a relatively short period of time. They are typically wide but short. Panel data sets are therefore more oriented toward cross-section analysis. In such an analysis the main focus of investigation is usually on the issue of heterogeneity across cross-sectional units. Heterogeneity implies that there are individual/unit effects present in the data. The focus of analysis is in capturing these individual effects. The individual effects are usually investigated either within a framework of a fixed effects model or a random effects model. In what follows we will briefly consider these two models.

Fixed effects model

It is often assumed that the differences across cross-sectional units can be captured in differences in the constant term. Within this framework, the intercept term of the regression model varies across the cross-sectional units. In order to fix some of the basic ideas we consider the following example.

Suppose in a panel data study of the neoclassical convergence hypothesis a researcher collects cross-sectional data on 20 countries over a five-year period. The relevant variables are assumed to be per capita annual growth rate of GDP, $\dot{GDP}$ and per capita GDP. The following simple regression model is specified:

$$\dot{GDP}_{it} = \alpha_i + \beta GDP_{it} + u_{it}; \; u_{it} \sim nid(0, \sigma_1^2)$$

$$i = 1, \ldots, 20; \; t = 1, \ldots, 5$$

α_i is the intercept term representing the fixed country effect. Within this framework, it is assumed that intercept term varies across 20 countries. To estimate the model it is customary to use dummy variables to allow for different country effects as follows:

$$GDP_{it} = \alpha_i + \beta GDP_{it} + u_{it}; \; t = 1, 2, 3, 4, \text{ and } 5$$

where:

$$D_i = 1 \text{ if } i\text{th country}$$

$$D_i = 0 \text{ otherwise}$$

The model is then estimated by the OLS to generate estimates of the country effects, α_i. This type of fixed effects model is usually referred to as the least squares dummy variables (LSDV) model. The assumption of the heterogeneity (different country effects) can be tested as follows:

$$H_0: \alpha_i = \ldots, = \alpha_{20} = \alpha \qquad \text{No specific effects}$$

$$H_1: H_0 \text{ is not true}$$

The appropriate test statistic is the familiar F test for parameter restrictions:

$$F = \frac{[RSS_R - RSS_u]/d}{RSS_u/n - k} \sim F(d, n - k)$$

where:

> RSS_R = residual sum of the restricted model (no country effect) corresponding to the regression model
>
> $GDP_{it} = \alpha + \beta GDP_{it} + u_{it}$
>
> RSS_U = residual sum of squares corresponding to the (LSDV) model
>
> d = no. of parameter restrictions (d = 19)
>
> $n = 20 \times 5 = 100$
>
> k = 21 (20 = country effects parameters + one slope parameter)

If the value of the F test statistic is greater than the corresponding theoretical value of F, at say level of significance of 5%, we reject H_0 and conclude that there are significant country effects presentation in the data. The fixed effect model is used when the number of cross-sectional units is relatively small and the sample exhausts all cross-sectional units. In this situation the LSDV method generates linear unbiased estimators. In the usual case when n is large as it typically is in a panel study of families, individuals etc, the LSDV method is not appropriate for two reasons: (i) degrees of freedom problem, due to a large number of cross-sectional units; and (ii) inference about the population from which the cross-sectional units are sampled is not possible, since LSDV method produces results conditional on the units in the data set. In a typical panel data study, because of the above reasons, researches tend to use random effects model, we will briefly consider this model below.

Random effect model

In a random effects model it is assumed that the individual effects are randomly distributed across cross-sectional units. To capture the individual effects, the regression model is specific with an intercept term representing an overall constant term. The error term is then assumed to have two components: $\epsilon_{it} + u_i$, where ϵ_{it} is the error term associated with each observation and u_i is an error term representing the extent to which the intercept term of the *i*th cross-sectional unit differs from the overall constant term. A random effect model may be specified as follows:

$$Y_{it} = \alpha + \beta_1 X_{it} + \beta_2 Z_{it} + + \epsilon_{it} + u_{it} \qquad i = 1, \ldots, n; t = 1, \ldots, T$$

where Y_{it} = dependent variable corresponding to the *i*th cross-section unit at time t and X_{it} and Z_{it} are cross-sectional units' characteristics, α is the overall constant term and $\epsilon_{it} + u_{it}$ is the composite error term with $\epsilon_{it} \sim nid(0, \sigma_\epsilon^2)$ and $u_i \sim nid(0, \sigma_{iu}^2)$, also $cov(\epsilon_{it} u_{it}) = 0$, for all i and t.

In the case of random effects model, the efficient estimator is generalised least squares. For estimation purpose, in practice, we use a two-step procedure. The various components are first estimated by using the residuals from OLS regression. Then, Feasible Generalised Least Squares (FGLS) estimates are computed using the estimated variances (for more details see for example W. H. Greene, 1993).

DISCUSSION QUESTIONS

1   Explain under what circumstances dummy variables are employed in regression models. Why is it important to test the significance of these variables in applied work?
2   What are the shortcomings of the LPM? Under what circumstances would you employ this model?
3   Explain how a logit model might be estimated. How would you interpret the regression results of a logit model?
4   Explain the main differences between a logit model and a probit model.
5   Explain under what circumstances you would use: (a) a fixed effects model; and (b) a random effects model, in a panel data study.

ADDITIONAL READING

See, D. Gujerati, 'Basic Econometrics'. McGraw Hill, 1997. Also see, D. W. Green. Econometric Analysis, 2000 for further aspects of the techniques used in this chapter.

4 Models with lagged variables

Key points

- Finite (D-L) models
- Infinite (D-L) models
- Partial adjustment and adaptive expectation models

1 INTRODUCTION

The classical normal linear multiple regression model is based on the following assumptions:

Linearity of the model

The general form of the linear in the parameters multiple regression model is:

$$Y_t = \beta_0 + \beta_1 X_{1t} + \beta_2 X_{2t} + \ldots + \beta_k X_{kt} + \epsilon_t, \text{ for } t = 1, 2, \ldots, N \tag{4.1}$$

where Y_t is the dependent variable; X_{jt} (for $j = 1, 2, \ldots, k$) are k independent, or explanatory, variables; ϵ_t is the stochastic disturbance, or error, term; β_j (for $j = 0, 1, \ldots, k$) are $k + 1$ unknown parameters to be estimated, the so called (partial) regression coefficients, with β_0 being the intercept; and t indicates the t-th observation, N being the size of the sample.

Assumptions involving the disturbance term

Zero mean: $E(\epsilon_t) = 0$, for all t. $\tag{4.2}$

Homoskedasticity: $\text{var}(\epsilon_t) = E(\epsilon_t^2) = \sigma^2$, for all t. $\tag{4.3}$

Serial independence, or no autocorrelation: $\text{cov}(\epsilon_t, \epsilon_s) = E(\epsilon_t \epsilon_s) = 0$, for $t \neq s$.

$$\tag{4.4}$$

Normality: $\epsilon_t \sim N(0, \sigma^2)$, for all t. $\tag{4.5}$

Assumptions involving the explanatory variables

Non-stochastic independent variables: Each independent variable X_j is fixed, or given, in repeated samples of size n, with its variance being a finite positive number. (4.6)

Adequate sample size: $n > k + 1$. (4.7)

No multicollinearity: There is no exact linear relationship between any of the independent variables. (4.8)

No specification bias: The regression model is correctly specified in terms of variables and functional form. (4.9)

Under these assumptions the OLS estimators b_j (for $j = 0, 1, \ldots, k$) of the corresponding unknown parameters β_j (for $j = 0, 1, \ldots, k$) are best linear unbiased estimators (BLUE).

In the case that t indicates time, or in other words the regression model involves time series variables, model (4.1) assumes that the current value of variable Y_t depends on the current values of all explanatory variables included in the model. However, this is not always true. In various economic phenomena the current value of a variable may depend on the current values and/or on past values of some explanatory variables as well. In the next section we will examine such cases.

2 CASES FROM ECONOMIC THEORY

In almost all economic phenomena the adjustment process between corresponding variables is rarely instantaneous. The process takes time to fully develop its effects depending on the characteristics and complexity of the phenomenon. Let us examine some simple economic cases.

Case 4.1 The consumption function

The simplest linear formulation of the consumption function is

$$C_t = \alpha_0 + \beta_0 Y_t + u_t \tag{4.10}$$

where C_t = private consumption, Y_t = personal disposable income, u_t = disturbance term, $\alpha_0 > 0$, and $0 < \beta_0 < 1$. The ratio C_t/Y_t is the 'average propensity to consume' and the first derivative $\partial C_t/\partial Y_t = \beta_0$ is the 'marginal propensity to consume'.

From equation (4.10) it is seen that the current value of consumption depends on the current value of income only and not on the current values of any other variable. However, this may be not true, taking into account that current consumption may also depend on the current level of savings. In such a case the consumption function could be written as

$$C_t = \alpha_0 + \beta_0 Y_t + \gamma_0 S_t + u_t \tag{4.11}$$

where S_t = private savings, and $\gamma_0 > 0$ parameter.

Following the theory that the current level of savings depends on the past levels of income, a savings function could be written as

$$S_t = \delta_0 + \delta_1 Y_{t-1} + \delta_2 Y_{t-2} + \ldots + v_t \tag{4.12}$$

where δ_j (for $j = 0, 1, 2, \ldots$) = parameters, and v_t = disturbance term.

Inserting equation (4.12) in equation (4.11) the following equation is produced

$$C_t = (\alpha_0 + \gamma_0 \delta_0) + \beta_0 Y_t + \gamma_0 \delta_1 Y_{t-1} + \gamma_0 \delta_2 Y_{t-2} + \ldots + (\gamma_0 v_t + u_t), \text{ or}$$

$$C_t = \alpha + \beta_0 Y_t + \beta_1 Y_{t-1} + \beta_2 Y_{t-2} + \ldots + \epsilon_t \tag{4.13}$$

where the substitution is obvious.

Although equation (4.13) states that the current level of consumption depends on current income and past levels of income, other models, referring to the so called 'habit persistence', postulate that the current level of consumption depends on past levels of consumption as well, as it is seen in the following equation.

$$C_t = \alpha + \beta_0 Y_t + \beta_1 Y_{t-1} + \beta_2 Y_{t-2} + \ldots + \alpha_1 C_{t-1} + \alpha_2 C_{t-2} + \ldots + \epsilon_t \tag{4.14}$$

Case 4.2 The accelerator model of investment

The accelerator model of investment in its simplest form asserts that there exists a fixed relationship between net investment and change in output. This relationship is written as

$$I_t = \beta_0 X_t + \epsilon_t \tag{4.15}$$

where I_t = net investment, $X_t = Q_t - Q_{t-1}$ with Q_t = output, and $\beta_0 > 0$ parameter.

This model predicts that investment will fluctuate considerably. Investment will be positive when the economy is in a recovery $(Q_t - Q_{t-1} > 0)$ and investment will be negative when the economy is in a recession $(Q_t - Q_{t-1} < 0)$. However, the fluctuation of investment depends on other factors such as the timing of investment decisions.

By assuming that investment decisions can be delayed for various reasons, the response of investment to change in output will spread out over time. In this case equation (4.15) is written as

$$I_t = \beta_0 X_t + \beta_1 X_{t-1} + \beta_2 X_{t-2} + \ldots + \beta_k X_{t-k} + \epsilon_t \tag{4.16}$$

Case 4.3 The quantity theory of money

The quantity theory of money asserts that the price level in an economy is proportional to the quantity of money of this economy. This can be derived from Irving Fisher's (1867–1947) equation:

$$M_t V_t = P_t Q_t \tag{4.17}$$

where M_t = nominal money stock, V_t = velocity of money, P_t = overall price level, and Q_t = real output, which can be written in natural logarithms as:

$$\ln M_t + \ln V_t = \ln P_t + \ln Q_t \tag{4.18}$$

By differentiating equation (4.18) with respect to time we get

$$m_t + v_t = p_t + q_t \tag{4.19}$$

where $x_t = \dfrac{d}{dt} \ln X_t = \dfrac{1}{X_t} \dfrac{dX_t}{dt}$, which in discrete time it corresponds to percentage

changes over time.

By rewriting equation (4.19) we get the following equation which decomposes inflation into three sources; nominal money changes, real output changes, and money velocity changes.

$$p_t = m_t - q_t + v_t \tag{4.20}$$

Monetarists argue that money changes is the major factor in determining inflation, treating the other two factors as being negligible. In cases when the combined effect of output and velocity changes is positive, zero, or negative inflation is less, equal, or higher than money changes respectively. Therefore, according to the monetarists view the 'money–inflation' link is written:

$$p_t = \alpha + \beta_0 m_t + \epsilon_t \tag{4.21}$$

Although equation (4.21) assumes that the change of money stock affects inflation instantaneously, in reality the response of inflation to changes in money stock is spread over time, and thus equation (4.21) is written as:

$$p_t = \alpha + \beta_0 m_t + \beta_1 m_{t-1} + \beta_2 m_{t-2} + \ldots + \beta_k m_{t-k} + \epsilon_t \tag{4.22}$$

Case 4.4 The Phillips Curve

The original Phillips (1958) Curve describes an empirical relationship between the rate of change in money wages and the rate of unemployment as a percentage of labour force. The higher the rate of unemployment, the lower the rate of change in money wages. This relationship is written as

$$w_t = \alpha U_t^\beta e^{\epsilon_t} \tag{4.23}$$

where $w_t = (W_t - W_{t-1})/W_{t-1}$ with $W_t =$ monetary wage level, $U_t =$ unemployment rate (percent), $\epsilon_t =$ disturbance term, $e =$ base of natural logarithm, and $\alpha_t > 0$ and $\beta < 0$ parameters.

Assuming that wages depend also on past prices, and allowing for some time adjustment between unemployment and wages, the wage determination equation (4.23) has been augmented to

$$w_t = \alpha U_t^{\beta_0} \ldots U_{t-m}^{\beta_m} p_{t-1}^{\gamma_1} \ldots p_{t-k}^{\gamma k} e^{\epsilon_t} \tag{4.24}$$

where $p_t = (P_t - P_{t-1})/P_{t-1}$ with P_t = overall price level, and $\beta_j < 0$ and $\gamma_h > 0$ parameters.

In all these cases we saw that the dependent variable of the economic phenomenon (function) was not only determined by variables of the current time point with the dependent variable (instantaneous adjustment) but from variables of past time points (continuous or dynamic adjustment) as well. The difference between the current time point and a past time point is called 'time lag', or simply 'lag', and the corresponding variable is called 'lagged variable'.

3 THE DEFINITIONS

Regression models including lagged variables can be distinguished as the following types:

Distributed lag models   In these models the explanatory variables include only current and lagged values of the independent variables. For example, if we consider one dependent and one independent variable the model is of the form:

$$Y_t = \alpha + \beta_0 X_t + \beta_1 X_{t-1} + \ldots + \beta_k X_{t-k} + \epsilon_t \tag{4.25}$$

Equations (4.13), (4.16), (4.22) and (4.24) are examples of distributed-lag models.

Autoregressive, or dynamic, models   In these models the explanatory variables include one or more lagged values of the dependent variable. For example:

$$Y_t = \alpha + \beta X_t + \gamma_1 Y_{t-1} + \gamma_2 Y_{t-2} + \epsilon_t \tag{4.26}$$

Equation (4.14) is an example of a (distributed lag) autoregressive model.

In distributed lag models the influence of the independent variable X_t on $E(Y_t)$ is distributed over a number, k, of past (lagged) values of X_t. According to the value of this number k, these models are distinguished into the following types:

Infinite lag models   This is the case when k is infinite. The model is written:

$$Y_t = \alpha + \sum_{i=0}^{k=\infty} \beta_i X_{t-i} + \epsilon_t \tag{4.27}$$

Finite lag models   This is the case when k is finite. The model is written:

$$Y_t = \alpha + \sum_{i=0}^{k<\infty} \beta_i X_{t-i} + \epsilon_t \tag{4.28}$$

In both cases, infinite or finite lag models, to avoid cases of explosive values of $E(Y_t)$, we make the assumption that the sum of the β_i coefficients is finite, i.e.:

$$\beta_0 + \beta_1 + \beta_2 + \ldots + \beta_k = \sum_{i=0}^{k} \beta_i = \beta < \infty \tag{4.29}$$

Let us explain the regression coefficients of a distributed lag model, say model (4.25). In this model, under the assumption of *ceteris paribus*, if the independent variable X_t is increased by one unit in period t, the impact of this change on $E(Y_t)$ will be β_0 in time t, β_1 in time $t+1$, β_2 in time $t+2$, and so on. We define this impact as follows:

Partial multipliers of order i   It is the marginal effect of X_{t-i} on Y_t; i.e. it is equal to $\partial Y_t / \partial X_{t-i} = \beta_i$. In other words, these multipliers show the effect on $E(Y_t)$ of a unit increase in X_t made i periods prior to period t.

Short-run, or impact multiplier   It is the partial multiplier of order $i=0$, i.e. it is equal to β_0. In words, this multiplier shows the effect on $E(Y_t)$ of a unit increase in X_t made in the same period t.

Interim, or intermediate, multipliers of order i   It is the sum of the first i partial multipliers, i.e. it is equal to $\beta_0 + \beta_1 + \ldots + \beta_i$. In other words, these multipliers show the effect on $E(Y_t)$ of a maintained unit increase in X_t for i periods prior to period t.

Long-run, or total, or equilibrium, multiplier   It is the sum of all partial multipliers of the distributed lag model, i.e. it is equal to β as defined in (4.29). In words, this multiplier shows the effect on $E(Y_t)$ of a maintained unit increase in X_t for all periods.

Since the partial multipliers are actually the corresponding partial regression coefficients, these coefficients depend on the units of measurement of the independent variable X_t. A method for expressing these coefficients of free of units of measurement values is their transformation into the following:

Standardised coefficients, or lag weights   They are the coefficients that are derived from the transformation,

$$
w_i = \frac{\beta_i}{\sum\limits_{i=0}^{k} \beta_i} = \frac{\beta_i}{\beta}, \text{ for } i = 0, 1, 2, \ldots, k \tag{4.30}
$$

and they show the proportion of the equilibrium multiplier realised by a specific time period.

Inserting (4.30) into (4.27) or (4.28) these models are written as

$$
Y_t = \alpha + \beta \sum_{i=0}^{k} w_i X_{t-i} + \epsilon_t \tag{4.31}
$$

Having defined the lag weights the following statistics that characterise the nature of the lags distribution can be also defined:

Mean, or average, lag   It is the weighted average of all lags involved, with the weights been the standardised coefficients,

$$\bar{w} = \sum_{i=1}^{k} i \cdot w_i \tag{4.32}$$

and it shows the average speed with which Y_t responds to a unit sustained change in X_t, provided that all regression coefficients are positive.

Median lag   It is the statistic that shows the time required for the 50 percent of the total change in Y_t is realised after a unit sustained change in X_t, and is given by

$$\text{median lag} = w, \text{ such that } \sum_{i=0}^{w-1} w_i = 0.50 \tag{4.33}$$

4 ESTIMATION OF DISTRIBUTED LAG MODELS

Let us consider for convenience the one dependent and one independent variable distributed lag model,

$$Y_t = \alpha + \sum_{i=0}^{k} \beta_i X_{t-i} + \epsilon_t \tag{4.34}$$

The approaches for estimating this model are usually grouped into the following two categories:

Unrestricted approaches   These approaches refer to the case where the lag length is

Table 4.1 GNP price deflator and nominal money stock for Greece

Year	Deflator 1970:100	Money (billion Drs)	Year	Deflator 1970:100	Money (billion Drs)
1960	73.6	13.8	1980	364.4	267.9
1961	74.8	16.1	1981	435.5	335.3
1962	78.1	18.5	1982	544.0	389.7
1963	79.2	21.3	1983	646.1	449.4
1964	82.0	25.4	1984	775.6	542.5
1965	85.3	28.4	1985	912.6	675.1
1966	89.4	32.1	1986	1073.9	738.8
1967	91.5	39.5	1987	1227.7	843.4
1968	93.2	40.5	1988	1419.4	973.2
1969	96.2	43.7	1989	1595.2	1265.0
1970	100.0	48.7	1990	1920.4	1583.5
1971	103.2	54.5	1991	2265.8	1742.9
1972	108.5	67.2	1992	2599.5	1968.2
1973	129.5	83.9	1993	2958.5	2223.7
1974	157.6	100.2	1994	3272.3	2793.5
1975	177.3	115.7	1995	3600.5	3149.0
1976	204.5	140.2			
1977	230.5	165.6			
1978	260.3	202.9			
1979	309.3	235.4			

Sources: Epilogi (1998). The Greek Economy in Figures.

finite and no specific restrictions about the nature of the lag pattern are imposed on the β coefficients of the model. We can distinguish two cases; known lag length, and unknown lag length.

Restricted approaches These approaches refer to the case where specific restrictions about the nature of the lag pattern are imposed on the β coefficients of the model. We can distinguish two cases; finite lag length, and infinite lag length.

4.1 Unrestricted estimation of distributed lag models when the lag length is known

According to this approach we consider that the model satisfies the classical assumptions presented in section 1. In principle, the ordinary least squares method can be applied to equation (4.34) and the corresponding estimator is the best linear unbiased estimator of the β coefficients of the model.

Example 4.1 The quantity theory of money for Greece, 1960–1995 (Known lag length)

Suppose that the response of inflation to changes in money stock in Greece is spread over three years. Therefore, equation (4.22) of the quantity theory of money is written:

$$p_t = \alpha + \beta_0 m_t + \beta_1 m_{t-1} + \beta_2 m_{t-2} + \beta_3 m_{t-3} + \epsilon_t \qquad (4.35)$$

Table 4.1 presents annually data on the implicit price deflator for GNP (P) and nominal money stock (M1) from 1960 to 1995. By computing first the rate of inflation as $p_t = (P_t - P_{t-1})/P_{t-1}$ and the rate of money change as $m_t = (M1_t - M1_{t-1})/M1_{t-1}$, OLS produces the following results. (The values in the parentheses below the estimated coefficients are asymptotic t ratios.)

$$\hat{p}_t = \quad -0.153 + 0.375m_t + 0.549m_{t-1} + 0.343m_{t-2} + 0.390m_{t-3} \qquad (4.36)$$
$$\quad (3.15) \quad (2.72) \qquad (3.92) \qquad (2.36) \qquad (2.68)$$

$$\bar{R}^2 = 0.5122 \; DW = 0.7980 \; F = 9.1381$$

where $\bar{R}^2$ = adjusted for degrees of freedom determination coefficient, DW = Durbin–Watson d statistic, and F = F-statistic.

Apart from the low Durbin–Watson statistic which suggests some autocorrelation, the results in (4.36) suggest that the response of inflation to changes in money stock in Greece is significantly spread over three years.

4.2 Unrestricted estimation of distributed lag models when the lag length is unknown

In most cases the true lag length k of model (4.34) is rarely known and thus it must be determined. Although there are various procedures for determining the lag length of a model, the approach adopted in this section is based on searches for model specification.

These approaches choose to optimise a criterion. Among the most common criteria used may be the following:

Maximise adjusted R^2: $\bar{R}^2 = 1 - \dfrac{n-1}{n-q}(1-R^2)$ (4.37)

Minimise Akaike's (1973) information criterion:

$$AIC = \ln\left(\frac{SSR}{n}\right) + \frac{2q}{n}$$ (4.38)

Minimise Schwartz (1978) criterion:

$$SC = \ln\left(\frac{SSR}{n}\right) + \frac{q}{n}\ln(n)$$ (4.39)

where n = sample size, q = total number of coefficients in the regression model, and SSR = sum of squared residuals.

Although these criteria reward good fit but place a penalty for extra coefficients to be included in the model, it is very possible conclusions based on these (and other) criteria to be different. A model could be ranked superior under one criterion and inferior under another.

Example 4.2 The quantity theory of money for Greece, 1960–1995 (Unknown lag length)

Suppose that the spread of the response of inflation to changes in money stock in Greece is not known. In this case we will try to estimate the lag length of the distributed lag model (4.34) by optimising the three criteria (4.37) to (4.39).

Table 4.2 presents the values of the three criteria by applying OLS to equation (4.34) for various lag lengths. According to these values it looks that criteria $\bar{R}^2$ and AIC favour the 5-period lag length, whilst SC favours the 4-period lag length. However the differences between these values are very small, thus, it looks that a 5-period lag length is acceptable for the distributed lag model. The estimation of this model is presented below:

$\hat{p}_t = \ -0.149 + 0.361m_t + 0.554m_{t-1} + 0.342m_{t-2} + 0.340m_{t-3} + 0.144m_{t-4} - 0.079m_{t-5}$
$\quad\quad\ (2.61)\ \ (2.71)\quad\quad (4.16)\quad\quad (2.45)\quad\quad (2.38)\quad\quad (1.03)\quad\quad (0.55)$

$\bar{R}^2 = 0.5369$ DW = 0.8335 F = 6.6041 (4.40)

Although the various criteria indicate that the 'best' model is that with a 5-period distributed lag pattern, the actual estimation of this model in (4.40) shows that as the lag length increases, the t-ratios of the estimated coefficients

Table 4.2 Estimation results for inflation – money growth equation for Greece

Lag length	$\bar{R}^2$	AIC	SC
0	0.0681	−5.3096	−5.2198
1	0.2797	−5.5748	−5.4401
2	0.3913	−5.7241	−5.5427
3	0.5122	−5.9712	−5.7422
4	0.5368	−6.0256	−5.7481*
5	0.5369*	−6.0308*	−5.7038
6	0.5174	−5.9919	−5.6147

Notes: Estimates from Eviews.
* Denotes 'best' model for the criterion.

corresponding to the lag variables decreases, indicating that these coefficients are statistically insignificant. This may possibly be due to the fact that as the lag length increases, the degrees of freedom decreases, and the introduction of more lagged variables will possibly introduce some multicollinearity between the independent variables.

Taking into consideration that most time series data are short, two of the most serious problems arise in estimation by the introduction of high lag lengths:

1 Low degrees of freedom: The higher the lag length, the lower the degrees of freedom. Lower degrees of freedom imply lower precision (lower efficiency) of the estimates, and therefore lower precision of the tests of hypotheses.
2 Multicollinearity: The higher the lag length, the higher the chance of successive lagged variables to be correlated. Multicollinearity leads to lower precision of the estimates (higher standard errors) and therefore lower precision of the tests of hypotheses.

Summarising, the introduction of higher lag lengths may erroneously lead to the rejection of estimated coefficients, being statistically insignificant, due to the lower precision of the tests of hypotheses. In order to bypass this problem of lower precision of estimates specific restrictions about the nature of the lag pattern must be imposed on the β coefficients of the distributed lag model. These specific restrictions are discussed in the following sections.

5 RESTRICTED ESTIMATION OF FINITE DISTRIBUTED LAG MODELS

Two of the most common methods for imposing a lag scheme on the β coefficients of a finite distributed lag model; are the arbitrary weights distributed lag schemes and the polynomial distributed lag schemes.

5.1 Arbitrary weights distributed lag schemes

This approach assigns arbitrary weights to each lagged variable according to a perceived influence of the lagged variable in determining the dependent variable. Various schemes can be used in this assignment exercise. The following are the most commonly used:

Arithmetic lag scheme

This is the case (Fisher, 1937) when the weights linearly decrease according to the scheme:

$$\beta_i = \begin{cases} (k+1-i)\beta & \text{for } i = 0, 1, 2, \ldots, k \\ 0 & \text{for } i > k \end{cases} \tag{4.41}$$

The rationale of this scheme is that more recent values of the independent variable have greater influence on the dependent variable than more remote values. Similarly, an increasing arithmetic scheme can be constructed. By inserting (4.41) into (4.34) the distributed lag model is written:

$$Y_t = \alpha + \beta \sum_{i=0}^{k} (k+1-i)X_{t-i} + \epsilon_t = \alpha + \beta Z_t + \epsilon_t \tag{4.42}$$

where the substitution for Z_t is obvious. By applying OLS in (4.42) the estimate b of β is obtained and therefore, the estimates $b_i = (k+1-i)$ b, for $i = 0, 1, 2, \ldots, k$, for the parameters β_i can be correspondingly calculated.

Inverted V lag scheme

According to this case (DeLeeuw, 1962) the weights linearly increase for most recent lags and then they decrease for more remote lags. The scheme, for k being even, is the following:

$$\beta_i = \begin{cases} (1+i)\beta & \text{for } 0 \leq i \leq k/2 \\ (k+1-i)\beta & \text{for } k/2 + 1 \leq i \leq k \\ 0 & \text{for } i > k \end{cases} \tag{4.43}$$

By inserting (4.43) into (4.34) the distributed lag model is written

$$Y_t = \alpha + \beta \left[\sum_{i=0}^{k/2} (1+i)X_{t-i} + \sum_{i=k/2+1}^{k} (k+1-i)X_{t-i} \right] + \epsilon_t = \alpha + \beta Z_t + \epsilon_t \tag{4.44}$$

where the substitution for Z_t is obvious. By applying OLS in (4.44) the estimate b of β is obtained and therefore, the estimates $b_i = (1+i)$ b, for $i = 0, 1, 2, \ldots, k/2$, and $b_i = (k+1-i)$ b, for $i = k/2+1, \ldots, k$, for the parameters β_i can be correspondingly calculated.

The approach of arbitrarily assigning weights suffers from the following limitations:

1 The actual scheme of the lag structure must be known, following a theoretical basis, i.e. if it increases, decreases, inverted V scheme, etc.
2 The mechanism for assigning specific weights to the lags must be also known, according to previous information, i.e. if the weights come from a linear, exponential, etc. mechanism.

The advantages of this approach might be the following:

1 The actual regression is a simple two variable regression.
2 The 'best' lag scheme and weights mechanism is determined by simply optimising statistical criteria, such as those of (4.37) to (4.39).

Example 4.3 The quantity theory of money for Greece, 1960–1995 (arbitrary weights)

Using the data of Table 4.1 the estimates for each arbitrary weights distributed lag scheme, say for a six period distributed lag equation, are the following:

Decreasing arithmetic lag scheme

Applying (4.41) the Z_t variable is:

$$z_t = 7m_t + 6m_{t-1} + 5m_{t-2} + 4m_{t-3} + 3m_{t-4} + 2m_{t-5} + m_{t-6}$$

The corresponding to (4.42) OLS estimates are the following:

$$\hat{p}_t = \ -0.169 + 0.064z_t \tag{4.45}$$
$$\phantom{\hat{p}_t = \ } (3.07) \quad (5.63)$$

$$\bar{R}^2 = 0.5226 \ DW = 1.1313 \ AIC = -6.1650 \ SC = -6.0707$$

Increasing arithmetic lag scheme

Applying (4.41) in an increasing order the Z_t variable is:

$$z_t = m_t + 2m_{t-1} + 3m_{t-2} + 4m_{t-3} + 5m_{t-4} + 6m_{t-5} + 7m_{t-6}$$

The corresponding to (4.42) OLS estimates are the following:

$$\hat{p}_t = \ -0.024 + 0.034z_t \tag{4.46}$$
$$\phantom{\hat{p}_t = \ } (0.32) \quad (2.18)$$

$$\bar{R}^2 = 0.1177 \ DW = 0.6380 \ AIC = -5.5509 \ SC = -5.4566$$

Inverted V lag-scheme

Applying (4.43) the Z_t variable is:

$$z_t = m_t + 2m_{t-1} + 3m_{t-2} + 4m_{t-3} + 3m_{t-4} + 2m_{t-5} + m_{t-6}$$

The corresponding to (4.44) OLS estimates are the following:

$$\hat{p}_t = \quad -0.114 + 0.092z_t \tag{4.47}$$
$$\quad\quad\quad (1.69)\quad (3.78)$$

$$\bar{R}^2 = 0.3215 \; DW = 0.7663 \; AIC = -5.8136 \; SC = -5.7193$$

Although all three estimates suffer from a considerable autocorrelation, the results of the decreasing arithmetic lag scheme (4.45) are preferable, according to all statistical criteria. Therefore, by applying the formula (4.41), and considering that

$$var(b_i) = (k + 1 - i)^2 var(b), \text{ for } i = 0, 1, 2, \ldots, k \tag{4.48}$$

the finally estimated model is the following:

$$\hat{p}_t = -0.169 + 0.448m_t + 0.384m_{t-1} + 0.320m_{t-2} + 0.256m_{t-3} + 0.192m_{t-4}$$
$$+ 0.128m_{t-5} + 0.064m_{t-6} \tag{4.49}$$

5.2 Polynomial distributed lag schemes

Given the lag length, k, the lag schemes in (4.41) and (4.43) assume that the β_i coefficients follow a specific function of the lag index i. However, this constitutes the major limitation of the arbitrary weights distributed lag schemes. The a priori restriction of the coefficients to follow a specific distribution function may be too arbitrary.

Shirley Almon (1965) proposed a method that generalises the arbitrary weights distributed lag schemes by relaxing the specific distribution function. Taking into consideration that a continuous function can generally be approximated by a polynomial, her method suggests that a true smooth distribution of the β_i coefficients can be traced and approximated by a polynomial, $\beta_i = f(i)$, of the lag index i of a fairly low degree r, such as the following:

$$\beta_i = f(i) = \alpha_0 + \alpha_1 i + \alpha_2 i^2 + \ldots + \alpha_r i^r, \text{ for } i = 0, 1, 2, \ldots, k > r \tag{4.50}$$

According to (4.50), a second degree polynomial could approximate the lag structure shown in Figure 4.1(a), and a third degree polynomial could approximate the lag structure shown in Figure 4.1(b). Generally, the degree of the polynomial is greater by one from the number of turning points shown by the lag structure.

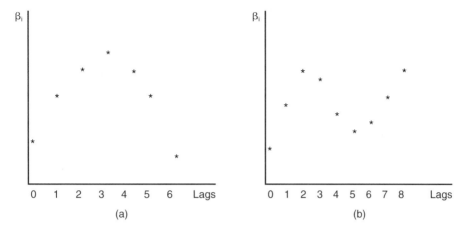

Figure 4.1 Second (a) and third (b) order polynomial approximation of distributed lag structures

Substituting (4.50) into (4.34) we obtain

$$Y_t = \alpha + \sum_{i=0}^{k} \beta_i X_{t-i} + \epsilon_t = \alpha + \sum_{i=0}^{k} (\alpha_0 + \alpha_1 i + \alpha_2 i^2 + \ldots + a_r i^r) X_{t-i} + \epsilon_t$$

or

$$Y_t = \alpha$$
$$+ \alpha_0 (X_t + X_{t-1} + X_{t-2} + X_{t-3} + \ldots + X_{t-k})$$
$$+ \alpha_1 (X_{t-1} + 2X_{t-2} + 3X_{t-3} + \ldots + kX_{t-k})$$
$$+ \alpha_2 (X_{t-1} + 2^2 X_{t-2} + 3^2 X_{t-3} + \ldots + k^2 X_{t-k})$$
$$+ \ldots$$
$$+ \alpha_r (X_{t-1} + 2^r X_{t-2} + 3^r X_{t-3} + \ldots + k^r X_{t-k})$$
$$+ \epsilon_t$$

or finally,

$$Y_t = \alpha + \alpha_0 Z_{0t} + \alpha_1 Z_{1t} + \alpha_2 Z_{2t} + \ldots + \alpha_r Z_{rt} + \epsilon_t \qquad (4.51)$$

where the substitution of the Zs is obvious. By considering that (4.51) satisfies the classical assumptions presented in section 1, OLS can be applied to this equation and the best linear unbiased estimates a and a_i of the α and α_i coefficients will correspondingly be obtained.

Having estimated with a_i the α_i coefficients, the actual estimates b_i of the β_i coefficients can be calculated by using (4.50), as follows:

$$b_i = f(i) = a_0 + a_1 i + a_2 i^2 + \ldots + a_r i^r, \text{ for } i = 0, 1, 2, \ldots, k \qquad (4.52)$$

Analytically, the estimates of the β_i coefficients are written:

$$b_0 = f(0) = a_0$$
$$b_1 = f(1) = a_0 + a_1 + a_2 + a_3 + \ldots + a_r$$
$$b_2 = f(2) = a_0 + 2a_1 + 2^2a_2 + 2^3a_3 + \ldots + 2^ra_r \tag{4.53}$$
$$b_3 = f(3) = a_0 + 3a_1 + 3^2a_2 + 3^3a_3 + \ldots + 3^ra_r$$

$$\ldots$$

$$b_k = f(k) = a_0 + ka_1 + k^2a_2 + k^3a_3 + \ldots + k^ra_r$$

The estimates in (4.52), or (4.53), are 'restricted least squares estimates' according to the Almon distributed lag approach, because there are restricted to fall on a polynomial of degree r.

Since the variances, $var(a_i)$, and the covariances, $cov(a_j, a_h)$, of the estimated coefficients a_i can be derived by applying the OLS method to equation (4.51), the variances, $var(b_i)$, of the estimated coefficients b_i, can be derived by the following formula:

$$var(b_i) = var(a_0 + a_1i + a_2i^2 + a_3i^3 + \ldots + a_ri^r)$$

$$= \sum_{j=0}^{r} i^{2j} var(a_j) + 2 \sum_{j<h} i^{(j+h)} cov(a_j, a_h) \text{ for } i = 0, 1, 2, \ldots, k \tag{4.54}$$

The last variances are used in the significance tests referring to the β_i coefficients. Therefore, it is very possible some of the β_i coefficients were significant although some of the α_i coefficients were insignificant.

Apart from the restriction of the β_i coefficients falling on the polynomial, it is common in some cases to impose extra restrictions on the coefficients. These extra restrictions, which are called 'endpoint restrictions' push the endpoint β_i coefficients to be equal to zero, i.e. $\beta_{-1} = 0$ and $\beta_{k+1} = 0$, or using (4.50) the endpoint restrictions are written:

$$\beta_{-1} = f(-1) = \alpha_0 - \alpha_1 + \alpha_2 + \ldots + (-1)^r\alpha_r = 0 \tag{4.55}$$

$$\beta_{k+1} = f(k+1) = \alpha_0 + \alpha_1(k+1) + \alpha_2(k+1)^2 + \ldots + \alpha_r(k+1)^r = 0 \tag{4.56}$$

By incorporating these restrictions (both, or either one) into equation (4.51) the OLS method on the 'endpoint restricted equation' can be applied. However, it must be noted here that the endpoint restrictions must be used with caution because these restrictions will have an impact not only on the endpoint coefficients but on all the coefficients of the model.

In the discussion up to now about the polynomial, or Almon distributed lag models, we assumed that we know both the lag length of the model and the degree of the polynomial. However, in most cases these two parameters are unknown, and thus have to be approximated.

Determining the lag length

This approach has been presented in section 4.2. In summary, searching for the lag length is a problem of testing nested hypotheses. We start by running a regression with a very large value of k. Then, by lowering the value of k by one at a time we run the corresponding regressions and seek to optimise a criterion, such as $\bar{R}^2$, AIC, or SC (see (4.37) to (4.39)).

Determining the degree of the polynomial

Similarly to determining the lag length, searching for the degree of the polynomial is a problem of testing nested hypotheses. Given the lag length k, we start by running a regression with a large value of r. Then, by lowering the value of r by one at a time we run the corresponding regressions and seek to optimise a criterion, such as $\bar{R}^2$, AIC, or SC. In practice, fairly low degrees of the polynomial (r = 2 or r = 3) give good results.

Misspecification problems

In cases when the lag length, or the degree of the polynomial, or both have been incorrectly determined specification problems arise. Although these problems depend of various specific cases they may be generally summarised as follows:

If k^* is the true lag length of the distributed lag model, then if $k > k^*$ (inclusion of irrelevant variables) the estimates are unbiased and consistent but inefficient; if $k < k^*$ (exclusion of relevant variables) the estimates are generally biased, inconsistent and inefficient.

If r^* is the true degree of the polynomial, then if $r < r^*$ (imposing invalid restrictions) the estimates are generally biased and inefficient; if $r > r^*$ (over-imposing restrictions) the estimates are unbiased but inefficient.

Example 4.4 The quantity theory of money for Greece, 1960–1995 (polynomial distributed lag model)

Using the data of Table 4.1 the steps involved in applying the polynomial distributed lag model are the following:

1 Determining the lag length

The lag length of this model has been determined in Example 4.2 to be equal to k = 5. Therefore, the distributed lag model is:

$$p_t = \alpha + \beta_0 m_t + \beta_1 m_{t-1} + \beta_2 m_{t-2} + \beta_3 m_{t-3} + \beta_4 m_{t-4} + \beta_5 m_{t-5} + \epsilon_t \qquad (4.57)$$

2 Calculating the Z variables

Using various assumed degrees of the polynomial, preferably $r \leq k$, say k = 5 in our case, calculate the corresponding Z variables, such as:

$$Z_{0t} = \sum_{i=0}^{5} m_{t-i}$$

$$Z_{1t} = \sum_{i=0}^{5} im_{t-i}$$

$$Z_{2t} = \sum_{i=0}^{5} i^2 m_{t-i} \tag{4.58}$$

$$\cdots$$

$$Z_{5t} = \sum_{i=0}^{5} i^5 m_{t-i}$$

3 Determining the degree of the polynomial

Starting from $r = 5$ and lowering this assumed degree of the polynomial by one at a time we run the corresponding regression equations, starting from

$$Y_t = \alpha + \alpha_0 Z_{0t} + \alpha_1 Z_{1t} + \alpha_2 Z_{2t} + \alpha_3 Z_{3t} + \alpha_4 Z_{4t} + \alpha_5 Z_{5t} + \epsilon_t \tag{4.59}$$

Table 4.3 presents the values of the three criteria by applying OLS to equation (4.59) for various degrees of the polynomial. All criteria agree that the degree of the polynomial is equal to 2.

Table 4.3 Determining the polynomial degree for inflation – money growth equation for Greece

Degree	$\bar{R}^2$	AIC	SC
1	0.5323	−6.1272	−5.9871
2	0.5754*	−6.1949*	−6.0081*
3	0.5604	−6.1326	−5.8991
4	0.5496	−6.0827	−5.8024
5	0.5369	−6.0308	−5.7038

Notes: Estimates from Eviews.
* Denotes 'best' model for the criterion.

4 Estimates of the α's

Having determined that the lag length is $k = 5$ and the degree of the polynomial is $r = 2$, the actual estimates are given by

$$\hat{p}_t = -0.1488 + 0.4207 Z_{0t} - 0.0681 Z_{1t} - 0.0346 Z_{2t}$$

$$\text{se} = (0.0545) \quad (0.0747) \quad (0.0345) \quad (0.0179)$$

$$\text{t} = (2.7292) \quad (5.6305) \quad (1.9705) \quad (1.9336) \tag{6.60}$$

$$\bar{R}^2 = 0.5754 \; DW = 0.9690 \; F = 14.0993 \; AIC = -6.1949 \; SC = -6.0081$$

5 Estimates of the β's

Having estimated the α coefficients as $a_0 = 0.4207$, $a_1 = -0.0681$ and $a_2 = -0.0346$, considering (4.53) the estimates of the β coefficients are the following:

$$b_0 = f(0) = a_0 = 0.4186$$

$$b_1 = f(1) = a_0 + a_1 + a_2 = 0.4542$$

$$b_2 = f(2) = a_0 + 2a_1 + 4a_2 = 0.4207$$

$$b_3 = f(3) = a_0 + 3a_1 + 9a_2 = 0.3181$$

$$b_4 = f(4) = a_0 + 4a_1 + 16a_2 = 0.1463$$

$$b_5 = f(5) = a_0 + 5a_1 + 25a_2 = -0.0946$$

6 The estimated model

The estimated distributed-lag model corresponding to (4.57), considering also the formula (4.54) for estimating the standard errors of the coefficients, is the following:

$$\hat{p}_t = -0.1488 + 0.4186m_t + 0.4542m_{t-1} + 0.4207m_{t-2} + 0.3181m_{t-3} + 0.1463m_{t-4} - 0.0946m_{t-5}$$
$$se = (0.0545) \quad (0.1056) \quad (0.0706) \quad (0.0747) \quad (0.0745) \quad (0.0715) \quad (0.1106)$$
$$t = (2.7292) \quad (3.9647) \quad (6.4370) \quad (5.6305) \quad (4.2725) \quad (2.0454) \quad (0.8556)$$

$$(6.61)$$

7 The endpoint restrictions

For comparison purposes, Table 4.4 presents the results from the unrestricted distributed lag estimation in (4.40), from the polynomial distributed lag estimation in (4.61) and for the polynomial distributed lag estimation where both endpoint restrictions (4.55) and (4.56) have been applied. Obviously other estimations could also been presented, such as polynomial distributed lag estimations with one endpoint restriction only, or with the sum of the distributed lag weights being equal to one, and so on.

Figure 4.2 shows the three estimates presented in Table 4.4. The distribution of the lag weights of the unrestricted distributed lag estimation (UNR) indicates one substantial turning point, and thus, the polynomial for the polynomial distributed lag estimation (PDL) has been determined as degree equal to two. Finally, the polynomial distributed-lag estimation with both endpoint restrictions restricts the distribution of the lag weights to a smooth symmetric quadratic curve.

Table 4.4 Various estimated models for inflation – money growth equation for Greece

Variable and Statistic	Unrestricted (UNR)	Polynomial lag (PDL)	Polynomial lag with endpoint restrictions (EPR)
Constant	−0.1487	−0.1488	−0.1548
m_t	0.3612	0.4186	0.1818
m_{t-1}	0.5539	0.4542	0.3030
m_{t-2}	0.3425	0.4207	0.3636
m_{t-3}	0.3396	0.3181	0.3636
m_{t-4}	0.1438	0.1463	0.3030
m_{t-5}	−0.0790	−0.0946	0.1818
Adjust. R^2	0.5369	0.5754	0.4312
DW	0.8335	0.9690	0.8489
F	6.6041	14.0993	22.9862
AIC	−6.0308	−6.1949	−5.9618
SC	−5.7038	−6.0081	−5.8684

Notes: Estimates from Eviews.
For brevity, standard errors or t ratios are not reported.

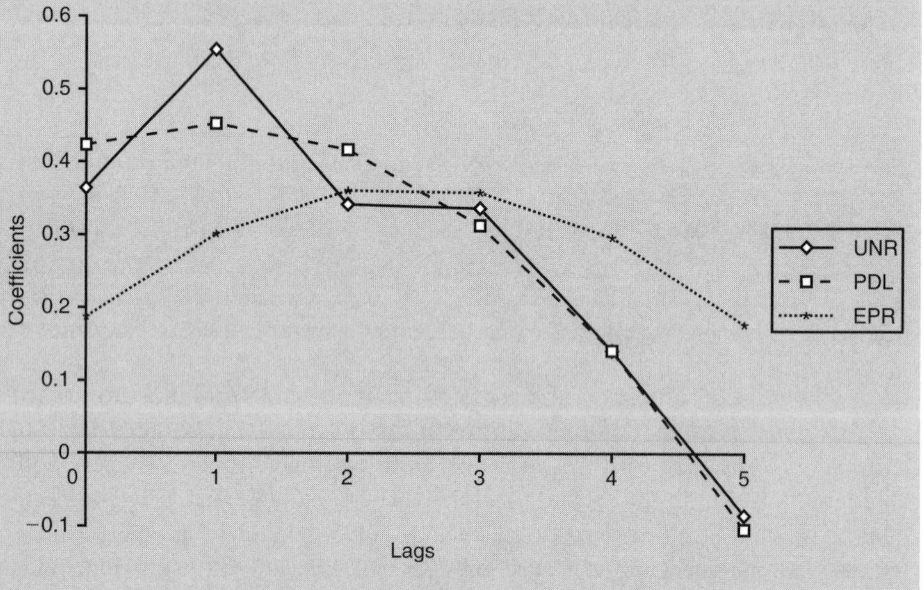

Figure 4.2 Distributed lag estimates for inflation – money growth equation for Greece

6 RESTRICTED ESTIMATION OF INFINITE DISTRIBUTED LAG MODELS

The major assumption in the finite distributed lag models is that the influence of the explanatory variable X_t on the dependent variable Y_t is negligible after a specific point in time and thus it should be treated as being zero. However, we saw in the previous section the problems in the estimation of the model which arise through by the wrong determination of that specific point in time, or the lag length, after which any influence is negligible. Of course, these problems will disappear if we accept that the influence of the explanatory variable X_t on the dependent variable Y_t after a specific point in time, in spite of being small, is still present in the model. The result of this consideration is the infinite distributed lag model met in (4.27), or the model

$$Y_t = \alpha + \sum_{i=0}^{\infty} \beta_i X_{t-i} + \epsilon_i \qquad (4.62)$$

In introducing model (4.62), although there is no problem in determining the lag length, there is now a new problem; the problem of estimating an infinite number of parameters, the β_i, using a finite number of observations. Therefore, methods should be employed in order to lower from infinite to finite the number of estimable parameters. In this section we will present several such methods.

6.1 The geometric, or, Koyck's, distributed lag model

This model, introduced by Koyck (1954), is possibly the most popular distributed lag model in empirical research. Koyck proposed that the influence of X_t on Y_t declines continuously with the lag coefficients (weights) following the geometric scheme

$$\beta_i = \beta_0 \lambda^i, 0 < \lambda < 1, \text{ for } i = 0, 1, 2, \ldots \qquad (4.63)$$

With the λ parameter, which is known as the 'rate of change', being positive and less than one it is guaranteed that the values of the β_i's corresponding to greater lags will be smaller than those corresponding to smaller lags, with infinite time lag weight tends to be zero. Taking into account that β_0 is common to all coefficients, the declining nature of the geometric scheme toward zero can be seen numerically and graphically in Table 4.5 and Figure 4.3 respectively. It can also be seen that the greater the value of λ the slower the decline of the series.

Substituting (4.63) into (4.62) we get

$$Y_t = \alpha + \beta_0 \sum_{i=0}^{\infty} \lambda^i X_{t-i} + \epsilon_i$$

or

$$Y_t = \alpha + \beta_0 X_t + \beta_0 \lambda X_{t-1} + \beta_0 \lambda^2 X_{t-2} + \beta_0 \lambda^3 X_{t-3} + \ldots + \epsilon_t \qquad (4.64)$$

By lagging (4.64) by one period we get

$$Y_{t-1} = \alpha + \beta_0 X_{t-1} + \beta_0 \lambda X_{t-2} + \beta_0 \lambda^2 X_{t-3} + \beta_0 \lambda^3 X_{t-4} + \ldots + \epsilon_{t-1} \qquad (4.65)$$

Table 4.5 Geometric distribution: successive values of λ^i

$Lag \rightarrow$ $\lambda \downarrow$	0	1	2	3	4	5	6	7	8	9	10
0.25	1	0.250	0.063	0.016	0.004	0.001	0.000	0.000	0.000	0.000	0.000
0.50	1	0.050	0.250	0.125	0.063	0.031	0.016	0.008	0.004	0.002	0.001
0.75	1	0.750	0.563	0.422	0.316	0.237	0.178	0.133	0.100	0.075	0.056

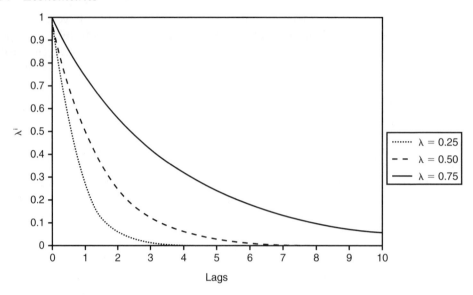

Figure 4.3 Geometric distribution: successive values of λ^i

Multiplying both terms of (4.65) by λ and subtracting the result from (4.64) we obtain

$$Y_t - \lambda Y_{t-1} = \alpha(1 - \lambda) + \beta_0 X_t + \epsilon_t - \lambda\epsilon_{t-1} \tag{4.66}$$

Finally, by rearranging (4.66) we get

$$Y_t = \alpha_0 + \beta_0 X_t + \lambda Y_{t-1} + \upsilon_t \tag{4.67}$$

where $\alpha_0 = \alpha(1 - \lambda)$ and $\upsilon_t = \epsilon_t - \lambda\epsilon_{t-1}$.

Ignoring here all the possible problems that the estimation of (4.67) creates, we could say that Koyck by introducing the geometric scheme (4.63) into model (4.62), managed to reduce the infinite number of coefficients of model (4.62) to three; α_0, β_0 and λ.

Model (4.67), by including the lagged dependent variable Y_{t-1} among its explanatory variables, is an autoregressive, or dynamic, model. Therefore, we see that the application of the geometric scheme to model (4.62) transformed this infinite distributed lag model into an autoregressive model. Furthermore, the disturbance term ϵ_t of model (4.62) has been transformed into the disturbance term $\upsilon_t = \epsilon_t - \lambda\epsilon_{t-1}$ of model (4.67), which is known as a 'moving average' of the first order.

The properties of the geometric distributed lag model (4.64) can be summarised as follows:

Mean lag

Using formulas (4.30) and (4.32) the mean lag is equal to:

$$\text{mean lag} = \bar{w} = \frac{\sum\limits_{i=0}^{\infty} i\beta_i}{\sum\limits_{i=0}^{\infty} \beta_i} = \frac{\beta_0(0 + \lambda + 2\lambda^2 + 3\lambda^3 + \ldots)}{\beta_0(1 + \lambda + \lambda^2 + \lambda^3 + \ldots)} = \frac{\lambda/(1-\lambda)^2}{1/(1-\lambda)} = \frac{\lambda}{1-\lambda} \quad (4.68)$$

Median lag

Using formula (4.33), and taking into account that

$$\sum_{i=0}^{k} \lambda^i = \frac{1 - \lambda^{k+1}}{1 - \lambda}$$

the median lag is equal to:

$$\text{median lag} = w = \frac{\ln(0.5)}{\ln(\lambda)} \quad (4.69)$$

Impact multiplier

This is equal to β_0.

Equilibrium multiplier

By using formula (4.29) the long-run multiplier is equal to:

$$\beta = \sum_{i=0}^{\infty} \beta_i = \beta_0(1 + \lambda + \lambda^2 + \lambda^3 + \ldots) = \beta_0\left(\frac{1}{1-\lambda}\right) \quad (4.70)$$

This is why some authors instead of the geometric scheme (4.63) use the following scheme:

$$\beta_i = \beta(1-\lambda)\lambda^i, 0 < \lambda < 1, \text{for } i = 0, 1, 2,\ldots \quad (4.71)$$

6.2 The Pascal distributed lag models

One characteristic of the geometric distributed lag model is that its lag weights continuously decrease with time. We also met this characteristic in the decreasing arithmetic lag schemes. To avoid the continuously decreasing lag weights Solow (1960) proposed the Pascal lag scheme which resembles the inverted V lag scheme. This lag scheme is the following:

$$\beta_i = \beta w_i, \text{for } i = 0, 1, 2,\ldots \quad (4.72)$$

where, for $i = 0, 1, 2,\ldots$ the weights are given by

$$w_i = \binom{i+r-1}{i}(1-\lambda)^r\lambda^i = \frac{(i+r-1)!}{i!(r-1)!}(1-\lambda)^r\lambda^i, r = \text{positive integer}, 0 < \lambda < 1 \quad (4.73)$$

Substituting (4.72) into (4.62) we get the Pascal distributed lag model

$$Y_t = \alpha + \beta(1-\lambda)^r \left[X_t + r\lambda X_{t-1} + \frac{r(r+1)}{2!} \lambda^2 X_{t-2} + \ldots \right] + \epsilon_t \qquad (4.74)$$

This model has four unknown parameters; α, β, λ, and r. In the case when $r = 1$, formula (4.73) is written $w_i = (1-\lambda)\lambda^i$ and thus, the Pascal distributed lag model is reduced to the geometric distributed lag model. This means that by giving various values for the r parameter we can shape the distribution of the lag weights in a way which seems more suitable for the case. Figure 4.4 shows the distribution of these weights for $\lambda = 0.4$ and $r = 1, 3, 5$ respectively.

In the case that $r = 2$ (4.74) is written

$$Y_t = \alpha + \beta(1-\lambda)^2 (X_t + 2\lambda X_{t-1} + 3\lambda^2 X_{t-2} + \ldots) + \epsilon_t \qquad (4.75)$$

Firstly, by multiplying with -2λ both terms of the once-lagged equation (4.75), secondly, by multiplying with λ^2 both terms of the twice-lagged equation (4.75), thirdly, by adding these two results to equation (4.75), and finally by rearranging we get

$$Y_t = \alpha(1-\lambda)^2 + \beta(1-\lambda)^2 X_t + 2\lambda Y_{t-1} - \lambda^2 Y_{t-2} + (\epsilon_t - 2\lambda\epsilon_{t-1} + \lambda^2\epsilon_{t-2}) \qquad (4.76)$$

Ignoring here all the possible problems that the estimation of (4.76) creates, including the problem of over-identification of the parameters, we could say that by introducing the Pascal scheme with $r = 2$ into model (4.62), this infinite distributed lag model has been transformed into an autoregressive model, where three parameters only have to be estimated; namely, α, β and λ.

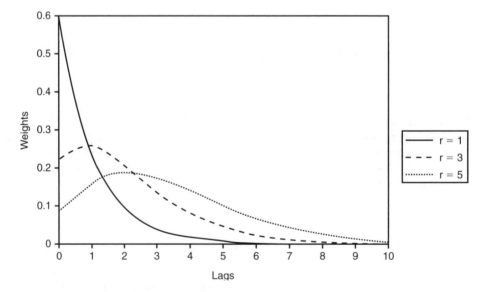

Figure 4.4 Pascal distribution of weights

Although not very practical for empirical research, generalising (4.76) to include any positive integer value of r, we get

$$Y_t = \alpha(1-\lambda)^r + \beta(1-\lambda)^r X_t - \sum_{i=1}^{r} \binom{r}{i}(-\lambda)^i Y_{t-i} + \sum_{i=0}^{r} \binom{r}{i}(-\lambda)^i \epsilon_{t-i} \tag{4.77}$$

6.3 The lag operator and the rational distributed lag models

The manipulations of complicate distributed lag models can be simplified by introducing the lag operator, L. This operator is defined by $Lx_t = x_{t-1}$.

Some useful algebraic operations with the lag operator are the following:

$$L(Lx_t) = L^2 x_t = x_{t-2}, L^p x_t = x_{t-p}, L^q(L^p x_t) = L^{p+q} x_t = x_{t-p-q}, L^0 x_t = 1x_t = x_t,$$

$$L^p(ax_t) = aL^p x_t = ax_{t-p} \text{ where } a = \text{constant}$$

Under the lag operator, the infinite distributed lag model (4.62) can be written

$$Y_t = \alpha + \sum_{i=0}^{\infty} \beta_i X_{t-i} + \epsilon_t = \alpha + \sum_{i=0}^{\infty} \beta_i L^i X_t + \epsilon_t = \alpha + \beta(L) X_t + \epsilon_t \tag{4.78}$$

where

$$\beta(L) = \sum_{i=0}^{\infty} \beta_i L^i = \beta_0 + \beta_1 L + \beta_2 L^2 + \beta_3 L^3 + \dots \tag{4.79}$$

is a polynomial in L.

Jorgenson (1963) approximated the infinite polynomial (4.79) by the ratio of two finite polynomials in L, as

$$\beta(L) = \frac{\gamma(L)}{\delta(L)} = \frac{\gamma_0 + \gamma_1 L + \gamma_2 L^2 + \dots + \gamma_p L^p}{\delta_0 + \delta_1 L + \delta_2 L^2 + \dots + \delta_q L^q} \tag{4.80}$$

which is known as the 'rational lag function'.

Substituting (4.80) into (4.78), putting $\delta_0 = 1$ for normalisation, and after rearranging we get the finite distributed lag autoregressive model

$$Y_t = \alpha_0 + \gamma_0 X_t + \gamma_1 X_{t-1} + \dots + \gamma_p X_{t-p} - \delta_1 Y_{t-1} - \dots - \delta_q Y_{t-q} + \upsilon_t \tag{4.81}$$

where

$$\alpha_0 = \alpha\delta(L) = \alpha(1 + \delta_1 + \dots + \delta_q) \text{ and } \upsilon_t = \delta(L)\epsilon_t = \epsilon_t + \delta_1\epsilon_{t-1} + \dots + \delta_q\epsilon_{t-q} \tag{4.82}$$

Equation (4.81) is the generalisation of all the distributed lag models we have seen till now. If for example, $\gamma(L) = \gamma_0$ and $\delta(L) = 1 - \lambda L$, then (4.81) becomes the geometric distributed lag model, and if $\gamma(L) = \gamma(1-\lambda)^r$ and $\delta(L) = (1-\lambda L)^r$, then (4.81) becomes the Pascal distributed lag model.

Chapter 4: Key terms

Distributed lag models   Models in which the impact of regressor/s on the dependent variable are distributed over time.

Infinite (D-L) models   The impact of regressor/s on the dependent variable is distributed over infinite time periods.

Autoregressive dynamic models   Lagged value/s of the dependent variable appears as a regressor/s.

The Almon's polynomial model   The influences of the regressors on the dependent variable are constrained to line on polynomials.

The Koyck D-L model   The impact of regressor on the dependent variable declines geometrically over time.

Partial adjustment models   Used in the (S-G) approach to generate a short-run dynamic model. The desired change is a fraction of actual change.

Adaptive expectations model   Used to replace expectational variables in the (S-G) approach. Expectations are formed by assuming agents make systematic mistakes. Agents revise expectations according to a fraction of previous period expectation errors.

The Sargan test   A test carried out to confirm instruments validity.

---

7 INFINITE DISTRIBUTED LAG MODELS AND ECONOMIC THEORY

We saw in the previous section that by imposing restrictions on the coefficients to fall on a specific distribution the infinite distributed lag model becomes an autoregressive model. In the case, for example, of the geometric distributed lag model (4.64) it becomes model (4.67), or

$$Y_t = \alpha(1 - \lambda) + \beta_0 X_t + \lambda Y_{t-1} + \epsilon_t - \lambda \epsilon_{t-1} \tag{4.83}$$

Although this transformation of the models was a mechanical one, seeking just to reduce the infinite number of the coefficients to be estimated, in this section we will try to connect the infinite distributed lag models with specific models of economic theory.

7.1 The partial adjustment model

According to the partial adjustment model, introduced by Marc Nerlove (1956, 1958), the current level of an explanatory variable, X_t, determines the 'desired' level of the dependent variable, Y_t^*, given by

$$Y_t^* = \alpha + \beta X_t \tag{4.84}$$

The desired level of inventories of a firm being a function of its sales, or the desired level of capital stock in an economy being a function of its output, might be examples of this model.

However, because a 'desired' level is not an 'observable' level, and thus cannot be used in estimation, Nerlove assumed that due to various reasons in the phenomenon there is a difference between the actual and the desired levels of the dependent variable. In fact he assumed that, apart from random disturbances, the actual change in the dependent variable, $Y_t - Y_{t-1}$, is only a fraction γ of the desired change, $Y_t^* - Y_{t-1}$, in any period t, i.e.

$$Y_t - Y_{t-1} = \gamma(Y_t^* - Y_{t-1}) + \epsilon_t, \ 0 < \gamma \le 1 \tag{4.85}$$

Equation (4.85) is known as the 'partial adjustment equation', and fraction γ is known as the 'adjustment coefficient'. The greater the value of γ, the greater the adjustment of the actual to the desired level of the dependent variable takes place in period t. In the extreme case where $\gamma = 1$ the adjustment is instantaneous, or in other words, all the adjustment takes place in the same time period.

Equation (4.85) can be written as

$$Y_t = \gamma Y_t^* + (1 - \gamma)Y_{t-1} + \epsilon_t \tag{4.86}$$

which expresses the actual value at time t of the dependent variable as a weighted average of its desired value at time t and its actual value at time $t - 1$, with γ and $(1 - \gamma)$ being respectively the weights. Substituting (4.84) into (4.86) and rearranging we get

$$Y_t = \alpha\gamma + \beta\gamma X_t + (1 - \gamma)Y_{t-1} + \epsilon_t \tag{4.87}$$

Equation (4.87) is similar to equation (4.83) which corresponds to the geometric distributed lag model. In fact, by using the lag operator, (4.87) can be written as:

$$Y_t = \alpha + \frac{\beta\gamma}{1 - (1 - \gamma)L} X_t + \frac{1}{1 - (1 - \gamma)L} \epsilon_t \tag{4.88}$$

or knowing that the inverse of the operator is,

$$\frac{1}{1 - (1 - \gamma)L} = 1 + (1 - \gamma)L + (1 - \gamma)^2 L^2 + (1 - \gamma)^3 L^3 + \dots \tag{4.89}$$

equation (4.88) expands to

$$Y_t = \alpha + \beta\gamma X_t + \beta\gamma(1-\gamma)X_{t-1} + \beta\gamma(1-\gamma)^2 X_{t-2} + \ldots + \upsilon_t \tag{4.90}$$

where $\upsilon_t = \sum_{i=0}^{\infty}(1-\gamma)^i \epsilon_{t-i}$. Equation (4.90) is nothing else but an infinite geometric distributed lag model.

7.2 Adaptive expectations models

According to the adaptive expectations model, introduced by Cagan (1956), the 'expected' level of an explanatory variable, X_t^*, determines the current level of the dependent variable, Y_t, given by

$$Y_t = \alpha + \beta X_t^* + \epsilon_t \tag{4.91}$$

The demand for money in an economy being a function of its expected long-run interest rate, the quantity demanded being a function of the expected price, or the level of consumption being a function of the expected, or permanent, income (Friedman, 1957) might be examples of this model.

Similarly to the partial adjustment model, since the 'expected' level is not an 'observable' level, and thus cannot be used in estimation, Cagan assumed that the interested agents revise their expectations according to the level from their earlier expectations. In fact he assumed that the change in expectations, $X_t^* - X_{t-1}^*$, is only a fraction δ of the distance between the actual level of the explanatory variable X_t and its expected level X_{t-1}^*, in any period t, i.e.

$$X_t^* - X_{t-1}^* = \delta(X_t - X_{t-1}^*), 0 < \delta \leq 1 \tag{4.92}$$

Equation (4.92) is known as the 'adaptive expectations equation', or due to its error searching nature, the 'error learning equation', and fraction δ is known as the 'expectation coefficient'. The greater the value of δ, the greater the realisation of expectations in period t. In the extreme case where $\delta = 1$ expectations are fully and instantaneously realised, or in other words, all expectations are realised in the same time period.

Equation (4.92) can be written as

$$X_t^* = \delta X_t + (1-\delta)X_{t-1}^* \tag{4.93}$$

which expresses the expected value at time t of the explanatory variable as a weighted average of its actual value at time t and its expected value at time $t-1$, with δ and $(1-\delta)$ being respectively the weights. Substituting (4.93) into (4.91) we get

$$Y_t = \alpha + \beta\delta X_t + \beta(1-\delta)X_{t-1}^* + \epsilon_t \tag{4.94}$$

By multiplying the lagged once equation (4.91) by $(1-\delta)$ and subtracting the result from equation (4.94), and after rearranging, we get

$$Y_t = \alpha\delta + \beta\delta X_t + (1-\delta)Y_{t-1} + (\epsilon_t - (1-\delta)\epsilon_{t-1}) \tag{4.95}$$

Equation (4.95) is similar to equation (4.83) which corresponds to the geometric distributed lag model. In fact, by using the lag operator, (4.95) can be written as

$$Y_t = \alpha + \frac{\beta\delta}{1 - (1 - \delta)L} X_t + \epsilon_t \qquad (4.96)$$

By using now (4.89) we can expand (4.96) as follows:

$$Y_t = \alpha + \beta\delta X_t + \beta\delta(1 - \delta)X_{t-1} + \beta\delta(1 - \delta)^2 X_{t-2} + \ldots + \epsilon_t \qquad (4.97)$$

Equation (4.97) is nothing else but an infinite geometric distributed lag model. So we saw that the adaptive expectations model, like the partial adjustment model, is a realisation of the geometric distributed lag model.

7.3 The mixed model: partial adjustment and adaptive expectations

By combining the partial adjustment and the adaptive expectations models, we obtain the following mixed model:

$$Y_t^* = \alpha + \beta X_t^* \qquad (4.98)$$

$$Y_t - Y_{t-1} = \gamma(Y_t^* - Y_{t-1}) + \epsilon_t, \, 0 < \gamma \leq 1 \qquad (4.99)$$

$$X_t^* - X_{t-1}^* = \delta(X_t - X_{t-1}^*), \, 0 < \delta \leq 1 \qquad (4.100)$$

where Y_t^* = desired level of the dependent variable, and X_t^* = expected level of the explanatory variable.

The most representative example of this model is Friedman's (1957) permanent income hypothesis, according to which hypothesis 'permanent consumption' depends on 'permanent income'.

Solving the system of the three equations (4.98) to (4.100), using the hints presented in the two previous sections, we obtain the following equation (Johnston, 1984):

$$Y_t = \alpha\gamma\delta + \beta\gamma\delta X_t + [(1 - \delta) + (1 - \gamma)]Y_{t-1} - (1 - \delta)(1 - \gamma)Y_{t-2} + [\epsilon_t - (1 - \delta)\epsilon_{t-1}]$$
$$(4.101)$$

Equation (4.101) is not similar to those of the partial adjustment and the adaptive expectations because it includes Y_{t-2} among its explanatory variables. This equation, ignoring the identification problems of the parameters involved, reminds us at equation (4.76) of the Pascal distributed lag models.

8 ESTIMATION METHODS OF INFINITE DISTRIBUTED LAG MODELS

We saw till now that an infinite distributed lag model

$$Y_t = \alpha + \beta_0 X_t + \beta_1 X_{t-1} + \beta_2 X_{t-2} + \ldots + \epsilon_t \qquad (4.102)$$

can be reduced into a dynamic model by imposing restrictions on its coefficients to fall on specific schemes. The most popular of such restrictions is that of the geometric lag scheme

$$\beta_i = \beta_0 \lambda^i, \, 0 < \lambda < 1, \text{ for } i = 0, 1, 2, \ldots \tag{4.103}$$

which substitutes model (4.102) into

$$Y_t = \alpha + \beta_0 \sum_{i=0}^{\infty} \lambda^i X_{t-i} + \epsilon_t \tag{4.104}$$

or reduces (4.102) into

$$Y_t = \alpha_0 + \beta_0 X_t + \lambda Y_{t-1} + \upsilon_t \tag{4.105}$$

where $\alpha_0 = \alpha(1 - \lambda)$ and $\upsilon_t = \epsilon_t - \lambda \epsilon_{t-1}$

In estimating (4.102) under the restriction of (4.103), i.e. in estimating the parameters α, λ, and the β's, we can use either equation (4.104) – transformed form estimation, or equation (4.105) – reduced form estimation. However, in both cases the specification of the disturbances must be considered.

8.1 Transformed form estimation of the geometric distributed lag model

Equation (4.104) as it stands cannot be used for direct estimation because it has an infinite number of coefficients to be estimated. Therefore, it has to be transformed into an equation suitable for estimation purposes. This transformation is the following:

$$\begin{aligned} Y_t &= \alpha + \beta_0 \sum_{i=0}^{\infty} \lambda^i X_{t-i} + \epsilon_t \\ &= \alpha + \beta_0(X_t + \lambda X_{t-1} + \ldots + \lambda^{t-1} X_1) + \beta_0 \lambda^t (X_0 + \lambda X_{-1} + \ldots) + \epsilon_t \end{aligned} \tag{4.106}$$

or

$$Y_t = \alpha + \beta_0 W_t + \theta_0 \lambda^t + \epsilon_t \tag{4.107}$$

where

$$W_t = X_t + \lambda X_{t-1} + \ldots + \lambda^{t-1} X_1 \tag{4.108}$$

and

$$\theta_0 = \beta_0 \sum_{i=0}^{\infty} \lambda^i X_{-i} = E(Y_0 - \alpha) \tag{4.109}$$

is the so-called 'truncation remainder', which is treated as an unknown parameter.

Equation (4.107) has four unknown parameters to be estimated, α, β_0, θ_0, and λ.

If we knew the specification of the error term ϵ_t we could apply various methods in estimating the equation. Without loss of generality we will consider the following specifications of the disturbances (for other specifications the reader is referred to Kmenta (1986), Judge *et al.* (1988), Greene (1993) and Griffiths *et al.* (1993)).

1 The error term ϵ_t is 'non-autocorrelated': That is it satisfies all the classical assumptions stated in section 1.
2 The error term ϵ_t is 'autoregressive of the first order': That is it has the form $\epsilon_t = \rho\epsilon_{t-1} + \eta_t$, or it is AR(1), where η_t satisfies all the classical assumptions and ρ is the correlation coefficient.
3 The error term ϵ_t is 'moving average of the first order': That is it has the form $\epsilon_t = \eta_t - \mu\eta_{t-1}$, or it is MA(1), where η_t satisfies all the classical assumptions, and μ is the moving average coefficient.

8.1.1 Transformed form estimation with non-autocorrelated disturbances

This is the case where for the disturbances we have:

$$E(\epsilon_t) = 0, \, var(\epsilon_t) = \sigma^2, \, cov(\epsilon_t, \epsilon_s) = 0 \text{ for } t \neq s, \, \epsilon_t \sim N(0, \sigma^2) \tag{4.110}$$

This means that if we knew the value of λ we could compute variables W_t and λ^t and then we could apply OLS to equation (4.107). For applied purposes the values of variable W_t could be computed recursively by

$$\begin{aligned} W_1 &= X_1 \\ W_t &= W_t + \lambda W_{t-1} \text{ for } t = 2, 3, \ldots, n \end{aligned} \tag{4.111}$$

However, the value of λ is not known and thus a search procedure similar to that of Hildreth and Lu (1960) could be used. According to this procedure the sum of squared residuals from the regression equation (4.107) is estimated for various values of λ between zero and one. The estimates of the parameters α, β_0, θ_0 and λ that correspond to the minimum sum of the squared residuals of the searching regressions have the maximum likelihood properties of consistency and asymptotic efficiency, because minimising the sum of the squared residuals is equivalent to maximising the logarithmic likelihood function $Y_1, Y_2, \ldots, Y_n$, with respect to α, β_0, θ_0 and λ, which is

$$L = -\frac{n}{2}\log(2\pi\sigma^2) - \frac{1}{2\sigma^2}\sum_{t=1}^{n}[Y_t - \alpha - \beta_0 W_t - \theta_0\lambda^t]^2 \tag{4.112}$$

8.1.2 Transformed form estimation with AR(1) disturbances

This is the case where for the disturbances we have:

$$\epsilon_t = \rho\epsilon_{t-1} + \eta_t$$
$$E(\eta_t) = 0, \, var(\eta_t) = \sigma_\eta^2, \, cov(\eta_t, \eta_s) = 0 \text{ for } t \neq s, \, \eta_t \sim N(0, \sigma_\eta^2) \tag{4.113}$$

$$E(\epsilon_t) = 0, \, var(\epsilon_t) = \sigma^2 = \frac{\sigma_\eta^2}{1 - \rho^2}, \, cov(\epsilon_t, \epsilon_{t-s}) = \rho^s\sigma^2, \, \epsilon_t \sim N(0, \sigma^2)$$

If we knew the value of the autocorrelation coefficient ρ, then equation (4.107) could be written as

$$Y_t - \rho Y_{t-1} = \alpha(1-\rho) + \beta_0(W_t - \rho W_{t-1}) + \theta_0(\lambda^t - \rho\lambda^{t-1}) + \eta_t \qquad (4.114)$$

Because the error term in this equation satisfies all the classical assumptions, the OLS method could be applied if of course we knew the value of λ. However, we do not know neither the value of ρ nor the value of λ. Therefore, a similar searching method with the method described in the previous section could be used, with the difference now that this method will be two-dimensional. In other words we will be searching for values of ρ between -1 and 1, and for values of λ between 0 and 1. The estimates of this methodology will have all the properties of the maximum likelihood estimation because minimising the sum of the squared residuals is equivalent to maximising the corresponding logarithmic likelihood function with respect to α, β_0, θ_0, λ and ρ, which is

$$L = -\frac{n-1}{2}\log(2\pi\sigma_\eta^2) - \frac{1}{2\sigma_\eta^2}\sum_{t=1}^{n}[Y_t^* - \alpha(1-\rho) - \beta_0 W_t^* - \theta_0\lambda_*^t]^2 \qquad (4.115)$$

where

$$Y_t^* = Y_t - \rho Y_{t-1}, \ W_t^* = W_t - \rho W_{t-1}, \ \lambda_*^t = \lambda^t - \rho\lambda^{t-1} \qquad (4.116)$$

8.1.3 Transformed form estimation with MA(1) disturbances

This is the case where for the disturbances we have:

$$\epsilon_t = \eta_t - \mu\eta_{t-1}$$
$$E(\eta_t) = 0, \ \text{var}(\eta_t) = \sigma_\eta^2, \ \text{cov}(\eta_t, \eta_s) = 0 \ \text{for} \ t \neq s, \ \eta_t \sim N(0, \sigma_\eta^2) \qquad (4.117)$$
$$E(\epsilon_t) = 0, \ \text{var}(\epsilon_t) = \sigma^2 = \sigma_\eta^2(1+\mu^2), \ \text{cov}(\epsilon_t, \epsilon_{t-s}) = \begin{cases} \mu \ (\text{for} \ s=1) \\ 0 \ (\text{otherwise}) \end{cases}, \ \epsilon_t \sim N(0, \sigma^2)$$

Taking into account that the error term is written as

$$\eta_t = \sum_{s=0}^{\infty}(-\mu)^s\epsilon_{t-s}^s = \epsilon_t - \mu\epsilon_{t-1} + \ldots + (-\mu)^{t-1}\epsilon_1 + (-\mu)^t\eta_0 \qquad (4.118)$$

and further assuming that the value of η_0 is negligible, i.e. $\eta_0 = 0$, then if we knew the value of the moving average coefficient μ equation (4.107) could be written as

$$Y_t^* = \alpha[1 - \mu + \ldots + (-\mu)^{t-1}] + \beta_0 W_t^* + \theta_0\lambda_*^t + \eta_t \qquad (4.119)$$

where

$$Y_t^* = Y_t - \mu Y_{t-1} + \ldots + (-\mu)^{t-1}Y_1$$
$$W_t^* = W_t - \mu W_{t-1} + \ldots + (-\mu)^{t-1}W_1 \qquad (4.120)$$
$$\lambda_*^t = \lambda^t - \mu\lambda^{t-1} + \ldots + (-\mu)^{t-1}\lambda$$

Because the error term in equation (4.119) satisfies all the classical assumptions, the OLS method could be applied if of course we knew the value of λ. However, we do not know either the value of μ or the value of λ. Therefore, a similar two-dimensional searching method with the method described in the previous section could be used. In other words we will be searching for values of μ between -1 and 1, and for values of λ between 0 and 1. The estimates of this methodology will have all the properties of the maximum likelihood estimation because minimising the sum of the squared residuals is equivalent to maximising the corresponding logarithmic likelihood function (under the assumption that $\eta_0 = 0$) with respect to α, β_0, θ_0, λ and μ, which is

$$L = -\frac{n}{2}\log(2\pi\sigma_\eta^2) - \frac{1}{2\sigma_\eta^2}\sum_{t=1}^{n}[Y_t^* - \alpha[1 - \mu + \ldots + (-\mu)^{t-1}] - \beta_0 W_t^* - \theta_0\lambda_*^t]^2 \tag{4.121}$$

The method just described is an 'approximate maximum likelihood method' because it involves assumption $\eta_0 = 0$. For the maximum likelihood method the interested reader is referred to Pesaran (1973) and Balestra (1980).

8.2 Reduced form estimation of the geometric distributed lag model

This is the case where estimation is applied to equation (4.105), i.e. to

$$Y_t = \alpha_0 + \beta_0 X_t + \lambda Y_{t-1} + \upsilon_t \tag{4.122}$$

where $\alpha_0 = \alpha(1 - \lambda)$ and $\upsilon_t = \epsilon_t - \lambda\epsilon_{t-1}$

In trying to estimate (4.122) we have to take into account two things; that the equation is 'dynamic', i.e. it contains a lagged dependent variable among its explanatory variables, and that its error term can take various specifications. These specifications can be summarised as follows:

1 The error term υ_t is 'non-autocorrelated': That is it has the form $\upsilon_t = \epsilon_t$, where ϵ_t satisfies all the classical assumptions. This is the case of the partial adjustment model (4.87).
2 The error term υ_t is 'autoregressive of the first order': That is it has the form $\upsilon_t = \rho\upsilon_{t-1} + \epsilon_t$, where ϵ_t satisfies all the classical assumptions and ρ is the correlation coefficient. This case is common in most models with time series data.
3 The error term υ_t is 'moving average of the first order': That is it has the form $\upsilon_t = \epsilon_t - \lambda\epsilon_{t-1}$, where ϵ_t satisfies all the classical assumptions. This is the case of the adaptive expectations model (4.95).

8.2.1 Reduced form estimation with non autocorrelated disturbances

This is the case of the partial adjustment model (4.87),

$$Y_t = \alpha\gamma + \beta\gamma X_t + (1 - \gamma)Y_{t-1} + \epsilon_t \tag{4.123}$$

or

$$Y_t = \alpha_0 + \beta_0 X_t + \gamma_1 Y_{t-1} + \epsilon_t \tag{4.124}$$

where

$$\alpha_0 = \alpha\gamma, \ \beta_0 = \beta\gamma, \ \gamma_1 = (1 - \gamma),$$
$$E(\epsilon_t) = 0, \ \text{var}(\epsilon_t) = \sigma^2, \ \text{cov}(\epsilon_t, \epsilon_s) = 0 \text{ for } t \neq s, \ \epsilon_t \sim N(0, \sigma^2) \tag{4.125}$$

In this model, the lagged dependent variable Y_{t-1} is a stochastic regressor which is uncorrelated with ϵ_t, but is correlated with past values of ϵ_t, i.e.

$$\text{cov}(Y_{t-1}, \epsilon_t) = 0 \text{ and } \text{cov}(Y_{t-1}, \epsilon_{t-s}) \neq 0 \text{ for } s = 1, 2, \ldots \tag{4.126}$$

or in words, the stochastic explanatory variable Y_{t-1} is distributed independently of the error term ϵ_t. In this case, equation (4.124) may be estimated by OLS. However, the OLS estimates of the coefficients and their standard errors will be consistent and asymptotically efficient, but will be biased in small samples. Thus, these estimates are not BLUE, and therefore the tests of hypotheses will be valid for large samples and invalid for small samples.

Summarising, we could apply the OLS methodology to equation (4.124) as long as the sample size is large enough. In this case the OLS estimates a_0, b_0 and c_1 of the coefficients α_0, β_0 and γ_1, respectively, could be used to estimate the initial parameters α, β and γ of the model, with the corresponding consistent estimates a, b and c, using (4.125), as follows:

$$c = 1 - c_1, \ b = \frac{b_0}{1 - c_1}, \ a = \frac{a_0}{1 - c_1} \tag{4.127}$$

8.2.2 *Reduced form estimation with AR(1) disturbances*

This is the most common case with time series data. This model could be written as

$$Y_t = \alpha_0 + \beta_0 X_t + \gamma_1 Y_{t-1} + v_t \tag{4.128}$$

where

$$v_t = \rho v_{t-1} + \epsilon_t$$
$$E(\epsilon_t) = 0, \ \text{var}(\epsilon_t) = \sigma^2, \ \text{cov}(\epsilon_t, \epsilon_s) = 0 \text{ for } t \neq s, \ \epsilon_t \sim N(0, \sigma^2) \tag{4.129}$$

$$E(v_t) = 0, \ \text{var}(v_t) = \sigma_v^2 = \frac{\sigma^2}{1 - \rho^2}, \ \text{cov}(v_t, v_{t-s}) = \rho^s \sigma^2, \ v_t \sim N(0, \sigma_v^2)$$

Given (4.129) it can be proved that

$$\text{cov}(Y_{t-1}, v_t) = \frac{\rho \sigma_v^2}{1 - \gamma_1 \rho} \neq 0 \tag{4.130}$$

which means that because v_t is correlated with v_{t-1} and the stochastic explanatory variable Y_{t-1} is correlated with v_{t-1}, then Y_{t-1} will be correlated with v_t. In other words, the application of OLS in equation (4.128) will give biased and inconsistent estimates of the coefficients and their standard errors. Thus, the corresponding tests of hypotheses will be invalid for small or even for large samples. Therefore, another alternative to OLS estimate is needed. In what it follows we will present such alternative methods.

(a) The method of instrumental variables (IV)

Having in mind that in equation (4.128) the cause of the problem was that the stochastic regressor Y_{t-1} was correlated with the error term v_t, Liviatan (1963) tried to eliminate this problem by introducing another variable Z_t, called 'instrumental variable', in the place of Y_{t-1} which had the following properties:

1   The instrumental variable Z_t is an approximation of variable Y_{t-1}.
2   Variable Z_t is not correlated with the error term v_t.
3   Variable Z_t is highly correlated with variable Y_{t-1}.

Liviatan used X_{t-1} as an instrumental variable for the estimation of (4.128). X_{t-1} being non stochastic was definitely not correlated with v_t, and being one of the explanatory variables of Y_{t-1}, was likely to be correlated with Y_{t-1}. The estimates a_0, b_0 and c_1 of the coefficients α_0, β_0 and γ_1 of equation (4.128) respectively, could be obtained by solving the system of the usual 'normal equations' shown below:

$$\left.\begin{array}{l} \Sigma Y_t = na_0 + b_0 \Sigma X_t + c_1 \Sigma Y_{t-1} \\[6pt] \Sigma Y_t X_t = a_0 \Sigma X_t + b_0 \Sigma X_t^2 + c_1 \Sigma Y_{t-1} X_t \\[6pt] \Sigma Y_t X_{t-1} = a_0 \Sigma X_{t-1} + b_0 \Sigma X_t X_{t-1} + c_1 \Sigma Y_{t-1} X_{t-1} \end{array}\right\} \qquad (4.131)$$

Although these estimates will be biased for small samples, as long as the sample increases the solution of system (4.131) will yield consistent estimates.

As another type of instrumental variable for Y_{t-1} in equation (4.128) could be used the lagged dependent predicted variable, $\hat{Y}_{t-1}$, obtained by the following regression:

$$\hat{Y}_t = d_0 + d_1 X_{t-1} + \ldots + d_h X_{t-h} \qquad (4.132)$$

i.e. by the regression of the dependent Y_t on lagged variables of X_t. The proper lag length in regression (4.132) could be determined by optimising a criterion, such as those in (4.37) to (4.39). Having obtained $\hat{Y}_{t-1}$ from (4.132), the OLS estimation applied to the following equation:

$$Y_t = \alpha_0 + \beta_0 X_t + \gamma_1 \hat{Y}_{t-1} + v_t \qquad (4.133)$$

which has been derived from equation (4.128) after substituting Y_{t-1} with $\hat{Y}_{t-1}$, will give, as before, consistent estimates. The method just described, i.e. the method where as a first step we get $\hat{Y}_{t-1}$ from (4.132) and as a second step we get the final estimates from (4.133), it is called the method of 'two stages least squares' (2SLS). Of course, $\hat{Y}_{t-1}$ could also be derived from the regression of Y_t on any other set of legitimate instruments and not just on the set of lagged values of X_t.

However, the instrumental variables methodology deals only with the problem of correlation between the stochastic regressor Y_{t-1} and the error term v_t and not with the problem of autocorrelation of the error term. In fact this method is free of the autocorrelation scheme. Therefore, the application of instrumental variables in equation (4.128) will give consistent but asymptotically inefficient estimates. In other words a method for dealing with the autocorrelation in the error term must be used. Such method could be the following Hatanaka's (1976) 'two-step' method, which has the same asymptotic properties as the maximum likelihood estimator (MLE) of normally distributed and serially correlated disturbances.

Step 1   Having estimated from the IV method consistent estimates a_0, b_0 and c_1 of the parameters α_0, β_0 and γ_1 respectively, estimate the residuals $u_t = Y_t - a_0 - b_0 X_t - c_1 Y_{t-1}$, and use them in order to get the estimate $\hat{\rho}$, for the autoregressive coefficient ρ, by regressing u_t against its own lagged value u_{t-1}.

Step 2   By regressing Y_t^* against a constant, X_t^*, Y_{t-1}^* and u_{t-1}, where $Y_t^* = Y_t - \hat{\rho} Y_{t-1}$, $X_t^* = X_t - \hat{\rho} X_{t-1}$, and $Y_{t-1}^* = Y_{t-1} - \hat{\rho} Y_{t-2}$, get the consistent and asymptotically efficient estimates a_0, b_0 and c_1, of the parameters α_0, β_0 and γ_1. If φ denotes the regression coefficient of u_{t-1}, then the two-step estimator of ρ is $\hat{\rho} + \varphi$.

Generally, the 'instrumental variable maximum likelihood with AR(1)' estimates could be obtained by maximising the function

$$L = -\frac{n-1}{2}\log(2\pi\sigma^2) - \frac{1}{2\sigma^2} \sum_{t=1}^{n} [Y_t^* - \alpha_0(1-\rho) - \beta_0 X_t^* - \gamma_1 \hat{Y}_{t-1}^*]^2 \qquad (4.134)$$

where

$$Y_t^* = Y_t - \rho Y_{t-1}, X_t^* = X_t - \rho X_{t-1}, \hat{Y}_{t-1}^* = \hat{Y}_{t-1} - \rho \hat{Y}_{t-2} \qquad (4.135)$$

with $\hat{Y}_{t-1}$ having obtained from (4.132), or any other similar estimation. This method is based on the philosophy of the Cochrane and Orcutt method that we will see next.

(b) The method of Cochrane and Orcutt (CORC)

The method of Cochrane and Orcutt (1949) refers to the application of OLS to equation

$$(Y_t - \rho Y_{t-1}) = \alpha_0(1-\rho) + \beta_0(X_t - \rho X_{t-1}) + \gamma_1(Y_{t-1} - \rho Y_{t-2}) + (v_t - \rho v_{t-1}) \qquad (4.136)$$

or

$$Y_t^* = \alpha^* + \beta_0 X_t^* + \gamma_1 Y_{t-1}^* + \epsilon_t \qquad (4.137)$$

where

$$Y_t^* = Y_t - \rho Y_{t-1}, X_t^* = X_t - \rho X_{t-1}, Y_{t-1}^* = Y_{t-1} - \rho Y_{t-2}, \epsilon_t = \upsilon_t - \rho \upsilon_{t-1},$$

and

$$\alpha^* = \alpha_0(1 - \rho))$$

which is a transformation of equation (4.128), under the assumption that ρ is known. The application of OLS to equation (4.137) is nothing else but the method of generalised least squares (GLS) under the assumption that the value of ρ is known. However, in most cases ρ is unknown and thus it has to be approximated by an estimate, $\hat{\rho}$. The iterative steps of the CORC methodology are in this case the following:

ITERATION 1

Step 1.1   Arrange for use the following variables: Y_t, X_t and Y_{t-1}.

Step 1.2   Apply OLS to equation (4.128) and get estimates a_0, b_0 and c_1, of the parameters α_0, β_0 and γ_1 respectively.

Step 1.3   Use the estimates from step 1.2, in order to estimate the corresponding residuals u_t, i.e. $u_t = Y_t - a_0 - b_0 X_t - c_1 Y_{t-1}$.

Step 1.4   Using the estimated residuals u_t, from step 1.3, regress u_t against its own lagged value u_{t-1}, and get the estimate $\hat{\rho}$, for the autoregressive coefficient ρ.

ITERATION 2

Step 2.1   Compute the following variables: $Y_t^* = Y_t - \hat{\rho} Y_{t-1}$, $X_t^* = X_t - \hat{\rho} X_{t-1}$, and $Y_{t-1}^* = Y_{t-1} - \hat{\rho} Y_{t-2}$.

Step 2.2   By regressing Y_t^* against a constant, X_t^*, Y_{t-1}^*, get new estimates a_0, b_0 and c_1, of the parameters α_0, β_0 and γ_1.

Step 2.3   Using the estimates from step 2.2, estimate a new set of residuals, u_t, i.e. $u_t = Y_t - a_0 - b_0 X_t - c_1 Y_{t-1}$.

Step 2.4   Using the estimated residuals u_t, from step 2.3, regress u_t against its own lagged value u_{t-1}, and get the new estimate $\hat{\rho}$, for the autoregressive coefficient ρ.

ITERATION X

By following the four steps x.1 to x.4 in each iteration x, continue this iterative procedure until two successive estimates $\hat{\rho}$ of the autoregressive coefficient ρ may not differ by more than a predetermined value, say 0.0001. The estimates a_0, b_0 and c_1 of this final iteration will be consistent although their standard errors will be inconsistent. To correct for this inconsistency the following two-step method may be followed (Harvey, 1990):

Step 1   Having estimated from the CORC method consistent estimates a_0, b_0, c_1 and $\hat{\rho}$ of the parameters α_0, β_0, γ_1 and ρ respectively, estimate the residuals $u_t = Y_t - a_0 - b_0 X_t - c_1 Y_{t-1}$ of the initial model (4.128) and the residuals $e_t = Y_t^* - a_0(1 - \hat{\rho}) - b_0 X_t^* - c_1 Y_{t-1}^*$ of the reduced model (4.137).

Step 2   By regressing e_t against a constant, X_t^*, Y_{t-1}^* and u_{t-1}, get the consistent and asymptotically efficient estimates a_0, b_0 and c_1 of the parameters α_0, β_0 and γ_1.

Equation (4.136), or (4.137), could also be used in obtaining consistent and asymptotically efficient estimates by maximising the corresponding logarithmic likelihood function with respect to α_0, β_0, γ_1 and ρ, which is

$$L = -\frac{n-1}{2}\log(2\pi\sigma^2) - \frac{1}{2\sigma^2}\sum_{t=1}^{n}[(Y_t - \rho Y_{t-1}) - \alpha_0(1 - \rho) -$$

$$\beta_0(X_t - \rho X_{t-1}) - \gamma_1(Y_{t-1} - \rho Y_{t-2})]^2 \tag{4.138}$$

This methodology is called the 'maximum likelihood with AR(1)'.

8.2.3 Reduced form estimation with MA(1) disturbances

This is the case of the Koyck infinite distributed lag model (4.102) to (4.105), or

$$Y_t = \alpha_0 + \beta_0 X_t + \lambda Y_{t-1} + \upsilon_t \tag{4.139}$$

where

$$\upsilon_t = \epsilon_t - \lambda\epsilon_{t-1}$$
$$E(\epsilon_t) = 0, \text{ var}(\epsilon_t) = \sigma^2, \text{ cov}(\epsilon_t, \epsilon_s) = 0 \text{ for } t \neq s, \epsilon_t \sim N(0, \sigma^2) \tag{4.140}$$

$$E(\upsilon_t) = 0, \text{ var}(\upsilon_t) = \sigma_\upsilon^2 = \sigma^2(1 + \lambda^2), \text{ cov}(\upsilon_t, \upsilon_{t-s}) = \begin{cases} \lambda \text{ (for } s = 1) \\ 0 \text{ (otherwise)} \end{cases}, \upsilon_t \sim N(0, \sigma_\upsilon^2)$$

We also met this case in the adaptive expectations model (4.95),

$$Y_t = \alpha\delta + \beta\delta X_t + (1 - \delta)Y_{t-1} + (\epsilon_t - (1 - \delta)\epsilon_{t-1}) \tag{4.141}$$

or

$$Y_t = \alpha_0 + \beta_0 X_t + \delta_1 Y_{t-1} + \upsilon_t \tag{4.142}$$

where

$$\alpha_0 = \alpha\delta, \beta_0 = \beta\delta, \delta_1 = (1 - \delta), \upsilon_t = \epsilon_t - (1 - \delta)\epsilon_{t-1} \tag{4.143}$$

Given (4.140) it can be proved that

$$\text{cov}(Y_{t-1}, v_t) = -\lambda\sigma^2 \tag{4.144}$$

which means that v_t is autocorrelated (for $\lambda \neq 0$), and taking into account that the stochastic explanatory variable Y_{t-1} is correlated with ϵ_{t-1}, this variable is also correlated with v_t, because v_t it includes ϵ_{t-1}. In other words, the autocorrelation of the error term and the correlation between an explanatory variable (Y_{t-1}) and the error term makes the application of OLS in equation (4.139) to give biased and inconsistent estimates of the coefficients and their standard errors. Thus, the corresponding tests of hypotheses will be invalid for small or even large samples.

In order to estimate (4.139) various methods could be used. The instrumental variable method, for example, could be used to obtain consistent estimates of the coefficients because this method does not require any specific assumptions regarding the error term. However, knowing that the error term in (4.139) is of the moving average scheme, the method that will incorporate this information into the estimation methodology may improve the asymptotic efficiency of its estimates. Such methods are the following:

(a) Zellner and Geisel search procedure

Zellner and Geisel (1970) starting from equation (4.139) defined $W_t = Y_t - \epsilon_t$, and using that

$$W_t - \lambda W_{t-1} = (Y_t - \epsilon_t) - \lambda(Y_{t-1} - \epsilon_{t-1}) = \alpha(1 - \lambda) + \beta_0 X_t$$

they obtained

$$W_t = \alpha(1 - \lambda) + \beta_0 X_t + \lambda W_{t-1} \tag{4.145}$$

Lagging repeatedly (4.145) and substituting the result into the same equation (4.145) we get

$$W_t = \alpha(1 - \lambda)(1 + \lambda + \lambda^2 + \ldots + \lambda^{t-1}) + \beta_0(X_t + \lambda X_{t-1} + \lambda^2 X_{t-2} + \ldots$$
$$+ \lambda^{t-1}X_1) + \lambda^t W_0$$

or

$$W_t = \alpha(1 - \lambda)\frac{1 - \lambda^t}{1 - \lambda} + \beta_0 Z_t + \lambda^t W_0$$

where

$$Z_t = X_t + \lambda X_{t-1} + \lambda^2 X_{t-2} + \ldots + \lambda^{t-1}X_1 \text{ or finally}$$

$$Y_t = \alpha + \beta_0 Z_t + (W_0 - \alpha)\lambda^t + \epsilon_t \tag{4.146}$$

If we knew the value of λ we could apply OLS in equation (4.146) because all the variables involved are either known or they could be computed, and the error term ϵ_t is white noise. However, the value of λ is not known and thus a search procedure similar to that of Hildreth and Lu could be used. According to this procedure the sum of squared residuals from the regression equation (4.146) is estimated for various values of λ between zero and one. The estimates of the parameters α, β_0 and λ that correspond to the minimum sum of the squared residuals of the searching regressions have the maximum likelihood properties of consistency and asymptotic efficiency, because minimising the sum of the squared residuals is equivalent to maximising the logarithmic likelihood function Y_1, Y_2,..., Y_n, with respect to α, β_0, W_0 and λ, which is

$$L = -\frac{n}{2}\log(2\pi\sigma^2) - \frac{1}{2\sigma^2}\sum_{t=1}^{n}[Y_t - \alpha - \beta_0 Z_t - (W_0 - \alpha)\lambda^t]^2 \tag{4.147}$$

(b) Instrumental variable maximum likelihood method with MA(1)

The procedure just described is based on the assumption that the moving average coefficient (μ) of the error term is equal to the Koyck coefficient (λ). In the case that these two coefficients were different ($\lambda \neq \mu$) function (4.115) could be used in the estimation instead of function (4.147). This would be the method of 'approximate maximum likelihood with MA(1)', as we saw in section 8.1.3.

However, by applying to equation

$$Y_t = \alpha_0 + \beta_0 X_t + \lambda \hat{Y}_{t-1} + \upsilon_t \tag{4.148}$$

the methodology described in section 8.1.3 we could get

$$Y_t^* = \alpha_0^* + \beta_0 X_t^* + \lambda \hat{Y}_{t-1}^* + \epsilon_t \tag{4.149}$$

where

$$\begin{aligned}
\alpha_0^* &= \alpha_0[1 - \mu + \ldots + (-\mu)^{t-1}] \\
X_t^* &= X_t - \mu X_{t-1} + \ldots + (-\mu)^{t-1}X_1 \\
\hat{Y}_{t-1}^* &= \hat{Y}_{t-1} - \mu\hat{Y}_{t-2} + \ldots + (-\mu)^{t-1}\hat{Y}_0
\end{aligned} \tag{4.150}$$

with $\hat{Y}_{t-1}$ having obtained from (4.132), or any other similar estimation.

The estimates of equation (4.148) could be obtained by maximising the corresponding logarithmic function which is the following:

$$L = -\frac{n-1}{2}\log(2\pi\sigma^2) - \frac{1}{2\sigma^2}\sum_{t=1}^{n}[Y_t^* - \alpha_0^* - \beta_0 X_t^* - \lambda\hat{Y}_{t-1}^*]^2 \tag{4.151}$$

The method just described is an 'instrumental variable approximate maximum likelihood method with MA(1)', because it involves assumptions about the 'zero' observations.

9 DIAGNOSTIC TESTS FOR MODELS WITH LAGGED DEPENDENT VARIABLES

We saw in the previous sections that the existence of lagged dependent variables among the explanatory variables of a model violates a basic assumption of the regression model, that the explanatory variables are not correlated with the error term. Furthermore, we saw that in trying to transform the infinite distributed lag model to a model with an estimable number of coefficients, no one guarantees us that the error term of the transformed model will not be autocorrelated. In this section we will present some diagnostic tests which refer to the preceding observations.

9.1 The Durbin h statistic for first-order autocorrelation

The most common statistic used for detecting first-order serial correlation in the error term of the classical regression model is the Durbin-Watson (1950, 1951) d statistic, given by

$$d = \frac{\sum\limits_{t=2}^{n} (u_t - u_{t-1})^2}{\sum\limits_{t=1}^{n} u_t^2} \qquad (4.152)$$

where u_t are the residuals of the disturbances v_t, given by $v_t = \rho v_{t-1} + \epsilon_t$, with ϵ_t being well behaved. This statistic takes the value of 2 when there is no autocorrelation in the disturbances, and the values of 0 and 4 when there exists a perfectly positive or a perfectly negative autocorrelation respectively.

The estimate $\hat{\rho}$ of the autocorrelation coefficient ρ in the disturbances is given by

$$\hat{\rho} = \frac{\sum\limits_{t=2}^{n} u_t u_{t-1}}{\sum\limits_{t=1}^{n} u_t^2} \qquad (4.153)$$

It can be shown that between the Durbin-Watson d statistic and $\hat{\rho}$ it approximately holds that

$$d \cong 2(1 - \hat{\rho}) \qquad (4.154)$$

However, in the presence of lagged dependent variables in the regression model the Durbin-Watson d statistic is biased because it tends toward 2, suggesting thus, that there is no autocorrelation although autocorrelation may be present. Therefore, in such cases another test for detecting autocorrelation should be employed.

Durbin (1970) developed a 'large-sample statistic' for detecting first-order serial correlation in models when lagged dependent variables are present. This statistic is called the 'h statistic' and is given by

$$h = \hat{\rho} \sqrt{\frac{n}{1 - n[var(c_1)]}} \cong \left(1 - \frac{1}{2}d\right) \sqrt{\frac{n}{1 - n[var(c_1)]}} \qquad (4.155)$$

where n = sample size, $\hat{\rho}$ = estimate of the autocorrelation coefficient according to (4.153), d = Durbin–Watson d statistic according to (4.154), and $var(c_1)$ = estimated variance of the coefficient of the lagged dependent variable Y_{t-1} in equation (4.128).

Durbin showed that for large sample sizes and if $\rho = 0$, then the h statistic follows the normal distribution with zero mean and unit variance (standardised normal distribution). Therefore, the actual h test for detecting first-order serial correlation in the disturbances of models with lagged dependent variables is the following:

H_0: $\rho = 0$, accept when h < |critical value| of N(0, 1)

H_a: $\rho \neq 0$, accept when h > |critical value| of N(0, 1)

(4.156)

Summarising, the steps for this test are the following:

Step 1 Estimate equation (4.128) and get the residuals and the estimated variance of the coefficient of the lagged dependent variable Y_{t-1}.

Step 2 Compute either d from (4.152) or $\hat{\rho}$ from (4.153).

Step 3 Compute h from (4.155).

Step 4 Apply the hypotheses testing of (4.156).

One could detect the following disadvantages in using the h test:

1 The h test is not applicable if $nvar(c_1) > 1$, because then the denominator in (4.155) becomes negative.
2 The test refers to large sample sizes only.
3 The test is not applicable for higher-order serial correlation than one.

9.2 The Breusch-Godfrey, or Lagrange multiplier, test for autocorrelation of any order

Breusch (1978) and Godfrey (1978) developed a test (BG) to detect higher-order autocorrelation, which can be also used when longer than one lags of the dependent variable are present in the model, and which is valid for large or small sample sizes. The steps of this test which follows a Lagrange multiplier test procedure can be constructed as follows:

Step 1 Define properly the model. The model could be of the following form:

$$Y_t = \alpha + \beta_1 X_{1t} + \beta_2 X_{2t} + \ldots + \beta_p X_{pt} + \gamma_1 Y_{t-1} + \gamma_2 Y_{t-2} + \ldots + \gamma_q Y_{t-q} + v_t$$

$$v_t = \rho_1 v_{t-1} + \rho_2 v_{t-2} + \ldots + \rho_m v_{t-m} + \epsilon_t$$

(4.157)

where the number of the regression coefficients is $k + 1$, with $k = p + q$ and $n > k + 1$, p is the number of the independent variables, q is the order of the lagged dependent variable, and m is the order of the autoregressive error term, with $m < q$.

Step 2 Apply OLS to equation (4.157) and obtain the corresponding residuals u_t.

Step 3 Regress u_t against all the regressors of the model, i.e. against the constant, $X_{1t}, \ldots, X_{pt}, Y_{t-1}, \ldots, Y_{t-q}$, plus all the lagged residuals till order m, i.e. $u_{t-1}, \ldots, u_{t-m}$, and obtain the corresponding R^2.

Step 4 Calculate the statistic BG as

$$BG = (n - q)R^2 \tag{4.158}$$

This statistic follows the χ^2 distribution with m degrees of freedom, i.e. it is $BG = (n - q) R^2 \sim \chi^2(m)$.

Step 5 Apply the following test of hypotheses:

H_0: all $\rho_i = 0$, accept if $BG <$ critical value of $\chi^2(m)$

H_a: not all $\rho_i = 0$, accept if $BG >$ critical value of $\chi^2(m)$ (4.159)

This test can be used also in the case when the error term follows a moving average process of order m, i.e. it is

$$v_t = \epsilon_t + \lambda_1 \epsilon_{t-1} + \lambda_2 \epsilon_{t-2} + \ldots + \lambda_m \epsilon_{t-m} \tag{4.160}$$

9.3 The Sargan test for testing instrument validity

We saw in section 8.2 that the general idea of the instrumental variable methodology was to take a set of instrumental variables which were not correlated with the error term and to use them in order to construct a new variable which will be used in place of the lagged dependent variable which was correlated with the error term. In estimating, for example, equation (4.128) we used instead equation (4.133). In equation (4.133), variable X_t was used as an instrument for itself, but variable $\hat{Y}_{t-1}$ was used as a 'proxy' for variable Y_{t-1}. The instruments used in order to construct $\hat{Y}_{t-1}$ were those shown in (4.132), i.e. they were $X_{t-1}, X_{t-2}, \ldots, X_{t-h}$.

Sargan (1964) developed a statistic (SARG) in order to test the validity of the instruments used in the IV estimation, or in other words to test if from the set of all possible instrumental variables the subset of instruments used in the estimation is independent of the error term. The steps of this test are the following:

Step 1 Divide the variables included in the structural equation to those which are independent of the error term, say, $X_{1t}, X_{2t}, \ldots, X_{pt}$, and to those which are not independent of the error term, say, $Z_{1t}, Z_{2t}, \ldots, Z_{qt}$.

Step 2 Define the set of instruments, say, $W_{1t}, W_{2t}, \ldots, W_{st}$, where $s \geq q$.

Step 3 Apply the IV estimation to the original equation and obtain the corresponding residuals u_t.

Step 4 Regress u_t against all the independent of the error term variables, i.e. the constant, $X_{1t}, \ldots, X_{pt}$, plus the instruments $W_{1t}, \ldots, W_{st}$, and obtain the corresponding R^2.

Step 5 Calculate the statistic SARG as

$$SARG = (n-k)R^2 \tag{4.161}$$

where n = number of observations, k = number of coefficients in the original structural model. This statistic follows the χ^2 distribution with $r = s - q$ degrees of freedom, i.e. it is $SARG = (n-k)R^2 \sim \chi^2(r)$.

Step 6 Apply the following test of hypotheses:

H_0: all instruments valid,   if SARG < critical value of $\chi^2(r)$

H_a: not all instruments valid,   if SARG > critical value of $\chi^2(r)$ (4.162)

In the case that the alternative hypothesis is accepted, at least one of the instrumental variables is correlated with the error term, and therefore not all instruments are valid. This means that the IV estimates of the coefficients are not also valid.

10 ILLUSTRATIVE EXAMPLES OF INFINITE DISTRIBUTED LAG MODELS

Having completed the theoretical presentation of infinite distributed lag models in the previous sections, we will present in this section some examples applying in real cases the theory presented.

Example 4.5 The private consumption function for Greece, 1960–1995

Assume that the expected, or permanent, private disposable income Y^* determines private consumption C in Greece according to the linear function

$$C_t = \alpha + \beta Y_t^* + \epsilon_t \tag{4.163}$$

Considering the adaptive expectations model, the mechanism used to transform the unobserved permanent level of the private disposable income into an observable level is

$$Y_t^* - Y_{t-1}^* = \delta(Y_t - Y_{t-1}^*), \, 0 < \delta \leq 1 \tag{4.164}$$

Following the methodology in 7.2, the private consumption function in observable levels is the following:

$$C_t = \alpha\delta + \beta\delta Y_t + (1-\delta)C_{t-1} + (\epsilon_t + (1-\delta)\epsilon_{t-1}) \tag{4.165}$$

or in estimable form

$$C_t = \alpha_0 + \beta_0 Y_t + \delta_1 C_{t-1} + v_t \qquad (4.166)$$

where

$$\alpha_0 = \alpha\delta, \beta_0 = \beta\delta, \delta_1 = (1-\delta), v_t = \epsilon_t - (1-\delta)\epsilon_{t-1} \qquad (4.167)$$

Finally, we saw that equation (4.165) can be expanded in a geometric (or Koyck) infinite distributed lag model as follows:

$$C_t = \alpha + \beta\delta Y_t + \beta\delta(1-\delta)Y_{t-1} + \beta\delta(1-\delta)^2 Y_{t-2} + \ldots + \epsilon_t \qquad (4.168)$$

Equation (4.168) is nothing else but equation (4.13) which we saw in the beginning of this chapter.

Table 4.6 presents annual data from 1960 to 1995 for private consumption and private disposable income of the Greek economy, for the estimation of equation (4.166). Furthermore, this table presents data for gross investment, gross national product, and long-term interest rates necessary for the estimation purposes of the next example. All nominal data are expressed at constant market prices of year 1970, and in million of drachmas. Private disposable income is deflated using the consumption price deflator.

Applying techniques described in section 8 to equation (4.166), Table 4.7 presents the obtained estimates. The instruments used for the IV estimation are a constant, Y_t, Y_{t-1} and Y_{t-2}. For the application of the Zellner-Geisel method corresponding to equation (4.146), we computed Z_t by the recursion

$$Z_1 = PDI_1$$
$$Z_t = PDI_t + \lambda Z_{t-1}, \text{ for } t = 2, 3, \ldots, n \qquad (4.169)$$

where

PDI = Private Disposable Income

Applying OLS to equation (4.146), having computed first Z_t using (4.169) and λ^t, for different values of λ between zero and one, the sum of the squared residuals (SSR) has been estimated. Figure 4.5 presents these results. It is seen in this Figure that the value of λ, or the value of $(1-\delta)$ in the adaptive expectations specification, that minimises the sum of squared residuals is 0.65. In our experimentation the steps used for the change of λ had the quite large value of 0.05. However, having found that the value of λ that globally minimises the SSR is 0.65, the precise value of λ could be found by decreasing the steps of change to very small values and repeating the pro cedure in the neighbourhood of 0.65. The final results are shown in Table 4.7 obtained by using the maximum likelihood routine of Microfit, where we can see that the exact estimate of λ is 0.65879, i.e. not different from the value of 0.65 found earlier.

Table 4.6 Data for the Greek economy referring to consumption and investment

Year	Private consumption	Private disposable income	Gross investment	Gross national product	Long-term interest rate (%)
1960	107,808	117,179	29,121	145,458	8.00
1961	115,147	127,599	31,476	161,802	8.00
1962	120,050	135,007	34,128	164,674	8.00
1963	126,115	142,128	35,996	181,534	8.25
1964	137,192	159,649	43,445	196,586	9.00
1965	147,707	172,756	49,003	214,922	9.00
1966	157,687	182,366	50,567	228,040	9.00
1967	167,528	195,611	49,770	240,791	9.00
1968	179,025	204,470	60,397	257,226	8.75
1969	190,089	222,638	71,653	282,168	8.00
1970	206,813	246,819	70,663	304,420	8.00
1971	217,212	269,249	80,558	327,723	8.00
1972	232,312	297,266	92,977	356,886	8.00
1973	250,057	335,522	100,093	383,916	9.00
1974	251,650	310,231	74,500	369,325	11.83
1975	266,884	327,521	74,660	390,000	11.88
1976	281,066	350,427	79,750	415,491	11.50
1977	293,928	366,730	85,950	431,164	12.00
1978	310,640	390,189	91,100	458,675	13.46
1979	318,817	406,857	99,121	476,048	16.71
1980	319,341	401,942	92,705	485,108	21.25
1981	325,851	419,669	85,750	484,259	21.33
1982	338,507	421,716	84,100	483,879	20.50
1983	339,425	417,930	83,000	481,198	20.50
1984	345,194	434,696	78,300	490,881	20.50
1985	358,671	456,576	82,360	502,258	20.50
1986	361,026	439,654	77,234	507,199	20.50
1987	365,473	438,454	73,315	505,713	21.82
1988	378,488	476,345	79,831	529,460	22.89
1989	394,942	492,334	87,873	546,572	23.26
1990	403,194	495,939	96,139	546,982	27.62
1991	412,458	513,173	91,726	566,586	29.45
1992	420,028	502,520	93,140	568,582	28.71
1993	420,585	523,066	91,292	569,724	28.56
1994	426,893	520,728	93,073	579,846	27.44
1995	433,723	518,407	98,470	588,691	23.05

Sources: Epilogi (1998). The Greek economy in figures.

Taking into account that autocorrelation in equation (4.166) is of the moving average scheme (although not significant in the actual estimates) the ML with MA(1) estimates in Table 4.7 have generally the highest asymptotic efficiency compared to the other two estimates in the same table, the estimated adaptive expectations model is written as

$$\hat{C}_t = 11199.7 + 0.26311Y_t + 0.65879C_{t-1} \qquad (4.170)$$

or using (4.167) as:

$$\hat{C}_t = 32823.5 + 0.77111Y_t^* \qquad (4.171)$$

with

$$Y_t^* - Y_{t-1}^* = 0.34121(Y_t - Y_{t-1}^*)$$

Table 4.7 Estimates of the consumption function for Greece, 1960–1995

	OLS	*Instrumental variables*	*Maximum likelihood with MA(1)*
Constant	11,282.2 (6.4971)	11,307.1 (5.8371)	11,199.7 (5.6583)
Income (Y_t)	0.24718 (5.6352)	0.30976 (5.6481)	0.26311 (5.9379)
Lagged Consumption(C_{t-1})	0.67864 (12.2495)	0.59902 (8.6738)	0.65879 (11.7678)
Adjusted R^2	0.99881	0.99861	0.99883
DW	1.5349	1.3383	1.9478
h	1.4563 [0.145]		
Breusch-Godfrey LM(1)	1.5630 [0.211]	2.6140 [0.106]	
SARGAN		0.0826 [0.774]	
MA(1)			0.24113[0.197]

Notes: Estimates from Microfit.
( ) t-ratios in parentheses.
[ ] significant levels in brackets.

Finally, the consumption function in the form of a geometric infinite distributed lag model, using (4.168), is written as

$$\hat{C}_t = 32823.5 + 0.26311X_t + 0.1333X_{t-1} + 0.11419X_{t-2} + \ldots \quad (4.172)$$

Estimates from (4.170) yield a marginal propensity to consume (MPC) equal to 0.26, implying that a 100 drachmas increase in the current income would increase current consumption by 26 drachmas. However, if this increase in income is sustained, then from equation (4.171) we see that the marginal propensity to consume out of permanent income will be 0.77, implying that a 100 drachmas increase in permanent income would increase current consumption by 77 drachmas. By comparing these two marginal propensities to consume (the short-run MPC = 0.26 and the long-run MPC = 0.77) and because the expectation coefficient is estimated 0.34, this implies that about one-third of the expectations of the consumers are realised in any given period.

Using finally the estimates in (4.172) and formulas (4.68) to (4.70) we find respectively the mean lag, the median lag, the impact multiplier, and the equilibrium multiplier as:

$$\text{mean lag} = \frac{\lambda}{1-\lambda} = \frac{0.65879}{1-0.65879} = 1.93$$

$$\text{median lag} = \frac{\ln(0.5)}{\ln(\lambda)} = \frac{\ln(0.5)}{\ln(0.65879)} = 1.66$$

$$\text{impact multiplier} = \beta_0 = 0.26$$

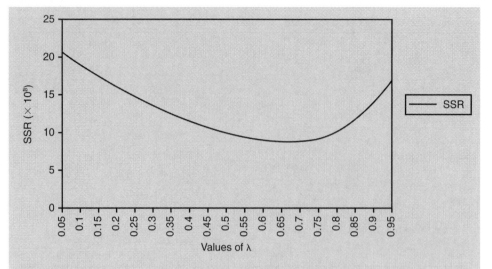

Figure 4.5 Sum of squared residuals with respect to λ

$$\text{equilibrium multiplier} = \beta_0 \frac{1}{1-\lambda} = 0.26311 \frac{1}{1-0.65879} = 0.77$$

We saw above that the short-run MPC is the impact multiplier and the long-run MPC is the equilibrium multiplier. Furthermore, the mean lag is 1.93, showing that it takes on average about two years for the effect of changes in income to be transmitted on consumption changes. Finally, the median lag is 1.66, meaning that 50% of the total change in consumption is accomplished in about one and a half years. Summarising, consumption adjusts to income within a relatively long time.

Example 4.6 The accelerator model of investment for Greece, 1960–1995

According to this model the desired level of capital stock K_t^* is a linear function of the level of output Q_t in the economy. This is written as

$$K_t^* = \alpha + \beta Q_t \tag{4.173}$$

Considering the partial adjustment model, the mechanism used to transform the unobserved level of the desired capital stock into an observed one is

$$K_t - K_{t-1} = \gamma(K_t^* - K_{t-1}) + \epsilon_t, \, 0 < \gamma \leq 1 \tag{4.174}$$

or

$$K_t = \gamma K_t^* + (1 - \gamma)K_{t-1} + \epsilon_t \tag{4.175}$$

meaning that the actual net investment is only a fraction of the investment required to achieve the desired capital stock. Following the methodology in 7.1, the capital stock function in observable levels is the following:

$$K_t = \alpha\gamma + \beta\gamma Q_t + (1 - \gamma)K_{t-1} + \epsilon_t \tag{4.176}$$

Although equation (4.176) is in an estimable form, in fact the partial adjustment form, this equation produces some estimation problems because the data for the capital stock variable is not usually very reliable. However, by taking into account that capital stock K_t at the end of a period is equal to the capital stock K_{t-1} at the beginning of the period plus gross investment I_t less depreciation D_t, i.e. it is

$$K_t = K_{t-1} + I_t - D_t \tag{4.177}$$

and assuming that depreciation in period t is proportional to the existing capital stock in period $t - 1$, i.e. it is

$$D_t = \delta K_{t-1} + \eta_t \tag{4.178}$$

where δ = depreciation rate and η_t = error term, then, equation (4.176) can be expressed in variables which may be more reliable in terms of data.

Substituting (4.178) into (4.177), and then substituting in the result (4.175), and then rearranging we obtain that

$$I_t = \gamma K_t^* - (\gamma - \delta)K_{t-1} + (\epsilon_t + \eta_t) \tag{4.179}$$

Multiplying both terms of the lagged once equation (4.179) times $(1 - \delta)$, adding the result to equation (4.179), and rearranging we get that

$$I_t = \gamma K_t^* - \gamma(1 - \delta)K_{t-1}^* + (1 - \gamma)I_{t-1} + \{[\epsilon_t - (1 - \delta)\epsilon_{t-1}] + [\eta_t - (1 - \delta)\eta_{t-1}]\} \tag{4.180}$$

Finally, substituting (4.173) into (4.180) we obtain

$$I_t = \alpha\gamma\delta + \beta\gamma Q_t - \beta\gamma(1 - \delta)Q_{t-1} + (1 - \gamma)I_{t-1} + \upsilon_t \tag{4.181}$$

where

$$\upsilon_t = \{[\epsilon_t - (1 - \delta)\epsilon_{t-1}] + [\eta_t - (1 - \delta)\eta_{t-1}]\}$$

or

$$I_t = \alpha_0 + \beta_0 Q_t + \beta_1 Q_{t-1} + \gamma_1 I_{t-1} + \upsilon_t \tag{4.182}$$

where

$$\alpha_0 = \alpha\gamma\delta, \; \beta_0 = \beta\gamma, \; \beta_1 = \beta\gamma(1 - \delta), \; \gamma_1 = 1 - \gamma \tag{4.183}$$

The good thing about the specification of equation (4.183) (for more specifications see Wallis (1973) and Desai (1976)) is that by estimating it we can estimate all the parameters of the partial adjustment model in equation (4.176) plus the depreciation rate without even having data on the capital stock and/or depreciation.

Using the data in Table 4.6, I_t = gross investment and Q_t = gross national product, Table 4.8 presents the estimates by applying to equation (4.182) techniques described in section 4.8. The instruments used for the IV estimation are a constant, Q_t, Q_{t-1}, r_t and r_{t-1}, where r_t = long-term interest rate.

Although all estimates in Table 4.8 are generally acceptable, with the coefficients having the proper a priori signs that equation (4.181) predicts, the question is which of these estimates is the 'best'. In order to answer this question we should look at the specification of equation (4.181). Because the error term in (4.181) is of the first-order moving average type it seems that the ML with MA(1) estimates in Table 4.8 are more appropriate. Therefore, using these estimates (a_0, b_0, b_1, c_1) of the corresponding regression coefficients (α_0, β_0, β_1, γ_1) the estimated parameters (a, b, c, d) of the corresponding parameters (α, β, γ, δ) of the accelerator model of investment for Greece are given by the formulas (4.183), i.e.

Table 4.8 Estimates of the investment function for Greece, 1960–1995.

	OLS	IV	CORC	Maximum likelihood with AR(1)	Maximum likelihood with MA(1)
Constant	565.7277 (0.1635)	−608.6250 (0.15899)	−280.8577 (0.09640)	−278.5964 (0.09564)	−945.8745 (2.2135)
Q_t	0.50091 (6.0105)	0.49435 (5.8521)	0.55709 (7.2828)	0.55702 (7.2823)	0.55788 (9.9300)
Q_{t-1}	−0.46599 (5.7579)	−0.46723 (5.7235)	−0.52420 (7.1150)	−0.52413 (7.1145)	−0.52697 (9.9701)
I_{t-1}	0.74733 (9.2742)	0.80591 (7.2558)	0.76090 (11.3840)	0.76088 (11.3861)	0.77908 (17.2404)
Adjust. R^2	0.94212	0.94113	0.93642	0.93642	0.94709
DW	2.4427	2.4942	2.0362	2.0368	1.8726
h	−1.4898 [0.136]				
BG	2.6856 [0.101]	2.1146 [0.146]			
SARGAN		0.44211 [0.506]			
AR(1)			−0.31511 [0.097]	−0.31462 [0.097]	
MA(1)					−0.44233 [0.037]

Notes: Estimates from Microfit.
( ) t-ratios in parentheses.
[ ] significant levels in brackets.

$$c_1 = 1 - c \Rightarrow c = 1 - c_1 \Rightarrow c = 0.22092$$

$$b_0 = bc \Rightarrow b = \frac{b_0}{c} \Rightarrow b = 2.52526 \qquad (4.184)$$

$$b_1 = -bc(1 - d) \Rightarrow d = 1 + \frac{b_1}{bc} = 0.05541$$

$$a_0 = acd \Rightarrow a = \frac{a_0}{cd} = -77269.9$$

These estimated parameters seem plausible, with the implicit desired capital stock–GNP ratio being 2.52526 (neglecting the constant term), the depreciation rate being 0.05541, and the adjustment coefficient being 0.22092. The adjustment coefficient is quite small showing that the adjustment of the current level of capital stock to the desired level of capital stock is rather slow.

DISCUSSION QUESTIONS

1   In a multiple linear regression model of your choice, show that a partial adjustment model implies a Koyck type lag structure. What are the implications of this result?

2   The impact of advertising expenditure on sales is assumed to increase first, reach a peak and then decline. Formulate an appropriate finite lag model of the relationship between sales and advertising expenditure assuming that the lag length is k. What are the consequences of an incorrect order of polynomial? How would you decide on the order of polynomial?

3   Consider the regression model $C_t = \beta_1 + \beta_2 Y^e_t + u_t$ where $u_t \sim \text{nid}(0, \sigma^2)$, and $(Y^e_t - Y^e_{t-1}) = \gamma(Y_t - Y^e_{t-1})$; $0 \le \gamma \le 1$. Explain how the regression model might be estimated. What are the problems with the OLS estimation?

Bibliography

Akaike, H. (1973) 'Information theory and an extension of the maximum likelihood principle' in B. Petrov and F. Csake (eds), *2nd International Symposium on Information Theory,* Budapest: Akademiai Kiado.

Almon, S. (1965) 'The distributed lag between capital appropriations and expenditures', *Econometrica,* 30, 178–196.

Balestra, P. (1980) 'A note on the exact transformation associated with the first-order moving average process', *Econometrica,* 14, 381–394.

Breusch, T. (1978) 'Testing for autocorrelation in dynamic linear models', *Australian Economic Papers,* 17, 334–355.

Cagan, P. (1956) 'The Monetary Dynamics of Hyper Inflations' in M. A. Friedman (ed.) *Studies in the Quantity Theory of Money,* Chicago: University Press.

Cochrane, D., and G. Orcutt (1949) 'Application of least squares regression to relationships containing autocorrelated error terms', *Journal of the American Statistical Association,* 44, 32-61.

DeLeeuw, F. (1962) 'The demand for capital goods by manufactures: A study of quarterly time series', *Econometrica,* 30, 407–423.

Desay, M. (1976) *Applied Econometrics,* London: Philip Allan.

Durbin, J. (1970) 'Testing for serial correlation in least squares regression when some of the regressors are lagged dependent variables', *Econometrica,* 38, 410–421.

Durbin, J. and G. Watson (1950) 'Testing for serial correlation in least squares regression-I', *Biometrica,* 37, 409–428.

Durbin, J. and G. Watson (1951) 'Testing for serial correlation in least squares regression-II', *Biometrica,* 38, 159–178.

Fisher, I. (1937) 'Note on a short-cut method for calculating distributed lags', *International Statistical Bulletin,* 29, 323–328.

Friedman, M. A. (1957) *Theory of the Consumption Function,* Princeton, N.J.: Princeton University Press.

Godfrey, L. (1978) 'Testing against general autoregressive and moving average error models when the regressors include lagged dependent variables', *Econometrica,* 46, 1293–1302.

Greene, W. H. (1993) *Econometric Analysis,* New York: Macmillan.

Griffiths, W. E., R. C. Hill, and G. G. Judge (1993) *Learning and Practicing Econometrics,* New York: John Wiley.

Harvey, A. C. (1990) *The Econometric Analysis of Time Series,* 2nd ed., Cambridge, Mass: MIT Press.

Hatanaka, M. (1976) 'Several efficient two-step estimators for dynamic simultaneous equations models with autoregressive disturbances', *Journal of Econometrics,* 4, 189–204.

Hildreth, G. and T. Lu (1960) 'Demand relations with autocorrelated disturbances', *Technical Bulletin 276,* Michigan State University Agricultural Experiment Station.

Johnston, J. (1984) *Econometric Methods,* 3nd ed., New York: McGraw-Hill.

Jorgenson, D. W. (1963) 'Capital theory and investment behaviour', *American Economic Review,* 53, 247–259.

Judge, G. G., W. E. Griffiths, R. C. Hill, H. Lutkepohl, and T. C. Lee (1985) *The Theory and Practice of Econometrics,* 2nd ed., New York: Wiley.

Kmenta, J. (1986) *Elements of Econometrics,* New York: Macmillan.

Koyck, L. M. (1954) *Distributed Lags and Investment Analysis,* Amsterdam: North-Holland.

Liviatan, N. (1963) 'Consistent estimation of distributed lags', *International Economic Review,* 4, 44–52.

Nerlove, M. (1956) 'Estimates of the elasticities of supply of selected agricultural commodities', *Journal of Farm Economics,* 38, 496–509.

Nerlove, M. (1958) *Distributed lags and Demand Analysis for Agricultural and other Commodities,* Agricultural Handbook No. 141, US Department of Agriculture.

Pesaran, M. H. (1973) 'The small sample problem of truncation remainders in the estimation of distribution lag models with autocorrelated errors', *International Economic Review,* 14, 120–131.

Phillips, A. W. (1958) 'The relation between unemployment and the rate of change of money wages in the United Kingdom, 1861–1957', *Economica,* XXV, 283–299.

Sargan, J. D. (1964) 'Wages and prices in the United Kingdom: a study in econometric methodology' in P. E. Hart, G. Mills, and J. K. Whitaker (eds) *Econometrics Analysis for National Economic Planning,* London: Butterworths.

Schwarz, R. (1978) 'Estimating the dimension of a model', *Annals of Statistics,* 6, 461–464.

Solow, R. M. (1960) 'On a family of lag distributions', *Econometrica,* 28, 393–406.

Wallis, K. F. (1973) *Topics in Applied Econometrics,* London: Gray-Mills.

Zellner, A. and Geisel, M. (1970) 'Analysis of distributed lag models with application to the consumption function', *Econometrica,* 38, 865–888.

5   Simultaneous equations models and econometric analysis

Key points

- Simultaneous equation bias;
- Identification;
- Estimation methods (single equation/system of equations)

1 INTRODUCTION

Most books on econometrics start, and exert considerable efforts investigating single equation models such as demand functions, consumption functions, wage rate functions, and so on. However, economic phenomena are not unique. In most cases the single equations under investigation are either related with other single equations or, they are part of a wider phenomenon which may be explained by a system of equations.

Case 5.1 A demand–supply model for a commodity

In many introductory textbooks in economics the market equilibrium model for a commodity is given by the following equations in systemic form.

$$\text{The demand function: } Q^d_t = \alpha_0 + \alpha_1 P_t + \alpha_2 Y_t + \epsilon_{dt}; \alpha_1 < 0, \alpha_2 > 0 \tag{5.1}$$

$$\text{The supply function: } Q^s_t = \beta_0 + \beta_1 P_t + \epsilon_{st}; \beta_1 > 0 \tag{5.2}$$

$$\text{The equilibrium condition: } Q^d_t = Q^s_t = Q_t \tag{5.3}$$

Where; Q^d_t = quantity demanded, Q^s_t = quantity supplied, Q_t = quantity sold, P_t = price of commodity, Y_t = income of consumers, α's and β's = parameters, ϵ's = random disturbances representing also other factors not included in each equation, and t = specific time period. The rationale of the system of equations (5.1) to (5.3) is that given income, the equilibrium quantity and the equilibrium price can be jointly and interdependently found by solving the system of three equations. This means that different pairs of equilibrium prices and equilibrium quantities correspond to different levels of income. In other words, although income is an explanatory variable for the demand equation only, changes in income cause effects in both the price and quantity sold.

Case 5.2   Income determination in a closed economy without government

The simple Keynesian model of income determination in an economy without transactions with the rest of the world (closed economy) and without any government activity can be written as:

The consumption function: $C_t = \alpha_0 + \alpha_1 Y_t + \epsilon_t; \; \alpha_1 > 0$ (5.4)

The income accounting identity: $Y_t = C_t + I_t$ (5.5)

where C_t = aggregate consumption, I_t = aggregate investment, Y_t = aggregate income, α's = parameters, ϵ_t = random disturbances representing other factors not included in the consumption function, and t = a specific time period. The rationale of the system of equations (5.4) and (5.5) is that given investment the levels of consumption and income can be jointly and interdependently found by the solution of this system. This means that different pairs of equilibrium consumption and equilibrium income correspond to different levels of investment. In other words, although investment enters the macroeconomic accounting income identity only, changes in investment cause effects in both equilibrium consumption and equilibrium income.

Case 5.3   The wages–prices link

The basic Phillips (1958) Curve can be augmented into the following model:

Wages inflation function: $w_t = \alpha_0 + \alpha_1 p_t + \alpha_2 U_t + \epsilon_{wt}; \; \alpha_1 > 0, \alpha_2 < 0$ (5.6)

Prices inflation function: $p_t = \beta_0 + \beta_1 w_t + \beta_2 Y_t + \epsilon_{pt}; \; \beta_1 > 0, \beta_2 > 0$ (5.7)

where $w_t = (W_t - W_{t-1})/W_{t-1}$ with W_t = monetary wage level, U_t = unemployment rate (percent), $p_t = (P_t - P_{t-1})/P_{t-1}$, with P_t = price level, Y_t = aggregate demand (income), α's and β's = parameters, ϵ's = random disturbances representing also other factors not included in each equation, and t = specific time period. The rationale of the system of equations (5.6) and (5.7) is that given the unemployment rate and aggregate demand the levels of monetary wage inflation, and price inflation can be jointly and interdependently found by solution of this system. This means that different pairs of money wage inflation and price inflation correspond to different levels of unemployment rate and aggregate demand. In other words, although the unemployment rate enters the wages function only and aggregate demand enters the prices function only, changes in both or any of these two explanatory variables cause effects in both the money wages inflation and prices inflation.

A common property of the three cases 5.1–5.3 is that, in each case, some of the variables included are *jointly and interdependently determined* by the corresponding system of equations that represents the economic phenomenon. In the first case, for example, price P_t and quantity Q_t are jointly and interdependently determined, whilst income Y_t and the disturbances ϵ_{dt} and ϵ_{st} affect P_t and Q_t but they are not affected by them. In the second case, consumption C_t and income Y_t are jointly and interdependently determined, whilst investment I_t and the disturbances ϵ_t affect C_t

and Y_t but they are not affected by them. In the third case, money wage inflation w_t and price inflation p_t are jointly and interdependently determined, whilst the unemployment rate U_t, aggregate demand Y_t and the disturbances ϵ_{wt} and ϵ_{pt} affect w_t and p_t but they are not affected by them.

Since some of the variables in these systems of equations are jointly and interdependently determined, changes in other variables in the same systems of equations that are not determined by these systems, but are determined outside the systems, are nonetheless transmitted to the jointly and interdependently determined variables. This transmission takes place as an *instantaneous adjustment*, or *feedback*, and we say that the system is a *system of simultaneous equations*. In the first case, for example, the initial result of an increase in income at the existing price level will be increases in demand. The result of a higher demand will be an increase in the price level which pushes demand to lower levels and supply to higher levels. This feedback between changes in price levels and changes in quantities continues until the system is again in equilibrium, i.e. until quantity demanded is equal to the quantity supplied.

Generally, we say that a set of equations is a simultaneous equations system, or model, if all its equations are needed for determining the level of at least one of its jointly and interdependently determined variables. In what follows we investigate models of simultaneous linear equations.

2 DEFINITIONS AND NOTATION

The variables in a simultaneous equations model may be classified as follows:

Endogenous variables

These are the jointly and interdependently determined random variables of the model. Consumption and income, for example, of case 5.2 are endogenous variables.

Exogenous variables

These are the variables that are determined outside the model and independently of the endogenous variables. Although these variables can affect the levels of the endogenous variables, there is no feedback effect. In other words, the endogenous variables cannot affect the levels of the exogenous variables. Unemployment rate and aggregate demand, for example, in case 5.3 are exogenous variables.

Lagged dependent variables

These are the lagged endogenous variables of any lag length. Because these values are predetermined due to lags for the current period t, their status is considered similar to the status of the exogenous variables.

Predetermined variables

These constitute the set of the exogenous variables and the lagged endogenous variables.

Structural disturbances

These variables refer to the usual random error variables, or disturbances, or, random shocks.

Because the equations in a simultaneous equations model describe the structure of an economic phenomenon they are called 'structural equations'. The parameters of a structural equation are called 'structural parameters'. Equations (5.1) to (5.3) are, for example, the structural equations representing the simultaneous equations model of case 5.1, with the α's and β's being the corresponding structural parameters of the model. The structural equations may be classified as follows:

Behavioural equations

These are the equations that are determined by the behaviour of economic agents. The demand and supply functions, for example, of case 5.1 are behavioural equations.

Technical equations

These are the equations that describe a technical situation. A production function, for example, that describes the relationship between inputs and outputs is a technical equation.

Identities

The identities, or definitional equations, express fixed relationships between various variables involved. Equation (5.5), for example, of case 5.2 is a macroeconomic accounting identity. Identities, being deterministic, do not contain either structural parameters or, error terms.

Equilibrium conditions

These are equations which specify the conditions under which some variables may be determined in the model. Equation (5.3), for example, in case 5.1, is an equilibrium condition specifying how the levels of price and quantity are determined. Equilibrium conditions, being also deterministic, do not contain either structural parameters or, error terms.

A system of simultaneous equations is *complete* if the total number of equations is equal to the total number of the endogenous variables it contains.

Given these definitions, the *structural form* of a simultaneous equations model may be written as follows:

$$\gamma_{11}Y_{1t} + \gamma_{12}Y_{2t} + \ldots + \gamma_{1G}Y_{Gt} + \beta_{11}X_{1t} + \beta_{12}X_{2t} + \ldots + \beta_{1K}X_{Kt} = \epsilon_{1t}$$

$$\gamma_{21}Y_{1t} + \gamma_{22}Y_{2t} + \ldots + \gamma_{2G}Y_{Gt} + \beta_{21}X_{1t} + \beta_{22}X_{2t} + \ldots + \beta_{2K}X_{Kt} = \epsilon_{2t}$$

$$\vdots$$

$$\gamma_{G1}Y_{1t} + \gamma_{G2}Y_{2t} + \ldots + \gamma_{GG}Y_{Gt} + \beta_{G1}X_{1t} + \beta_{G2}X_{2t} + \ldots + \beta_{GK}X_{Kt} = \epsilon_{Gt}$$

$$(5.8)$$

where the Y's are G endogenous variables, the X's are K predetermined variables (exogenous and lagged dependent variables), the ϵ's are the disturbances, the γ's and β's are the structural parameters, and t = 1, 2,..., n. This structural model is complete because the number of equations is G as it is the number of endogenous variables. Of course, the equations in this model may contain any type of equations. Furthermore, some of the structural coefficients may equal zero, indicating that not all equations involve the same exact variables. The inclusion of a constant term is indicated by the unitary values of one for the X variables. Normally, the γ_{ii} coefficients (diagonal coefficients) are set equal to one to indicate the dependent variable of the corresponding equation.

Model (5.8) may be written in matrix form as

$$
\begin{bmatrix} \gamma_{11} & \gamma_{12} & \cdots & \gamma_{1G} \\ \gamma_{21} & \gamma_{22} & \cdots & \gamma_{2G} \\ \vdots & \vdots & \vdots & \vdots \\ \gamma_{G1} & \gamma_{G2} & \cdots & \gamma_{GG} \end{bmatrix} \begin{bmatrix} Y_{1t} \\ Y_{2t} \\ \vdots \\ Y_{Gt} \end{bmatrix} + \begin{bmatrix} \beta_{11} & \beta_{12} & \cdots & \beta_{1K} \\ \beta_{21} & \beta_{22} & \cdots & \beta_{2K} \\ \vdots & \vdots & \vdots & \vdots \\ \beta_{G1} & \beta_{G2} & \cdots & \beta_{GK} \end{bmatrix} \begin{bmatrix} X_{1t} \\ X_{2t} \\ \vdots \\ X_{Kt} \end{bmatrix} = \begin{bmatrix} \epsilon_{1t} \\ \epsilon_{2t} \\ \vdots \\ \epsilon_{Gt} \end{bmatrix} \tag{5.9}
$$

or

$$
\mathbf{\Gamma Y}_t + \mathbf{B X}_t = \mathbf{\epsilon}_t, \, t = 1, 2, \dots, n \tag{5.10}
$$

where $\mathbf{\Gamma}$ is a $G \times G$ matrix of the γ coefficients, $\mathbf{B}$ is a $G \times K$ matrix of the β coefficients, $\mathbf{Y}_t$ is a $G \times 1$ vector of the G endogenous variables for time t, $\mathbf{X}_t$ is a $K \times 1$ vector of the K predetermined variables for time t, and $\mathbf{\epsilon}_t$ is a $G \times 1$ vector of the structural disturbances for time t.

The assumptions underlying structural disturbances are those of the classical normal linear regression model. These assumptions are written as

$$\epsilon_{it} \sim N(0, \sigma_{ii}), \quad \text{for all t, and i = 1, 2,..., G, where } \sigma_{ii} = var(\epsilon_{it})$$

$$E(\epsilon_{it}\epsilon_{is}) = 0, \quad \text{for t} \neq s, \text{ and i = 1, 2,..., G} \tag{5.11}$$

$$E(\epsilon_{it}\epsilon_{jt}) = \sigma_{ij}, \quad \text{for all t, and i, j = 1, 2,..., G, where } \sigma_{ij} = cov(\epsilon_{it}, \epsilon_{jt})$$

or in matrix form

$$
\mathbf{\epsilon}_t \sim N(\mathbf{0}, \mathbf{\Sigma}), \text{ with } E(\mathbf{\epsilon}_t\mathbf{\epsilon}_s') = \mathbf{0}, \text{ and } E(\mathbf{\epsilon}_t\mathbf{\epsilon}_t') = \mathbf{\Sigma} = \begin{bmatrix} \sigma_{11} & \sigma_{12} & \cdots & \sigma_{1G} \\ \sigma_{21} & \sigma_{22} & \cdots & \sigma_{2G} \\ \vdots & \vdots & \vdots & \vdots \\ \sigma_{G1} & \sigma_{G2} & \cdots & \sigma_{GG} \end{bmatrix} \tag{5.12}
$$

where symbol (´) indicates 'transpose', and $\mathbf{\Sigma}$ is the so-called variance-covariance matrix of disturbances.

Since model (5.8) is complete, it may be generally solved for endogenous variables. This solution is called 'reduced form model' and it is written as

$$Y_{1t} = \pi_{11}X_{1t} + \pi_{12}X_{2t} + \dots + \pi_{1K}X_{Kt} + \upsilon_{1t}$$

$$Y_{2t} = \pi_{21}X_{1t} + \pi_{22}X_{2t} + \dots + \pi_{2K}X_{Kt} + \upsilon_{2t}$$

$$\vdots \tag{5.13}$$

$$Y_{Gt} = \pi_{G1}X_{1t} + \pi_{G2}X_{2t} + \dots + \pi_{GK}X_{Kt} + \upsilon_{Gt}$$

where the π's are the *reduced form coefficients*, and the v's are the *reduced form disturbances*. The reduced form coefficients show the effects on the equilibrium values of the endogenous variables from a change in the corresponding exogenous variables after all feedbacks have taken place.

Model (5.13) may be written in matrix form as

$$
\begin{bmatrix} Y_{1t} \\ Y_{2t} \\ \vdots \\ Y_{Gt} \end{bmatrix} = \begin{bmatrix} \pi_{11} & \pi_{12} & \cdots & \pi_{1K} \\ \pi_{21} & \pi_{22} & \cdots & \pi_{2K} \\ \vdots & \vdots & \vdots & \vdots \\ \pi_{G1} & \pi_{G2} & \cdots & \pi_{GK} \end{bmatrix} \begin{bmatrix} X_{1t} \\ X_{2t} \\ \vdots \\ X_{Kt} \end{bmatrix} + \begin{bmatrix} v_{1t} \\ v_{2t} \\ \vdots \\ v_{Gt} \end{bmatrix}
\tag{5.14}
$$

or

$$
\mathbf{Y}_t = \mathbf{\Pi}\mathbf{X}_t + \mathbf{v}_t, t = 1, 2, \ldots, n
\tag{5.15}
$$

where $\mathbf{\Pi}$ is a $G \times K$ matrix of the π coefficients, and $\mathbf{v}_t$ is a $G \times 1$ vector of the reduced form disturbances for time t.

If now we return to the structural model (5.10), this can be written as

$$
\mathbf{Y}_t = -\mathbf{\Gamma}^{-1}\mathbf{B}\mathbf{X}_t + \mathbf{\Gamma}^{-1}\boldsymbol{\epsilon}_t
\tag{5.16}
$$

which is the explicit solution for the endogenous variables it contains, under the assumption that the inverse matrix $\mathbf{\Gamma}^{-1}$ exists, or that matrix $\mathbf{\Gamma}$ is non-singular. In other words, solution (5.16) is the *reduced form model* corresponding to the structural form model (5.10).

By comparing (5.16) with (5.15) we obtain

$$
\mathbf{\Pi} = -\mathbf{\Gamma}^{-1}\mathbf{B}
\tag{5.17}
$$

and

$$
\mathbf{v}_t = \mathbf{\Gamma}^{-1}\boldsymbol{\epsilon}_t
\tag{5.18}
$$

From (5.17) it is seen that the reduced form coefficients are functions of the structural coefficients. Furthermore, from (5.18) it is seen that each reduced form disturbance is a linear function of all structural disturbances, and therefore the stochastic properties of the reduced form disturbances depend on the stochastic properties of the structural disturbances. These properties are written as

$$
v_{it} \sim N(0, \omega_{ii}), \text{ for all t, and } i = 1, 2, \ldots, G, \text{ where } \omega_{ii} = \text{var}(v_{it})
$$

$$
E(v_{it}v_{is}) = 0, \text{ for } t \neq s, \text{ and } i = 1, 2, \ldots, G
\tag{5.19}
$$

$$
E(v_{it}v_{jt}) = \omega_{ij}, \text{ for all t, and } i, j = 1, 2, \ldots, G, \text{ where } \omega_{ij} = \text{cov}(v_{it}, v_{jt})
$$

or in matrix form

$$
\mathbf{v}_t \sim N(\mathbf{0},\mathbf{\Omega}), \text{ with } E(\mathbf{v}_t\mathbf{v}_s') = \mathbf{0}, \text{ and}
$$

$$E(\upsilon_t\upsilon_t')= E[\Gamma^{-1}\epsilon_t\epsilon_t'(\Gamma^{-1})'] = \Omega = \Gamma^{-1}\Sigma(\Gamma^{-1})' = \begin{bmatrix} \omega_{11} & \omega_{12} & \cdots & \omega_{1G} \\ \omega_{21} & \omega_{22} & \cdots & \omega_{2G} \\ \vdots & \vdots & \vdots & \vdots \\ \omega_{G1} & \omega_{G2} & \cdots & \omega_{GG} \end{bmatrix} \quad (5.20)$$

Having established the notation and the properties of the general structural model and the corresponding reduced form model, we distinguish two important cases of specific models according to the values of the γ coefficients in matrix Γ.

(a) Seemingly unrelated equations

This is the case where matrix Γ is diagonal, i.e. it has the following form.

$$\Gamma = \begin{bmatrix} \gamma_{11} & 0 & \cdots & 0 \\ 0 & \gamma_{22} & \cdots & 0 \\ \vdots & \vdots & \vdots & \vdots \\ 0 & 0 & \cdots & \gamma_{GG} \end{bmatrix} \quad (5.21)$$

In this case each endogenous variable appears in one and only one equation. In fact we do not have a system of simultaneous equations but instead we have a set of seemingly unrelated equations.

(b) Recursive equations model

This is the case where matrix Γ is triangular, i.e. it has the following form.

$$\Gamma = \begin{bmatrix} \gamma_{11} & 0 & \cdots & 0 \\ \gamma_{21} & \gamma_{22} & \cdots & 0 \\ \vdots & \vdots & \vdots & \vdots \\ \gamma_{G1} & \gamma_{G2} & \cdots & \gamma_{GG} \end{bmatrix} \quad (5.22)$$

In this case the first equation contains one only endogenous variable, let us say the first endogenous variable. The second equation contains the first endogenous variable plus a new one, let us say the second endogenous variable, and so on. The final equation contains all the endogenous variables of the system. In other words, the solution for the first endogenous variable is completely determined by the first equation of the system. The solution for the second endogenous variable is completely determined by the first and the second equations of the system, and so on. The solution for the final endogenous variable is completely determined by all equations of the system.

3 SIMULTANEOUS EQUATIONS BIAS

Let us return to the simple macroeconomic model of case 5.2 presented in section 1, i.e. to the model

The consumption function: $\qquad C_t = \alpha_0 + \alpha_1 Y_t + \epsilon_t;\, \alpha_1 > 0$ $\qquad\qquad$ (5.23)

The income accounting identity: $\quad Y_t = C_t + I_t$ $\qquad\qquad\qquad\qquad$ (5.24)

Chapter 5: Key terms

Simultaneous equations bias OLS estimators are biased and inconsistent due to interrelationships between variables.

Identification problem A Problem concerning simultaneous equation models where one or more equations may not have unique statistical forms (not identified).

Order condition A necessary but not sufficient condition for an equation to be identified.

Rank condition A necessary and sufficient condition for an equation to be identified.

Indirect least squares A single method of estimation based on the OLS method used on an exactly identified equation. The reduced form model is estimated by the OLS.

Two stage least squares (2SLS) A method of estimation for exactly or over identified equations.

According to the notation established in section two, and having classified variables C_t and Y_t as being endogenous and variable I_t as being exogenous, the structural model of equations (5.23) and (5.24) is written in standard form as

$$C_t - \alpha_1 Y_t - \alpha_0 + 0 I_t = \epsilon_t$$
$$- C_t + Y_t + 0 - I_t = 0 \tag{5.25}$$

or, in matrix form as:

$$\begin{bmatrix} 1 & -\alpha_1 \\ -1 & 1 \end{bmatrix} \begin{bmatrix} C_t \\ Y_t \end{bmatrix} + \begin{bmatrix} -\alpha_0 & 0 \\ 0 & -1 \end{bmatrix} \begin{bmatrix} i_t \\ I_t \end{bmatrix} = \begin{bmatrix} \epsilon_t \\ 0 \end{bmatrix} \tag{5.26}$$

where i, is the unitary variable (a $n \times 1$ vector, all values being one) and the disturbance term ϵ_t follows the assumptions of the classical normal linear model, i.e.

$$E(\epsilon_t) = 0, \text{var}(\epsilon_t) = \sigma^2, \text{cov}(\epsilon_t, \epsilon_s) = 0 \text{ for } t \neq s, \epsilon_t \sim N(0, \sigma^2) \tag{5.27}$$

According to (5.10), (5.26) can be written

$$\mathbf{\Gamma Y}_t + \mathbf{B X}_t = \epsilon_t, t = 1, 2, \ldots, n \tag{5.28}$$

where

$$\mathbf{\Gamma} = \begin{bmatrix} 1 & -\alpha_1 \\ -1 & 1 \end{bmatrix}, \mathbf{B} = \begin{bmatrix} -\alpha_0 & 0 \\ 0 & -1 \end{bmatrix}, \mathbf{Y}_t = \begin{bmatrix} C_t \\ Y_t \end{bmatrix}, \mathbf{X}_t = \begin{bmatrix} i_t \\ I_t \end{bmatrix}, \boldsymbol{\epsilon}_t = \begin{bmatrix} \epsilon_t \\ 0 \end{bmatrix} \tag{5.29}$$

Solving the model of equations (5.23) and (5.24) for endogenous variables we obtain the explicit solution

$$C_t = \frac{\alpha_0}{1-\alpha_1} + \frac{\alpha_1}{1-\alpha_1} I_t + \frac{1}{1-\alpha_1} \epsilon_t$$

$$Y_t = \frac{\alpha_0}{1-\alpha_1} + \frac{\alpha_1}{1-\alpha_1} I_t + \frac{1}{1-\alpha_1} \epsilon_t \tag{5.30}$$

which, according to (5.14) and (5.15), in matrix form is written as:

$$\begin{bmatrix} C_t \\ Y_t \end{bmatrix} = \begin{bmatrix} \dfrac{\alpha_0}{1-\alpha_1} & \dfrac{\alpha_1}{1-\alpha_1} \\ \dfrac{\alpha_0}{1-\alpha_1} & \dfrac{1}{1-\alpha_1} \end{bmatrix} \begin{bmatrix} i_t \\ I_t \end{bmatrix} + \begin{bmatrix} \dfrac{1}{1-\alpha_1}\epsilon_t \\ \dfrac{1}{1-\alpha_1}\epsilon_t \end{bmatrix} \tag{5.31}$$

or,

$$\mathbf{Y}_t = \mathbf{\Pi}\mathbf{X}_t + \boldsymbol{v}_t, \, t = 1, 2, \ldots, n \tag{5.32}$$

where:

$$\mathbf{\Pi} = \begin{bmatrix} \pi_{11} & \pi_{12} \\ \pi_{21} & \pi_{22} \end{bmatrix} = \begin{bmatrix} \dfrac{\alpha_0}{1-\alpha_1} & \dfrac{\alpha_1}{1-\alpha_1} \\ \dfrac{\alpha_0}{1-\alpha_1} & \dfrac{1}{1-\alpha_1} \end{bmatrix} \text{ and } \boldsymbol{v}_t = \begin{bmatrix} v_{1t} \\ v_{2t} \end{bmatrix} = \begin{bmatrix} \dfrac{1}{1-\alpha_1}\epsilon_t \\ \dfrac{1}{1-\alpha_1}\epsilon_t \end{bmatrix} \tag{5.33}$$

Solution (5.30), or equivalently (5.32), is the reduced form model of the structural model of the two simultaneous equations (5.23) and (5.24). If fact, this solution can be verified by applying (5.17) and (5.18) respectively as:

$$\mathbf{\Pi} = \begin{bmatrix} \pi_{11} & \pi_{12} \\ \pi_{21} & \pi_{22} \end{bmatrix} = -\mathbf{\Gamma}^{-1}\mathbf{B} = -\frac{1}{1-\alpha_1}\begin{bmatrix} 1 & \alpha_1 \\ 1 & 1 \end{bmatrix}\begin{bmatrix} -\alpha_0 & 0 \\ 0 & -1 \end{bmatrix} = \begin{bmatrix} \dfrac{\alpha_0}{1-\alpha_1} & \dfrac{\alpha_1}{1-\alpha_1} \\ \dfrac{\alpha_0}{1-\alpha_1} & \dfrac{1}{1-\alpha_1} \end{bmatrix} \tag{5.34}$$

and

$$\boldsymbol{v}_t = \begin{bmatrix} v_{1t} \\ v_{2t} \end{bmatrix} = \mathbf{\Gamma}^{-1}\boldsymbol{\epsilon}_t = \frac{1}{1-\alpha_1}\begin{bmatrix} 1 & \alpha_1 \\ 1 & 1 \end{bmatrix}\begin{bmatrix} \epsilon_t \\ 0 \end{bmatrix} = \begin{bmatrix} \dfrac{1}{1-\alpha_1}\epsilon_t \\ \dfrac{1}{1-\alpha_1}\epsilon_t \end{bmatrix} \tag{5.35}$$

Of course, for the existence of the inverse of matrix Γ it must be $1 - \alpha_1 \neq 0$. Furthermore, from (5.19) it can be derived that the properties of the reduced form disturbances $v_{1t} = v_{2t} = v_t$ are given by

$$E(v_t) = 0, \text{var}(v_t) = \sigma_v^2 = \frac{1}{(1 - \alpha_1)^2} \sigma^2, \text{cov}(v_t, v_s) = 0 \text{ for } t \neq s, v_t \sim N(0, \sigma_v^2) \quad (5.36)$$

We said in section 2 that the reduced form coefficients show the effects on the equilibrium values of the endogenous variables from a change in the corresponding exogenous variables after all feedbacks have taken place. Coefficients $\pi_{22} = 1/(1 - \alpha_1)$ and $\pi_{12} = \alpha_1/(1 - \alpha_1)$ for example in (5.33), show the 'total effects' on Y_t and on C_t respectively of a change in I_t. By analysing, for example, coefficient π_{22}, we obtain

$$\pi_{22} = \frac{1}{1 - \alpha_1} = 1 + \frac{\alpha_1}{1 - \alpha_1} \quad (5.37)$$

In other words (5.37) shows that the 'total effect' $1/(1 - \alpha_1)$ on Y_t of a change in I_t, i.e. the investment multiplier, may be divided between a 'direct effect' equal to 1, and an 'indirect effect' equal to $\alpha_1/(1 - \alpha_1)$. The direct effect, 1, is the direct increase in Y_t, shown in the income accounting identity (5.24), as the coefficient of I_t, and the indirect effect is the indirect increase in Y_t through the increase in consumption C_t, $\alpha_1/(1 - \alpha_1)$, in equation (5.30), transmitted to the income accounting identity.

In the structural model of equations (5.23) and (5.24) it is assumed that the exogenous variable I_t is independent of the error term ϵ_t. However, this is not true for the endogenous variable Y_t, although this variable appears as an explanatory variable in equation (5.23). This is because this variable is jointly and interdependently determined by variables I_t and ϵ_t, as shown in the reduced form model in (5.30) or (5.32), and therefore it can be proved that:

$$\text{cov}(Y_t, \epsilon_t) = \frac{1}{1 - \alpha_1} \sigma^2 \neq 0 \quad (5.38)$$

(5.38) shows that the explanatory variable Y_t in equation (5.23) is not independent of the error term ϵ_t, and therefore this equation does not satisfy the assumptions of the classical regression model. Thus, the application of OLS to this equation would yield biased and inconsistent estimates.

Generally, because the endogenous variables of a simultaneous equations model are all correlated with the disturbances, the application of OLS to equations in which endogenous variables appear as explanatory variables, yields biased and inconsistent estimates. This failure of OLS is called 'simultaneous equations bias'.

This simultaneous equations bias does not disappear by increasing the size of the sample, because it can be proved that

$$\text{plim}(a_1) = \alpha_1 + \frac{1}{1 - \alpha_1} \left(\frac{\sigma^2}{\text{var}(Y_t)} \right) \quad (5.39)$$

where a_1 is the OLS estimator of α_1. (5.39) shows that the probability limit of a_1 is not equal to the true population parameter α_1, and thus the estimator is inconsistent. In fact (5.39) shows that the plim(a_1) will be always greater than α_1 because $\alpha_1 < 1$ and the variances are positive numbers.

4 COPING WITH SIMULTANEOUS EQUATIONS BIAS

We saw earlier that the problem of the simultaneous equations bias arises when we apply the OLS method to the structural model because, due to the correlation between the endogenous variables of the system and its error terms, the OLS estimator is biased and inconsistent. But we also saw in the previous section that the reduced form model contains only predetermined variables on the right hand side of its equations which are not correlated with the reduced form disturbances. Therefore, by assuming that the reduced form disturbances satisfy the classical linear regression model assumptions, the application of the OLS method to the structural form equations generally give unbiased and consistent estimates.

The question that now arises is, can we estimate the structural parameters of the model by using the consistent estimates of the reduced form parameters? The answer to this question is that may be there is a solution to this problem by employing (5.17) which connects the parameters of the reduced form model with the parameters of the structural model. Let us consider the following examples.

Example 5.1 Estimating a simple income determination model

Assume a structural model of equations (5.23) and (5.24). The corresponding reduced form model in its specific form is that given in (5.30), and in general form is that given in (5.32). From the specific and general structural forms we saw in the previous section that:

$$\pi_{11} = \pi_{21} = \frac{\alpha_0}{1 - \alpha_1}, \; \pi_{12} = \frac{\alpha_1}{1 - \alpha_1}, \; \pi_{22} = \frac{1}{1 - \alpha_1} \tag{5.40}$$

This means that if we had the consistent estimates of the π's, then we could get consistent estimates of the α's by solving the system of equations in (5.40). By denoting with $\wedge$ the estimates of the π's and with 'a' the estimates of the α's the estimates of the structural coefficients are:

$$\frac{\hat{\pi}_{21}}{\hat{\pi}_{22}} = \frac{a_0/(1 - a_1)}{1/(1 - a_1)} = a_0 \text{ and } \frac{\hat{\pi}_{12}}{\hat{\pi}_{22}} = \frac{a_1/(1 - a_1)}{1/(1 - a_1)} = a_1 \tag{5.41}$$

This technique by which we first estimate the reduced form parameters and then we obtain through the system (5.40) estimates of the structural parameters it is called 'indirect least squares (ILS)'.

Table 4.6 in Chapter 4 contains data referring to the private consumption

(C_t) and to private disposable income (Y_t) for an EU member. Furthermore, private savings ($S_t = Y_t - C_t$) it assumed that all are invested (I_t). In other words we have the simple structural model of equations (5.23) and (5.24) and the corresponding reduced form model (5.30), or (5.33).

Using the data in Table 4.6, the OLS estimates of the reduced form equations of the model are (s.e. = standard errors and t = t ratio in absolute values):

$$\hat{C}_t = 74218.71 + 3.22239 I_t \qquad \bar{R}^2 = 0.833$$

s.e.   (14013.2)   (0.19763)   (5.42)

t   [5.29635]   [16.30515]

and

$$\hat{Y}_t = 74218.71 + 4.22239 I_t \qquad \bar{R}^2 = 0.929$$

s.e.   (14013.2)   (0.19763)   (5.43)

t   [5.29635]   [21.36510]

Corresponding to (5.40) we have:

$$\hat{\pi}_{11} = \hat{\pi}_{21} = \frac{a_0}{1 - a_1} = 74218.71, \ \hat{\pi}_{12} = \frac{a_1}{1 - a_1} = 3.22239, \ \hat{\pi}_{22} = \frac{1}{1 - a_1} = 4.22239$$

and therefore, from (5.41) we obtain the estimates of the structural coefficients, using the method of indirect least squares, as:

$$a_0 = \frac{\hat{\pi}_{21}}{\hat{\pi}_{22}} = 17577.39 \text{ and } a_1 = \frac{\hat{\pi}_{12}}{\hat{\pi}_{22}} = 0.76317 \qquad (5.44)$$

Summarising we have managed to get *unique estimates* of the structural coefficients by applying the method of indirect least squares.

Ignoring the problem of the simultaneous equations bias, if we apply the method of OLS directly to the structural equation of private consumption we get the following estimates:

$$\hat{C}_t = 11907.23 + 0.77958 Y_t, \qquad \bar{R}^2 = 0.994$$

s.e.   (3812.87)   (0.01032)   (5.45)

t   [3.12291]   [75.5662]

By comparing (5.45) with (5.44) we see that $a_{1,OLS} = 0.77958 > a_{1,ILS} = 0.76317$. This result verifies, in a sense, the theoretical result stated in (5.39).

Example 5.2 Estimating a demand–supply model for meat

Considering the market equilibrium condition $Q_t^d = Q_t^s$ a model describing the demand and supply equations for meat could be:

The demand function: $Q_t = \alpha_0 + \alpha_1 P_t + \alpha_2 P_{ft} + \alpha_3 Y_t + \epsilon_{1t}$ (5.46)

The supply function: $Q_t = \beta_0 + \beta_1 P_t + \beta_2 w_t + \epsilon_{2t}$ (5.47)

where Q_t = the quantity of meat, P_t = the price of meat, P_{ft} = the price of fish as a substitute for meat, Y_t = income, w_t = cost of labour as a main factor of production, and ϵ_t's are the error terms. According to the laws of demand and supply we expect $\alpha_1 < 0$ and $\beta_1 > 0$ respectively. We expect $\alpha_2 > 0$ because fish is a substitute for meat and $\alpha_3 > 0$ because the higher the income the greater the meat demanded. Finally, we expect $\beta_2 < 0$ because the higher the prices of the factors of production the lower the production.

By solving the system of two structural equations (5.46) and (5.47) for the two endogenous variables P_t and Q_t we get the following reduced form model.

$$P_t = \pi_{11} + \pi_{12} w_t + \pi_{13} P_{ft} + \pi_{14} Y_t + \upsilon_{1t}$$ (5.48)

$$Q_t = \pi_{21} + \pi_{22} w_t + \pi_{23} P_{ft} + \pi_{24} Y_t + \upsilon_{2t}$$ (5.49)

where:

$$\pi_{11} = \frac{\beta_0 - \alpha_0}{\alpha_1 - \beta_1}, \ \pi_{12} = \frac{\beta_2}{\alpha_1 - \beta_1}, \ \pi_{13} = -\frac{\alpha_2}{\alpha_1 - \beta_1}, \ \pi_{14} = -\frac{\alpha_3}{\alpha_1 - \beta_1}$$ (5.50)

$$\pi_{21} = \frac{\beta_0 \alpha_1 - \alpha_0 \beta_1}{\alpha_1 - \beta_1}, \ \pi_{22} = \frac{\alpha_1 \beta_2}{\alpha_1 - \beta_1}, \ \pi_{23} = -\frac{\alpha_2 \beta_1}{\alpha_1 - \beta_1}, \ \pi_{24} = -\frac{\alpha_3 \beta_1}{\alpha_1 - \beta_1}$$ (5.51)

and

$$\upsilon_{1t} = \frac{\epsilon_{2t} - \epsilon_{1t}}{\alpha_1 - \beta_1}, \ \upsilon_{2t} = \frac{\alpha_1 \epsilon_{2t} - \beta_1 \epsilon_{1t}}{\alpha_1 - \beta_1}$$ (5.52)

If we make use of the data in Table 5.1 where Q_t = meat consumption, Y_t = personal disposable income, P_t = the meat consumption price index, P_t = the fish consumption price index, and w_t = real unit labour cost, the OLS estimates of the two reduced form equations (5.48) and (5.49) are:

$$\hat{P}_t = -4.08811 + 3.15485 w_t + 0.46503 P_{ft} + 5.97E - 06 Y_t \qquad \bar{R}^2 = 0.996$$

s.e.   (1.8931) (1.6274)   (0.0082)   (1.37E − 06)   (5.53)

t   [2.1595] [1.9385]   [56.5939]   [4.3701]

Table 5.1 Meat consumption data for an EU member

Year	Meat consumption (constant 1970 prices, million drs)	Personal disposable income (constant 1970 prices, million drs)	Meat consumption price index (base year: 1970)	Fish consumption price index (base year: 1970)	Real unit labour cost* (total economy, base year: 1970)
1960	6978.00	117179.2	0.730582	0.631294	1.245935
1961	7901.00	127598.9	0.687128	0.643396	1.160569
1962	8660.00	135007.1	0.668129	0.664978	1.153455
1963	9220.00	142128.3	0.705857	0.686165	1.096545
1964	9298.00	159648.7	0.825554	0.723415	1.092480
1965	11718.00	172755.9	0.844342	0.793698	1.070122
1966	12858.00	182365.5	0.874242	0.854387	1.073171
1967	13686.00	195611.0	0.886746	0.878201	1.074187
1968	14468.00	204470.4	0.871924	0.890306	1.074187
1969	15041.00	222637.5	0.888305	0.913162	1.032520
1970	16273.00	246819.0	1.000000	1.000000	1.000000
1971	17538.00	269248.9	1.088836	1.109464	0.980691
1972	18632.00	297266.0	1.139330	1.181059	0.969512
1973	19789.00	335521.7	1.421598	1.398077	0.895325
1974	20333.00	310231.1	1.642847	1.600000	0.917683
1975	21508.00	327521.3	1.809792	1.993345	0.927846
1976	22736.00	350427.4	2.100325	2.158964	0.942073
1977	25379.00	366730.0	2.294023	2.390215	0.991870
1978	26151.00	390188.5	2.498719	3.168778	1.018293
1979	26143.00	406857.2	3.243239	3.794650	1.021341
1980	26324.00	401942.8	3.861875	6.849719	1.000000
1981	24324.00	419669.1	5.269651	9.000263	1.065041
1982	26155.00	421715.6	6.708813	10.17698	1.073171
1983	27098.00	417930.3	7.873865	11.63759	1.101626
1984	27043.00	434695.7	8.948711	13.95143	1.077236
1985	27047.00	456576.2	10.54561	18.22986	1.104675
1986	27136.00	439654.1	12.26179	23.00331	1.046748
1987	28616.00	438453.5	13.43940	27.15239	1.026423
1988	29624.00	476344.7	15.09249	29.46009	1.033537
1989	30020.00	492334.4	18.44444	33.97218	1.081301
1990	29754.00	495939.2	21.63215	40.39124	1.116870
1991	29332.00	513173.0	23.91613	45.39971	1.016260
1992	30665.00	502520.1	27.24601	51.81363	0.969512
1993	31278.00	523066.1	30.08261	57.15542	0.965447
1994	35192.00	520727.5	31.82757	63.53983	0.977642
1995	36505.00	518406.9	33.60107	72.49126	0.997967

Sources: Epilogi (1998).
* EC (1997). Annual Economic Report for 1997, *European Economy*.

and

$$\hat{Q}_t = 8501.874 - 6367.821w_t + 19.37733P_{ft} + 0.05726Y_t \qquad \bar{R}^2 = 0.974$$

s.e.   (3991.87)   (3431.64)   (17.3266)   (0.0029)   (5.54)

t   [2.1298]   [1.8556]   [1.1184]   [19.8658]

Having obtained the reduced form estimates in (5.53) and (5.54) let us try, following the methodology of the previous example, to find the indirect least squares estimates of the structural form model. Let us try, for example, to find an estimate b_1 of β_1. From (5.50) and (5.51) we get:

$$\frac{\hat{\pi}_{23}}{\hat{\pi}_{13}} = \frac{-a_2 b_1/(a_1 - b_1)}{-a_2/(a_1 - b_1)} = b_1 = \frac{19.3773}{0.46503} = 41.6689$$

However, from the same system of (5.50) and (5.51) we can also get that

$$\frac{\hat{\pi}_{24}}{\hat{\pi}_{14}} = \frac{-a_3 b_1/(a_1 - b_1)}{-a_3/(a_1 - b_1)} = b_1 = \frac{0.05726}{5.97\text{E} - 06} = 9591.62$$

By comparing the two values of the estimate b_1 we see that these values are very different. This means that by applying in this example the same methodology of indirect least squares as we did in example 5.1 we got two consistent but different estimates of β_1. This result is due to the fact that, in estimating equations (5.48) and (5.49) we did not take into account the restriction of $\pi_{23}/\pi_{13} = \pi_{24}/\pi_{14}$. In other words the use of the indirect least squares method does not give always unique estimates. This creates a problem which is known as the 'problem of identification'.

5 IDENTIFICATION

Let us summarise the simultaneous equations models used until now as follows:
Structural equations model:

$$\mathbf{\Gamma Y}_t + \mathbf{B X}_t = \boldsymbol{\epsilon}_t \tag{5.55}$$

with $\boldsymbol{\epsilon}_t \sim N(\mathbf{0}, \boldsymbol{\Sigma})$ (5.56)

where $\mathbf{Y}_t$ = is a $G \times 1$ vector of the endogenous variables in time t.

$\mathbf{X}_t$ = is a $K \times 1$ vector of the predetermined variables in time t.

$\boldsymbol{\epsilon}_t$ = is a $G \times 1$ vector of the structural disturbances in time t.

$\mathbf{\Gamma}$ = is a $G \times G$ nonsingular matrix of the γ's structural coefficients.

$\mathbf{B}$ = is a $G \times K$ matrix of the β's structural coefficients.

$\boldsymbol{\Sigma}$ = is a $G \times G$ symmetric positive definite matrix of the variance–covariance parameters of the structural disturbances.

The specific solution, or restricted reduced form model; being

$$\mathbf{Y}_t = -\mathbf{\Gamma}^{-1}\mathbf{B X}_t + \mathbf{\Gamma}^{-1}\boldsymbol{\epsilon}_t \tag{5.57}$$

The general solution, or unrestricted reduced form model; being

$$\mathbf{Y}_t = \mathbf{\Pi}\mathbf{X}_t + \mathbf{v}_t \tag{5.58}$$

with $\mathbf{v}_t \sim \mathrm{N}(\mathbf{0}, \mathbf{\Omega})$ (5.59)

where: $\mathbf{\Omega} = \mathbf{\Gamma}^{-1}\mathbf{\Sigma}(\mathbf{\Gamma}^{-1})'$ (5.60)

And, $\mathbf{v}_t$ = is a $G \times 1$ vector of the reduced form disturbances in time t.

$\quad$ $\mathbf{\Pi}$ = is a $G \times K$ matrix of the π's reduced form coefficients.

$\quad$ $\mathbf{\Omega}$ = is a $G \times G$ matrix of the variance–covariance parameters of structural disturbances.

The coefficients system:

$$\mathbf{\Pi} = -\mathbf{\Gamma}^{-1}\mathbf{B} \tag{5.61}$$

The variance–covariance system:

$$\mathbf{v}_t = \mathbf{\Gamma}^{-1}\mathbf{\epsilon}_t \tag{5.62}$$

Furthermore, we saw that if we apply an OLS estimator to the structural model, then we have the problem of simultaneous equation bias, i.e. we will get inconsistent estimates of the structural coefficients $\mathbf{B}$ and $\mathbf{\Gamma}$, because the endogenous variables of the model, $\mathbf{Y}_t$, are not independent of the structural error term, $\mathbf{\epsilon}_t$. We also saw that we can apply the OLS estimator to the unrestricted reduced form model, in order to obtain consistent estimates of the reduced form coefficients $\mathbf{\Pi}$, because the predetermined variables, $\mathbf{X}_t$, are independent of the reduced form error term, $\mathbf{v}_t$. Finally, we saw that using the indirect least squares method, i.e. having estimated first the reduced form coefficients consistently with OLS and secondly obtaining consistent estimates of the structural coefficients through the coefficients system, there are cases where we can get unique estimates. Of course, there are other cases where we cannot obtain unique estimates of the structural coefficients, or, estimates at all. This brings the problem of identification.

Equations in a simultaneous equations model may be grouped into the following two categories:

Identified equations   These are the equations for which estimates of the structural coefficients can be obtained from the estimates of the reduced form coefficients.

Unidentified, or underidentified, equations   These are the equations for which estimates of the structural coefficients cannot be obtained from the estimates of the reduced form coefficients.

The identified equations may further be grouped into the following two categories:

Exactly, or just, or fully, identified equations   These are the identified equations for which unique estimate of the structural coefficients can be obtained.

Overidentified equations   These are the identified equations for which more than one estimate of at least one of their structural coefficients can be obtained.

Generally, the problem of identification arises because the same reduced form model may be compatible with more than one structural model, or in other words, with more than one theory. We say then that we have 'observationally equivalent relations' and we cannot distinguish them without more information.

5.1 Observationally equivalent relations

We shall investigate the problem of identification by using the concept of the observationally equivalent relations in various cases.

Case 5.4 Underidentified equations

If Q = quantity and P = price, assume the simple structural demand-supply model under equilibrium conditions.
   Where:

$$\text{Demand} = Q_t = \alpha_0 + \alpha_1 P_t + \epsilon_{1t} \tag{5.63}$$

$$\text{Supply} = Q_t = \beta_0 + \beta_1 P_t + \epsilon_{2t} \tag{5.64}$$

The reduced form model is:

$$P_t = \frac{\beta_0 - \alpha_0}{\alpha_1 - \beta_1} + \frac{\epsilon_{2t} - \epsilon_{1t}}{\alpha_1 - \beta_1} = \pi_1 + \upsilon_{1t} \tag{5.65}$$

$$Q_t = \frac{\alpha_1 \beta_0 - \alpha_0 \beta_1}{\alpha_1 - \beta_1} + \frac{\alpha_1 \epsilon_{2t} - \beta_1 \epsilon_{1t}}{\alpha_1 - \beta_1} = \pi_2 + \upsilon_{2t} \tag{5.66}$$

where:

$$\pi_1 = \frac{\beta_0 - \alpha_0}{\alpha_1 - \beta_1} \text{ and } \pi_2 = \frac{\alpha_1 \beta_0 - \alpha_0 \beta_1}{\alpha_1 - \beta_1} \tag{5.67}$$

By counting the coefficients of the structural and reduced form models we see that the structural model has four coefficients and the reduced form model has two coefficients. Furthermore, each of the two equations in the coefficients system (5.67) contains all four structural coefficients. Therefore, it is not possible from system (5.67) to find solutions for the structural coefficients and thus, both structural equations are underidentified.

The problem of underidentification of both equations in this model is shown in Figure 5.1. Figure 5.1(a) represents some scatter points of pairs of equilibrium prices and quantities data. Figure 5.1(b) shows the scatter points plus some different structures of demand (D, d) and supply (S, s) curves. The reduced scatter points of Figure 5.1(a) may be the result of the intersection of the demand (D) and supply (S) curves, or of the demand (d) and supply (s) curves, or of any other demand or supply curves

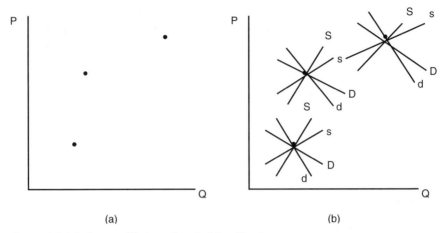

Figure 5.1 Market equilibria and underidentification

in Figure 5.1(b). In other words, having only the information included in the scatter points, it is impossible to distinguish which structure of demand and supply curves corresponds to these scatter points. Thus, more than one structure (theories) is consistent with the same scatter points (data) and there is no way of distinguishing them without further information. These structures are underidentified.

Let us see the same problem presented in Figure 5.1 from another point of view. If we multiply equation (5.63) by λ, where $1 \geq \lambda \geq 0$, and equation (5.64) by $(1 - \lambda)$ and add the two results we obtain the following combined equation which is a linear combination of the two original equations:

$$Q_t = \gamma_0 + \gamma_1 P_t + \epsilon_t \tag{5.68}$$

where

$$\gamma_0 = \lambda \alpha_0 + (1 - \lambda)\beta_0, \ \gamma_1 = \lambda \alpha_1 + (1 - \lambda)\beta_1, \ \epsilon_t = \lambda \epsilon_{1t} + (1 - \lambda)\epsilon_{2t} \tag{5.69}$$

Observing the combined equation (5.68) we see that it has the same exact form as the forms of the original equations (5.63) and (5.64). We say that these three equations are 'observationally equivalent'. Therefore, if we just have pairs of data on Q_t and P_t, we do not know that by regressing Q_t on P_t the underlying relation is that of the demand equation (5.63), or of the supply equation (5.64), or of the combined equation (5.68). In other words, more than one structure (relation) are consistent with the same data and there is no way of distinguishing them without further information. These structures are underidentified.

Case 5.5 Exactly, or just, identified equations

Assume a structural demand-supply model under the equilibrium conditions as:

$$\text{The demand} = Q_t = \alpha_0 + \alpha_1 P_t + \alpha_2 Y_t + \epsilon_{1t} \tag{5.70}$$

$$\text{The supply} = Q_t = \beta_0 + \beta_1 P_t + \epsilon_{2t} \tag{5.71}$$

This model differs from that of case 5.4 because the demand equation now contains more information. The quantity demanded depends now not only on prices, but on income, Y_t, as well. The reduced form model is

$$P_t = \frac{\beta_0 - \alpha_0}{\alpha_1 - \beta_1} - \frac{\alpha_2}{\alpha_1 - \beta_1} Y_t + \frac{\epsilon_{2t} - \epsilon_{1t}}{\alpha_1 - \beta_1} = \pi_{11} + \pi_{12} Y_t + \upsilon_{1t} \tag{5.72}$$

$$Q_t = \frac{\alpha_1 \beta_0 - \alpha_0 \beta_1}{\alpha_1 - \beta_1} - \frac{\alpha_2 \beta_1}{\alpha_1 - \beta_1} Y_t + \frac{\alpha_1 \epsilon_{2t} - \beta_1 \epsilon_{1t}}{\alpha_1 - \beta_1} = \pi_{21} + \pi_{22} Y_t + \upsilon_{2t} \tag{5.73}$$

where:

$$\pi_{11} = \frac{\beta_0 - \alpha_0}{\alpha_1 - \beta_1}, \ \pi_{12} = -\frac{\alpha_2}{\alpha_1 - \beta_1}, \ \pi_{21} = \frac{\alpha_1 \beta_0 - \alpha_0 \beta_1}{\alpha_1 - \beta_1}, \ \pi_{22} = -\frac{\alpha_2 \beta_1}{\alpha_1 - \beta_1} \tag{5.74}$$

By counting the coefficients of the structural and of the reduced form models we see that the structural model has five coefficients and the reduced form model has four coefficients. Therefore, it is not possible from the coefficients system (5.74) to find solutions for all of the structural coefficients. In fact we see that from system (5.74) we can find a unique solution for the coefficients of the supply function as:

$$\beta_1 = \pi_{22}/\pi_{12} \text{ and } \beta_0 = \pi_{21} - \beta_1 \pi_{11} \tag{5.75}$$

Unfortunately, we cannot find from the same system solutions for all coefficients of the demand function. Therefore, the supply function is exactly identified, whilst the demand function is not identified.

The exact identification of the supply equation and the non-identification of the demand equation in this model is shown in Figure 5.2. Figure 5.2(a) represents some scatter points of pairs of equilibrium prices and quantities. Figure 5.2(b) shows the scatter points plus the information that the demand curve shifts over time to the right, from D_1 to D_3, because of increasing income, and the supply curve remains

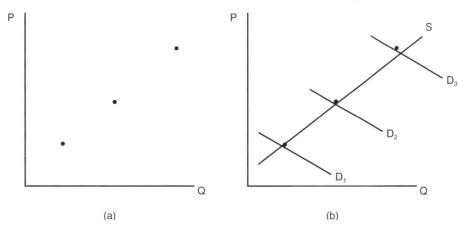

(a)   (b)

Figure 5.2 Market equilibria and exact identification of the supply function

stable, or relatively stable. Therefore, the introduction of this new information in this model, with respect to the model in case 5.4, distinguishes a unique supply curve but it is impossible to distinguish a unique demand curve. The observed values of Q_t and P_t, i.e. the intersection points of demand and supply curves, trace (identify) the supply curve. Thus, the supply equation is exactly identified, whilst the demand equation is not identified.

This can be seen also by the following combined equation which is a linear combination of the two original equations (5.70) and (5.71):

$$Q_t = \gamma_0 + \gamma_1 P_t + \gamma_2 Y_t + \epsilon_t \tag{5.76}$$

where

$$\gamma_0 = \lambda\alpha_0 + (1-\lambda)\beta_0,\ \gamma_1 = \lambda\alpha_1 + (1-\lambda)\beta_1,\ \gamma_2 = \lambda\alpha_2,\ \epsilon_t = \lambda\epsilon_{1t} + (1-\lambda)\epsilon_{2t} \tag{5.77}$$

Observing the combined equation (5.76) we see that it has the same exact form to that of the demand equation (5.70) and different form from that of the supply equation (5.71). We say therefore, that the combined and the demand equations are 'observationally equivalent', meaning that the demand function is not identified, and the combined and the supply equations are 'not observationally equivalent', meaning that the supply function is identified.

Case 5.6   An exactly, or just, identified model

Assume a structural demand–supply model under the equilibrium conditions as:

$$\text{The demand} = Q_t = \alpha_0 + \alpha_1 P_t + \alpha_2 Y_t + \epsilon_{1t} \tag{5.78}$$

$$\text{The supply} = Q_t = \beta_0 + \beta_1 P_t + \beta_2 W_t + \epsilon_{2t} \tag{5.79}$$

This model differs from the model in case 5.5 because the supply equation now contains more information. The quantity supplied depends now not only on prices, but on wages, W_t, as well, depicting the negative effect of the cost of a major factor of production to the quantity produced and supplied. The reduced form model is:

$$P_t = \frac{\beta_0 - \alpha_0}{\alpha_1 - \beta_1} - \frac{\alpha_2}{\alpha_1 - \beta_1}Y_t + \frac{\beta_2}{\alpha_1 - \beta_1}W_t + \frac{\epsilon_{2t} - \epsilon_{1t}}{\alpha_1 - \beta_1} = \pi_{11} + \pi_{12}Y_t + \pi_{13}W_t + \upsilon_{1t} \tag{5.80}$$

$$Q_t = \frac{\alpha_1\beta_0 - \alpha_0\beta_1}{\alpha_1 - \beta_1} - \frac{\alpha_2\beta_1}{\alpha_1 - \beta_1}Y_t + \frac{\alpha_1\beta_2}{\alpha_1 - \beta_1}W_t + \frac{\alpha_1\epsilon_{2t} - \beta_1\epsilon_{1t}}{\alpha_1 - \beta_1}$$

$$= \pi_{21} + \pi_{22}Y_t + \pi_{23}W_t + \upsilon_{2t} \tag{5.81}$$

where:

$$\pi_{11} = \frac{\beta_0 - \alpha_0}{\alpha_1 - \beta_1},\ \pi_{12} = -\frac{\alpha_2}{\alpha_1 - \beta_1},\ \pi_{13} = \frac{\beta_2}{\alpha_1 - \beta_1},\ \pi_{21} = \frac{\alpha_1\beta_0 - \alpha_0\beta_1}{\alpha_1 - \beta_1},$$

$$\pi_{22} = -\frac{\alpha_2\beta_1}{\alpha_1 - \beta_1}, \ \pi_{23} = \frac{\alpha_1\beta_2}{\alpha_1 - \beta_1} \tag{5.82}$$

By counting the coefficients of the structural and the reduced form models we see that the structural model has six coefficients and the reduced form model has six coefficients also. Therefore, it is generally possible from the coefficients system (5.82) to find a unique solution for all structural coefficients, and thus both equations are exactly identified. In this case the model as a whole is exactly identified.

We can derive the same results using the following combined equation which is a linear combination of the two original equations (5.78) and (5.79):

$$Q_t = \gamma_0 + \gamma_1 P_t + \gamma_2 Y_t + \gamma_3 W_t + \epsilon_t \tag{5.83}$$

where

$$\gamma_0 = \lambda\alpha_0 + (1 - \lambda)\beta_0, \ \gamma_1 = \lambda\alpha_1 + (1 - \lambda)\beta_1, \ \gamma_2 = \lambda\alpha_2, \ \gamma_3 = (1 - \lambda)\beta_2,$$

$$\epsilon_t = \lambda\epsilon_{1t} + (1 - \lambda)\epsilon_{2t} \tag{5.84}$$

Observing the combined equation (5.84) we see that it has a different form from the forms of both the demand equation (5.78) and the supply equation (5.79). We say therefore, that the combined and the demand and supply equations are 'not observationally equivalent', meaning that both the demand and the supply functions are identified.

Case 5.7 Overidentified equations

Assume a structural demand–supply model under the equilibrium conditions as

$$\text{The demand} = Q_t = \alpha_0 + \alpha_1 P_t + \alpha_2 Y_t + \alpha_3 P_{ft} + \epsilon_{1t} \tag{5.85}$$

$$\text{The supply} = Q_t = \beta_0 + \beta_1 P_t + \beta_2 W_t + \epsilon_{2t} \tag{5.86}$$

This model differs from the model of case 5.6 because the demand equation now contains even more information. The quantity demanded depends now not only on prices and income, but on the prices of a substitute good, P_{ft}, as well, depicting thus the positive effect on the quantity demanded by an increase in the price of the substitute good. We met this case in example 5.2. The reduced form model is

$$P_t = \frac{\beta_0 - \alpha_0}{\alpha_1 - \beta_1} - \frac{\alpha_2}{\alpha_1 - \beta_1}Y_t - \frac{\alpha_3}{\alpha_1 - \beta_1}P_{ft} + \frac{\beta_2}{\alpha_1 - \beta_1}W_t + \frac{\epsilon_{2t} - \epsilon_{1t}}{\alpha_1 - \beta_1}$$

$$= \pi_{11} + \pi_{12}Y_t + \pi_{13}P_{ft} + \pi_{14}W_t + \upsilon_{1t} \tag{5.87}$$

$$Q_t = \frac{\alpha_1\beta_0 - \alpha_0\beta_1}{\alpha_1 - \beta_1} - \frac{\alpha_2\beta_1}{\alpha_1 - \beta_1}Y_t - \frac{\alpha_3\beta_1}{\alpha_1 - \beta_1}P_{ft} + \frac{\alpha_1\beta_2}{\alpha_1 - \beta_1}W_t + \frac{\alpha_1\epsilon_{2t} - \beta_1\epsilon_{1t}}{\alpha_1 - \beta_1}$$

$$= \pi_{21} + \pi_{22}Y_t + \pi_{23}P_{ft} + \pi_{14}W_t + \upsilon_{2t} \tag{5.88}$$

where:

$$\pi_{11} = \frac{\beta_0 - \alpha_0}{\alpha_1 - \beta_1}, \ \pi_{12} = -\frac{\alpha_2}{\alpha_1 - \beta_1}, \ \pi_{13} = -\frac{\alpha_3}{\alpha_1 - \beta_1}, \ \pi_{14} = \frac{\beta_2}{\alpha_1 - \beta_1}$$

$$\pi_{21} = \frac{\alpha_1\beta_0 - \alpha_0\beta_1}{\alpha_1 - \beta_1}, \ \pi_{22} = -\frac{\alpha_2\beta_1}{\alpha_1 - \beta_1}, \ \pi_{23} = -\frac{\alpha_3\beta_1}{\alpha_1 - \beta_1}, \ \pi_{24} = \frac{\alpha_1\beta_2}{\alpha_1 - \beta_1}$$

(5.89)

By counting the coefficients of the structural and of the reduced form models we see that the structural model has seven coefficients and the reduced form model has eight coefficients. Therefore, it is not possible from the coefficients system (5.89) to find a unique solution for all of the structural coefficients. In fact we see that from system (5.89) we can obtain that

$$\beta_1 = \pi_{23}/\pi_{13} \text{ or } \beta_1 = \pi_{22}/\pi_{12} \tag{5.90}$$

meaning that from the same data we can get two generally different estimates of β_1. Furthermore, because β_1 appears in all the equations of system (5.89), for each estimate of β_1 in (5.90) we get a set of estimates of the other structural coefficients as well. Therefore, the two equations are overidentified.

We can derive the same results by using the following combined equation which is a linear combination of the two original equations (5.85) and (5.86):

$$Q_t = \gamma_0 + \gamma_1 P_t + \gamma_2 Y_t + \gamma_3 W_t + \gamma_4 P_{ft} + \epsilon_t \tag{5.91}$$

where

$$\gamma_0 = \lambda\alpha_0 + (1-\lambda)\beta_0, \ \gamma_1 = \lambda\alpha_1 + (1-\lambda)\beta_1, \ \gamma_2 = \lambda\alpha_2, \ \gamma_3 = (1-\lambda)\beta_2, \ \gamma_4 = \lambda\alpha_3,$$

$$\epsilon_t = \lambda\epsilon_{1t} + (1-\lambda)\epsilon_{2t}$$

(5.92)

Observing the combined equation (5.91) we see that it has a different form from that of the demand equation (5.85) and the supply equation (5.86). We say therefore, that the combined and the demand and supply equations are 'not observationally equivalent', meaning that both the demand and supply functions are identified.

Summarising the cases, we can say that if an equation has an omitted variable, in a two simultaneous equations system, then this equation is identified. This 'condition' is the general topic of investigation in the next section.

5.2 The conditions for identification

Counting the parameters in the structural model (5.55) and (5.56) we see that there are $G \times G = G^2$ parameters in matrix $\mathbf{\Gamma}$, $G \times K$ parameters in matrix $\mathbf{B}$, and $G \times (G+1)/2$ in matrix $\mathbf{\Sigma}$. Similarly, counting the parameters in the reduced form model (5.58) and (5.59) we see that there are $G \times K$ parameters in matrix $\mathbf{\Pi}$, and $G \times (G+1)/2$ parameters in matrix $\mathbf{\Omega}$. Comparing the number of parameters of the structural and of the reduced form models we see that the structural model has

$$G^2 + G \times K + G \times (G+1)/2 - G \times K - G \times (G+1)/2 = G^2 \tag{5.93}$$

more parameters than the reduced from model. Therefore, the indirect least squares method is not generally successful in providing unique estimates of the structural parameters, because the parameters of the structural model are in excess of G^2 with respect to the parameters of the reduced form model. For obtaining unique estimates, i.e. for identification, we need further prior information in the form of G^2 a priori restrictions on the parameters.

This prior information referring to the parameters of the $\mathbf{\Gamma}$, $\mathbf{B}$ and $\mathbf{\Sigma}$ matrices comes in several forms as follows:

Normalisation   This is the case in each equation when the coefficient of one of the G endogenous variables is normalised to be equal to one. Usually it is set that $\gamma_{ii} = 1$, for $i = 1, 2, \ldots, G$. It corresponds in other words to the endogenous variable that appears on the left hand side of the equation. Therefore, with normalisation the G^2 a priori restrictions reduce to $G(G - 1)$ restrictions.

Zero restrictions   This is the case when some of the endogenous or predetermined variables do not appear in each of the equations. The corresponding coefficients of the *excluded variables* are set equal to zero.

Identity restrictions   This refers to the inclusion of identities in the model. The identities do not carry any coefficients to be estimated but they aid in the identification of other equations.

Parameter restrictions within equations   Refers to certain relationships among parameters within equations. If the Cobb-Douglas, for example, production function is of constant returns of scale, the sum of the elasticities of the factors of production must equal one.

Parameter restrictions across equations   Refers to certain relationships among parameters across equations. If consumers behaviour, for example, in the EU Member States is the same, then the marginal propensity to consume across the member states consumption functions must be set equal across equations.

Disturbance variance–covariance parameter restrictions   By assuming, for example, that the structural disturbances among equations are not correlated, then the off diagonal elements of matrix $\mathbf{\Sigma}$ are set equal to zero.

Let us now write the coefficients system (5.61) as

$$\mathbf{\Gamma}\mathbf{\Pi} = -\mathbf{B} \tag{5.94}$$

For a single equation of the system (5.55), say equation i, (5.94) becomes

$$\gamma_i \mathbf{\Pi} = -\boldsymbol{\beta}_i \tag{5.95}$$

where

$$\gamma_i = [\gamma_{i1}\ \gamma_{i2} \cdots \gamma_{iG}],\ \boldsymbol{\beta}_i = [\beta_{i1}\ \beta_{i2} \cdots \beta_{iK}], \tag{5.96}$$

For this new writing of the coefficients system let us introduce the following notation:

G = number of all the endogenous variables of the system.

g = number of the endogenous variables appearing in the ith equation
(i.e. number of nonzero elements in γ_i).

$g^* = G - g$ = number of endogenous variables not appearing in the ith equation.

K = number of all the predetermined variables of the system.

k = the number of the predetermined variables appearing in the ith equation.
(i.e. number of non zero elements in β_i).

$k^* = K - k$ = the number of predetermined variables not appearing in the ith equation.

Under this notation, and without loss of generality, (5.96) is written:

$$\gamma_i = \left[\underbrace{\gamma_{i1} \quad \gamma_{i2} \quad \cdots \quad \gamma_{ig}}_{g} \quad \underbrace{0 \quad 0 \quad \cdots \quad 0}_{g^*} \right] = \left[\underbrace{\gamma_g}_{1 \times g} \quad \underbrace{0_{g^*}}_{1 \times g^*} \right] \tag{5.97}$$

and

$$\beta_i = \left[\underbrace{\beta_{i1} \quad \beta_{i2} \quad \cdots \quad \beta_{ik}}_{k} \quad \underbrace{0 \quad 0 \quad \cdots \quad 0}_{k^*} \right] = \left[\underbrace{\beta_k}_{1 \times k} \quad \underbrace{0_{k^*}}_{1 \times k^*} \right] \tag{5.98}$$

By substituting (5.97) and (5.98) into (5.95) and by partitioning matrix Π appropriately, (5.95) is written as follows:

$$[\gamma_g \ 0_{g^*}] \begin{bmatrix} \underbrace{\Pi_{gk}}_{g \times k} & \underbrace{\Pi_{gk^*}}_{g \times k^*} \\ \underbrace{\Pi_{g^*k}}_{g^* \times k} & \underbrace{\Pi_{g^*k^*}}_{g^* \times k^*} \end{bmatrix} = -[\beta_k \ 0_{k^*}] \tag{5.99}$$

System (5.99) can be analysed into the following two subsystems:

$$\gamma_g \Pi_{gk} = -\beta_k \tag{5.100}$$

$$\gamma_g \Pi_{gk^*} = 0_{k^*} \tag{5.101}$$

In these two subsystems we observe that (5.101) contains γ coefficients only, which are g in number and correspond to the coefficients of all the endogenous variables included in equation i. By normalisation we can put one of these coefficients equal to 1 and therefore, the number of the unknown coefficients in (5.101) reduces to

$g - 1$. Subsystem (5.100) contains both the γ coefficients that are also contained in subsystem (5.101) plus the β coefficients which are k in number and correspond to the coefficients of all the predetermined variables included in equation i. Therefore, if we estimate first the γ coefficients from subsystem (5.101) then we can estimate the β coefficients by substituting the estimated γ coefficients in subsystem (5.100).

Because now subsystem (5.101) contains $g - 1$ unknown coefficients, therefore, for the existence of a solution this subsystem must have at least $g - 1$ equations, or it must be that:

$$k* \geq g - 1 \tag{5.102}$$

In fact, (5.102) says that in order for equation i to be identified the number of predetermined variables excluded (not appearing) from this equation i (k*) must be greater or equal to the number of the endogenous variables included (appearing) in this equation i less one ($g - 1$). (5.102) is known as the 'order condition'.

However, it is possible some of the equations in subsystem (5.101) are not independent with respect to the γ and π coefficients. Therefore, the order condition is only a *necessary* condition for identification and not a sufficient condition. Thus, a *necessary and sufficient* condition for identification requires that the number of independent equations in subsystem (5.101) to be equal to $g - 1$. This of course will happen if and only if by forming all square submatrices of matrix $\mathbf{\Pi}_{gk*}$, the order of the corresponding to the submatrices largest non-zero determinant is equal to $g - 1$, or in other words, if and only if it is true that:

$$\text{rank}(\mathbf{\Pi}_{gk*}) = g - 1 \tag{5.103}$$

(5.103) is known as the 'rank condition'. Because it can be proved (see Kmenta, 1971) that:

$$\text{rank}(\mathbf{\Pi}_{gk*}) = \text{rank}(\mathbf{\Delta}) - g* \tag{5.104}$$

where $\mathbf{\Delta}$ is the matrix which consists of all the structural coefficients for the variables of the system excluded from the *i*th structural equation but included in the other structural equations, the rank condition (5.103) may be also written as:

$$\text{rank}(\mathbf{\Delta}) = G - 1 \tag{5.105}$$

Let us now summarise the order and rank conditions for identification of an equation in a simultaneous equations model.

The order condition of identification

In a model of G simultaneous equations with G endogenous variables and K predetermined variables, an equation which includes g endogenous variables and k predetermined variables is identified if the number of predetermined variables excluded from the equation ($K - k$) is not less than the number of endogenous variables included in that equation less one ($g - 1$), i.e. it is

$$K - k \geq g - 1 \tag{5.106}$$

The rank condition of identification

In a model of G simultaneous equations an equation is identified if and only if matrix Δ, which is constructed from the coefficients of all the variables excluded from that specific equation but included in the other equations of the model, has a rank equal to the number of equations less one, i.e. it is

$$\text{rank}(\Delta) = G - 1 \tag{5.107}$$

The properties of the conditions for identification

The order condition for identification is a necessary condition, whilst the rank condition for identification is a necessary and sufficient condition. In other words, for an equation to be identified the rank condition must be satisfied, but if the equation is exactly identified or overidentified this is determined by the order condition according to whether is $K - k = g - 1$ or $K - k > g - 1$ respectively.

Furthermore, let us state the identification possibilities for an equation.

1   *Overidentified equation:* If $K - k > g - 1$ and $\text{rank}(\Delta) = G - 1$.
2   *Exactly identified equation:* If $K - k = g - 1$ and $\text{rank}(\Delta) = G - 1$.
3   *Underidentified equation:* If $K - k \geq g - 1$ and $\text{rank}(\Delta) < G - 1$, or if $K - k < g - 1$.

In what it follows we apply the preceding theory in investigating identification in two simple models of simultaneous equations.

Example 5.3 Investigating identification of the demand–supply model of case 5.5

Consider the demand-supply model of case 5.5 which under equilibrium conditions is written in equations (5.70) and (5.71), or

$$The\ Demand = Q_t = \alpha_0 + \alpha_1 P_t + \alpha_2 Y_t + \epsilon_{1t}$$
$$The\ Supply = Q_t = \beta_0 + \beta_1 P_t + \epsilon_{2t} \tag{5.108}$$

Where Q_t = quantity, P_t = prices, Y_t = income, and ϵ_t's = disturbances. This model is complete because it has two equations and two endogenous variables. The endogenous variables are Q_t, and P_t, and the exogenous variable is Y_t.

According to (5.55) model (5.108) can be written in the following form:

$$\text{Demand} = Q_t - \alpha_1 P_t - \alpha_0 - \alpha_2 Y_t = \epsilon_{1t}$$
$$\text{Supply} = Q_t - \beta_1 P_t - \beta_0 - 0 Y_t = \epsilon_{2t} \tag{5.109}$$

Investigating the demand function   In this function there are $k = 1$ exogenous variables (Y_t) and $g = 2$ endogenous variables (Q_t, P_t). Therefore, the order condition $K - k = 1 - 1 = 0 < g - 1 = 2 - 1 = 1$ is 'satisfied' for underidentification.

Investigating the supply function In this function there are $k = 0$ exogenous variables and $g = 2$ endogenous variables (Q_t, P_t). Therefore, the order condition $K - k = 1 - 0 = 1 = g - 1 = 2 - 1 = 1$ is satisfied for exact identification.

The rank condition can be investigated by forming matrix Δ from (5.109), which is

$$\Delta = [-\alpha_2]$$

Evaluating the $(G - 1) \times (G - 1) = (2 - 1) \times (2 - 1) = 1 \times 1$ determinants of matrix Δ we get that

$$|\Delta_1| = |-\alpha_2| = -\alpha_2 \neq 0$$

Therefore, the rank condition is satisfied since we can construct at least one non-zero determinant of order 1, i.e. $\text{rank}(\Delta) = G - 1 = 1$. Summarising, by taking into account both the order and rank conditions, the supply function is exactly identified.

Example 5.4 Investigating identification in a simple macroeconomic model

Given the following macroeconomic model:

Consumption: $C_t = \alpha_0 + \alpha_1 Y_t + \alpha_2 C_{t-1} + \epsilon_{1t}$

Investment: $I_t = \beta_0 + \beta_1 Y_t + \beta_2 Y_{t-1} + \beta_3 I_{t-1} + \epsilon_{2t}$

Taxes: $T_t = \gamma_0 + \gamma_1 Y_t + \epsilon_{3t}$ $\qquad$ (5.110)

Income: $Y_t = C_t + I_t + G_t$

Where C_t = consumption, Y_t = income, I_t = investment, T_t = taxes, G_t = government spending, and ϵ_t's = disturbances. This model is complete because it has four equations and four endogenous variables. The endogenous variables are C_t, I_t, T_t, and Y_t and the predetermined variables are $C_{t-1}, Y_{t-1}, I_{t-1}$, and G_t.

According to (5.55) model (5.110) can be written in the following way.

Consumption: $C_t - \alpha_1 Y_t + 0I_t + 0T_t - \alpha_0 - \alpha_2 C_{t-1} + 0Y_{t-1} + 0I_{t-1} + 0G_t = \epsilon_{1t}$

Investment: $\quad 0C_t - \beta_1 Y_t + I_t + 0T_t - \beta_0 + 0C_{t-1} - \beta_2 Y_{t-1} - \beta_3 I_{t-1} + 0G_t = \epsilon_{2t}$

Taxes: $\quad 0C_t - \gamma_1 Y_t + 0I_t + T_t - \gamma_0 + 0C_{t-1} + 0Y_{t-1} + 0I_{t-1} + 0G_t = \epsilon_{3t}$

Income: $\quad -C_t + Y_t - I_t + 0T_t + 0 + 0C_{t-1} + 0Y_{t-1} + 0I_{t-1} - G_t = 0$

$$(5.111)$$

Investigating the consumption function   In this function there are $k = 1$ predetermined variables (C_{t-1}) and $g = 2$ endogenous variables (C_t, Y_t). Therefore, the order condition $K - k = 4 - 1 = 3 > g - 1 = 2 - 1 = 1$ is satisfied for overidentification.

The rank condition can be investigated by forming matrix Δ from (5.111), which is

$$\Delta = \begin{bmatrix} 1 & 0 & -\beta_2 & -\beta_3 & 0 \\ 0 & 1 & 0 & 0 & 0 \\ -1 & 0 & 0 & 0 & -1 \end{bmatrix}$$

Evaluating the $(G-1) \times (G-1) = (4-1) \times (4-1) = 3 \times 3$ determinants of matrix Δ we get that

$$|\Delta_1| = \begin{vmatrix} 1 & 0 & -\beta_2 \\ 0 & 1 & 0 \\ -1 & 0 & 0 \end{vmatrix} = -\beta_2 \neq 0$$

Therefore, the rank condition is satisfied since we can construct at least one nonzero determinant of order 3, i.e. rank$(\Delta) = G - 1 = 3$. Summarising, by taking into account both the order and rank conditions, the consumption function is overidentified.

Investigating the investment function   In this function there are $k = 2$ predetermined variables (Y_{t-1}, I_{t-1}) and $g = 2$ endogenous variables (I_t, Y_t). Therefore, the order condition $K - k = 4 - 2 = 2 > g - 1 = 2 - 1 = 1$ is satisfied for overidentification.

The rank condition can be investigated by forming matrix Δ from (5.111), which is

$$\Delta = \begin{bmatrix} 1 & 0 & -\alpha_2 & 0 \\ 0 & 1 & 0 & 0 \\ -1 & 0 & 0 & -1 \end{bmatrix}$$

Evaluating the $(G-1) \times (G-1) = (4-1) \times (4-1) = 3 \times 3$ determinants of matrix Δ we get that

$$|\Delta_1| = \begin{vmatrix} 1 & 0 & -\alpha_2 \\ 0 & 1 & 0 \\ -1 & 0 & 0 \end{vmatrix} = -\alpha_2 \neq 0$$

Therefore, the rank condition is satisfied since we can construct at least one nonzero determinant of order 3, i.e. rank$(\Delta) = G - 1 = 3$. Summarising, by taking into account both the order and rank conditions, the investment function is overidentified.

Investigating the taxes function   In this function there are $k = 0$ predetermined variables and $g = 2$ endogenous variables (T_t, Y_t). Therefore, the order condition $K - k = 4 - 0 = 4 > g - 1 = 2 - 1 = 1$ is satisfied for overidentification.

The rank condition can be investigated by forming matrix Δ from (5.111), which is

$$\Delta = \begin{bmatrix} 1 & 0 & -\alpha_2 & 0 & 0 & 0 \\ 0 & 1 & 0 & -\beta_2 & -\beta_3 & 0 \\ -1 & -1 & 0 & 0 & 0 & -1 \end{bmatrix}$$

Evaluating the $(G - 1) \times (G - 1) = (4 - 1) \times (4 - 1) = 3 \times 3$ determinants of matrix Δ we get that

$$|\Delta_1| = \begin{vmatrix} 1 & 0 & -\alpha_2 \\ 0 & 1 & 0 \\ -1 & -1 & 0 \end{vmatrix} = -\alpha_2 \neq 0$$

Therefore, the rank condition is satisfied since we can construct at least one nonzero determinant of order 3, i.e. $\text{rank}(\Delta) = G - 1 = 3$. Summarising, by taking into account both the order and rank conditions, the taxes function is overidentified.

6 METHODS OF ESTIMATION

In the beginning of section 5 we noted that if we apply the OLS method to the structural equations of a simultaneous equations system we will get inconsistent estimates because the endogenous variables of the systems' equations are not independent by the structural error terms. Furthermore, we saw that there are cases where we can obtain indirectly consistent estimates of the structural coefficients of the system by applying the OLS methodology to the reduced form equations of the system because the predetermined variables of the system are independent of the reduced form error terms. However, there are specific simultaneous equations models where the direct application of OLS to the structural equations leads to consistent estimates of the structural parameters.

The methods for estimating the structural parameters of a simultaneous equations model may be grouped into the following two categories:

Single equation methods of estimation   These are the methods where each equation in the model is estimated individually, taking into account only the information included in the specific equation and without considering all the other information included in the rest equations of the system. For this reason, these methods are known as 'limited information methods'. We present the following such methods:

1   Ordinary least squares method (OLS), for fully recursive models.
2   Indirect least squares (ILS) method, for exactly identified equations.

3   Instrumental variables (IV) method.
4   Two-stage least squares (2SLS) method.
5   Limited information maximum likelihood (LIML) method.

System methods of estimation   These are the methods where all the equations of the model are estimated simultaneously taking into account all the information included in all equations simultaneously. For this reason, these methods are known as 'full information methods'. We present the following methods:

1   Three-stage least squares (3SLS) method.
2   Full information maximum likelihood (FIML) method.

6.1 Single equation methods of estimation

6.1.1 Ordinary least squares method (OLS), for fully recursive models

We saw in section 2 that a recursive equations model is that where matrix Γ, in the structural model (5.10), is triangular, i.e. it is of the form shown in (5.22). Specifically, by normalising with respect to the coefficient of each endogenous dependent variable in the corresponding equation, the recursive equations model may be written as follows:

$$Y_{1t} = \beta_{11}X_{1t} + \beta_{12}X_{2t} + \ldots + \beta_{1K}X_{Kt} + \epsilon_{1t}$$

$$Y_{2t} = \gamma_{21}Y_{1t} + \beta_{21}X_{1t} + \beta_{22}X_{2t} + \ldots + \beta_{2K}X_{Kt} + \epsilon_{2t}$$

$$Y_{3t} = \gamma_{31}Y_{1t} + \gamma_{32}Y_{2t} + \beta_{31}X_{1t} + \beta_{32}X_{2t} + \ldots + \beta_{3K}X_{Kt} + \epsilon_{3t}$$

$$\ldots$$

$$Y_{Gt} = \gamma_{G1}Y_{1t} + \gamma_{G2}Y_{2t} + \ldots + \gamma_{G,G-1}Y_{G-1,t} + \beta_{G1}X_{1t} + \beta_{G2}X_{2t} + \ldots + \beta_{GK}X_{Kt} + \epsilon_{Gt}$$

$$(5.112)$$

If now we assume that the variance-covariance matrix Σ of the structural disturbances is diagonal, i.e. if it is

$$E(\epsilon_t\epsilon_t') = \Sigma = \begin{bmatrix} \sigma_{11} & \sigma_{12} & \ldots & \sigma_{1G} \\ \sigma_{21} & \sigma_{22} & \ldots & \sigma_{2G} \\ \vdots & \vdots & \vdots & \vdots \\ \sigma_{G1} & \sigma_{G2} & \ldots & \sigma_{GG} \end{bmatrix} = \begin{bmatrix} \sigma_{11} & 0 & \ldots & 0 \\ 0 & \sigma_{22} & \ldots & 0 \\ \vdots & \vdots & \vdots & \vdots \\ 0 & 0 & \ldots & \sigma_{GG} \end{bmatrix} \qquad (5.113)$$

then the model is called 'fully recursive model'. In this case we can apply the OLS estimation in the first equation of the model because the predetermined variables in this equation are uncorrelated with the error term ϵ_{1t}. We can also apply the OLS estimation in the second equation of the model because both the predetermined variables in this equation are uncorrelated with the error term ϵ_{2t}, and the endogenous variable Y_{1t} is also uncorrelated with ϵ_{2t}. This is because the endogenous variable Y_{1t}, which is also an explanatory variable in the second equation, although it is a function of ϵ_{1t}, still is not correlated with ϵ_{2t} due to (5.113), which assumes that ϵ_{1t}

and ϵ_{2t} are not correlated. For exactly the same reasons we can apply OLS to all equations of the fully recursive model. In other words, the structural parameters of the fully recursive model can be estimated consistently and asymptotically efficiently with OLS. Of course, in cases where the structural disturbances are correlated among equations, then the OLS approach is not applicable and therefore other methods have to be used.

Example 5.5 The wages–prices link for an EU member

Assume the following revised version of the wages–prices link model of case 5.3. The prices inflation function:

$$p_t = \alpha_0 + \alpha_1 w_{t-1} + \alpha_2 Y_t + \alpha_3 p_{t-1} + \epsilon_{1t}, \ \alpha_1 > 0, \ \alpha_2 > 0, \ 0 < \alpha_3 < 1 \qquad (5.114)$$

The wages inflation function:

$$w_t = \beta_0 + \beta_1 p_t + \beta_2 U_t + \beta_3 w_{t-1} + \epsilon_{2t}, \ \beta_1 > 0, \ \beta_2 < 0, \ 0 < \beta_3 < 1 \qquad (5.115)$$

where w_t = percentage change of money wage level, p_t = percentage change of the general level of prices, Y_t = aggregate demand, and U_t = the unemployment rate. The model of equations (5.114) and (5.115) is a recursive model because the dependent (endogenous) variable p_t of the first equation is a function of exogenous (Y_t) and lagged endogenous (w_{t-1}, p_{t-1}) variables only. Furthermore, the dependent (endogenous) variable w_t of the second equation is a function of the dependent variable p_t of the first equation plus a function of exogenous (U_t) and lagged endogenous (w_{t-1}) variables. By assuming that the variance–covariance matrix of the structural disturbances is diagonal, we can apply the OLS procedure to the two equations individually.

Using data in Table 5.2 where w_t = the percentage change of nominal compensation per employee, p_t = the annual percentage change of the GNP price deflator, U_t = the unemployment rate as a percentage of civilian labour force, and Y_t = GNP at constant prices, the OLS estimates of the two equations (5.114) and (5.115) are:

$$\hat{p}_t = -3.29432 + 0.33264 w_{t-1} + 0.01391 Y_t + 0.39320 p_{t-1} \qquad \bar{R}^2 = 0.7575$$

s.e. (2.3883) (0.1482)   (0.0074)   (0.1755)   (5.116)

t   [1.3794] [2.2443]   [1.8847]   [2.2402]

and

$$\hat{w}_t = 7.59395 + 0.38277 p_t - 0.58217 U_t + 0.40607 w_{t-1} \qquad \bar{R}^2 = 0.6941$$

s.e. (2.1741) (0.1319)   (0.2714)   (0.1466)   (5.117)

t   [3.4929] [2.9017]   [2.1449]   [2.7700]

Table 5.2 Wages–prices link data of an EU member

Year	GNP Price deflator (annual percentage change)	Nominal compensation per employee (annual percentage change)	Unemployment rate (percentage of civilian labour force)	GNP (in constant 1970 prices, billion drs)*
1960	1.2	4.0	6.1	145.458
1961	1.5	4.6	5.9	161.802
1962	4.6	6.6	5.1	164.674
1963	1.4	7.7	5.0	181.534
1964	3.7	13.3	4.6	196.586
1965	4.0	12.2	4.8	214.922
1966	4.9	12.6	5.0	228.040
1967	2.4	9.5	5.4	240.791
1968	1.7	9.8	5.6	257.226
1969	3.4	9.6	5.2	282.168
1970	3.9	8.8	4.2	304.420
1971	3.2	8.0	3.1	327.723
1972	5.0	12.6	2.1	356.886
1973	19.4	17.2	2.0	383.916
1974	20.9	19.3	2.1	369.325
1975	12.3	20.3	2.3	390.000
1976	15.4	23.2	1.9	415.491
1977	13.0	22.0	1.7	431.164
1978	12.9	23.1	1.8	458.675
1979	18.6	22.1	1.9	476.048
1980	17.7	15.7	2.7	485.108
1981	19.8	21.3	4.0	484.259
1982	25.1	27.6	5.8	483.879
1983	19.1	21.5	7.1	481.198
1984	20.3	20.5	7.2	490.881
1985	17.7	23.2	7.0	502.258
1986	17.5	12.8	6.6	507.199
1987	14.3	11.5	6.7	505.713
1988	15.6	18.6	6.8	529.460
1989	14.4	24.0	6.7	546.572
1990	20.6	23.1	6.4	546.982
1991	19.8	14.3	7.0	566.586
1992	14.6	8.2	7.9	568.582
1993	14.1	10.1	8.6	569.724
1994	10.9	11.9	8.9	579.846
1995	9.3	12.7	9.1	588.691

Sources: EC (1997) Annual Economic Report for 1997, *European Economy.*
*Epilogi (1998).

6.1.2 Indirect least squares method (ILS), for exactly identified equations

Previously we used many times this method in trying to obtain unique estimates of the structural coefficients using consistent estimates of the reduced form coefficients. In fact, we said there that, if an equation is exactly identified then the indirect least squares method yields unique estimates of the structural coefficients of this equation. To summarise, the steps used for this method are:

Step 1

Solve the structural equations system in order to get the reduced form system, i.e. the system of (5.57), or

$$\mathbf{Y}_t = -\mathbf{\Gamma}^{-1}\mathbf{B}\mathbf{X}_t + \mathbf{\Gamma}^{-1}\boldsymbol{\epsilon}_t \tag{5.118}$$

Step 2

Apply the OLS method to each reduced form equation and get the consistent and asymptotically efficient estimates of the corresponding reduced form coefficients, i.e. apply the OLS method to (5.58), or

$$\mathbf{Y}_t = \mathbf{\Pi}\mathbf{X}_t + \boldsymbol{\upsilon}_t \tag{5.119}$$

Step 3

Substituting the estimates of the reduced form coefficients to the coefficients system, solve it in order to get consistent and asymptotically efficient estimates of the structural coefficients, i.e. solve the system (5.95), or the two subsystems (5.100) and (5.101), or finally

$$\mathbf{c}_g\mathbf{P}_{gk} = -\mathbf{b}_k \tag{5.120}$$

$$\mathbf{c}_g\mathbf{P}_{gk^*} = \mathbf{0}_{k^*} \tag{5.121}$$

where c's are the estimates of γ's, b's are the estimates of β's, and $\mathbf{P}$'s are the estimates of $\mathbf{\Pi}$'s.

Example 5.6 A wages–prices link model for an EU member

Assume the following revised version of the wages–prices link model of Example 5.5.
A prices inflation function:

$$p_t = \alpha_0 + \alpha_1 w_t + \alpha_2 Y_t + \epsilon_{1t}, \, \alpha_1 > 0, \alpha_2 > 0 \tag{5.122}$$

A wages inflation function:

$$w_t = \beta_0 + \beta_1 p_t + \beta_2 U_t + \epsilon_{2t}, \, \beta_1 > 0, \beta_2 < 0 \tag{5.123}$$

This model is a simultaneous equations model with $G = 2$ endogenous variables (p_t, w_t) and $K = 2$ exogenous variables (Y_t, U_t). Furthermore, according to the order and rank conditions both equations are exactly identified. Therefore, the model of equations (5.122) and (5.123) is an exactly identified model (see case 5.6).

According to step 1, the reduced form model is

$$p_t = \frac{\alpha_0 + \alpha_1\beta_0}{1 - \alpha_1\beta_1} + \frac{\alpha_2}{1 - \alpha_1\beta_1} Y_t + \frac{\alpha_1\beta_2}{1 - \alpha_1\beta_1} U_t + \frac{\epsilon_{1t} + \alpha_1\epsilon_{2t}}{1 - \alpha_1\beta_1}$$

$$w_t = \frac{\beta_0 + \alpha_0\beta_1}{1 - \alpha_1\beta_1} + \frac{\alpha_2\beta_1}{1 - \alpha_1\beta_1} Y_t + \frac{\beta_2}{1 - \alpha_1\beta_1} U_t + \frac{\beta_1\epsilon_{1t} + \alpha_1\epsilon_{2t}}{1 - \alpha_1\beta_1}$$

$$(5.124)$$

According to step 2 and using the data in Table 5.2, the results of the OLS application to the unrestricted reduced form model

$$p_t = \pi_{11} + \pi_{12}Y_t + \pi_{13}U_t + \upsilon_{1t}$$

$$w_t = \pi_{21} + \pi_{22}Y_t + \pi_{23}U_t + \upsilon_{2t}$$

$$(5.125)$$

are the following:

$$\hat{p}_t = -1.68241 + 0.04479Y_t - 0.87508U_t \qquad \bar{R}^2 = 0.6520$$

s.e. (2.4202) (0.0054) (0.3485) $\qquad\qquad\qquad\qquad$ (5.126)

t [0.6951] [8.2165] [2.5106]

$$\hat{w}_t = 9.37057 + 0.03250Y_t - 1.42523U_t \qquad \bar{R}^2 = 0.4923$$

s.e. (2.5689) (0.0058) (0.3700) $\qquad\qquad\qquad\qquad$ (5.127)

t [3.6476] [5.6174] [3.8522]

According to step 3, the estimates of the structural coefficients will be obtained from the solution of the following corresponding coefficients system:

$$\pi_{11} = \frac{\alpha_0 + \alpha_1\beta_0}{1 - \alpha_1\beta_1}, \ \pi_{12} = \frac{\alpha_2}{1 - \alpha_1\beta_1}, \ \pi_{13} = \frac{\alpha_1\beta_2}{1 - \alpha_1\beta_1}$$

$$\pi_{21} = \frac{\beta_0 + \alpha_0\beta_1}{1 - \alpha_1\beta_1}, \ \pi_{22} = \frac{\alpha_2\beta_1}{1 - \alpha_1\beta_1}, \ \pi_{23} = \frac{\beta_2}{1 - \alpha_1\beta_1} \qquad (5.128)$$

These estimates are given by:

$$a_1 = \frac{\hat{\pi}_{13}}{\hat{\pi}_{23}} = 0.61399, \qquad\qquad\qquad b_1 = \frac{\hat{\pi}_{22}}{\hat{\pi}_{12}} = 0.72569$$

$$a_2 = (1 - a_1b_1)\hat{\pi}_{12} = 0.02483, \qquad b_2 = (1 - a_1b_1)\hat{\pi}_{23} = -0.79019 \quad (5.129)$$

$$a_0 = (1 - a_1b_1)\hat{\pi}_{11} - a_1b_0 = -7.43590, \quad b_0 = \hat{\pi}_{21} - b_1\hat{\pi}_{11} = 10.59149$$

6.1.3 *Instrumental variables (IV) method*

Assuming that one equation of the structural model, say the first equation, has g endogenous variables and k predetermined variables, and normalising the coefficient of its dependent (endogenous) variable to be equal to one, this equation may generally be written as

$$Y_1 = \mathbf{Y}_1\boldsymbol{\gamma}_1 + \mathbf{X}_1\boldsymbol{\beta}_1 + \boldsymbol{\epsilon}_1 \tag{5.130}$$

where

$$Y_1 = \begin{bmatrix} Y_{11} \\ Y_{12} \\ \vdots \\ Y_{1n} \end{bmatrix} \mathbf{Y}_1 = \begin{bmatrix} Y_{21} & Y_{31} & \cdots & Y_{g1} \\ Y_{22} & Y_{32} & \cdots & Y_{g2} \\ \vdots & \vdots & \vdots & \vdots \\ Y_{2n} & Y_{3n} & \cdots & Y_{gn} \end{bmatrix} \mathbf{X}_1 = \begin{bmatrix} X_{11} & X_{21} & \cdots & X_{k1} \\ X_{12} & X_{22} & \cdots & X_{k2} \\ \vdots & \vdots & \vdots & \vdots \\ X_{1n} & X_{2n} & \cdots & X_{kn} \end{bmatrix} \boldsymbol{\epsilon}_1 = \begin{bmatrix} \epsilon_{11} \\ \epsilon_{12} \\ \vdots \\ \epsilon_{1n} \end{bmatrix}$$

$$\boldsymbol{\gamma}_1 = \begin{bmatrix} \gamma_{12} \\ \gamma_{13} \\ \vdots \\ \gamma_{1g} \end{bmatrix} \boldsymbol{\beta}_1 = \begin{bmatrix} \beta_{11} \\ \beta_{12} \\ \vdots \\ \beta_{1k} \end{bmatrix} \tag{5.131}$$

Equation (5.130) may also be written as

$$Y_1 = \mathbf{Z}_1\boldsymbol{\delta}_1 + \boldsymbol{\epsilon}_1 \tag{5.132}$$

where

$$\mathbf{Z}_1 = [\mathbf{Y}_1 \ \mathbf{X}_1] \text{ and } \boldsymbol{\delta}_1' = [\boldsymbol{\gamma}_1 \ \boldsymbol{\beta}_1] \tag{5.133}$$

If $\mathbf{d}_1$ denotes an estimator of $\boldsymbol{\delta}_1$, the OLS estimator for equation (5.132) is

$$\mathbf{d}_{1,\text{OLS}} = (\mathbf{Z}_1'\mathbf{Z}_1)^{-1}\mathbf{Z}_1'Y_1 \tag{5.134}$$

and

$$\text{var--cov}(\mathbf{d}_{1,\text{OLS}}) = s_1^2(\mathbf{Z}_1'\mathbf{Z}_1)^{-1} \tag{5.135}$$

where

$$s_1^2 = (Y_1 - \mathbf{Z}_1\mathbf{d}_{1,\text{OLS}})'(Y_1 - \mathbf{Z}_1\mathbf{d}_{1,\text{OLS}})/(n - g - k + 1) \tag{5.136}$$

We have argued till now that this estimator yields inconsistent estimates because of the correlation of $\mathbf{Z}_1$ with $\boldsymbol{\epsilon}_1$, or because of the correlation of $\mathbf{Y}_1$ with $\boldsymbol{\epsilon}_1$.

If now $\mathbf{W}_1$ is an $n \times (g - 1 + k)$ matrix that satisfies all the requirements to be considered as an instrumental variables matrix, then the instrumental variables estimator, which gives consistent estimates, is given by

$$\mathbf{d}_{1,\text{IV}} = (\mathbf{W}_1'\mathbf{Z}_1)^{-1}\mathbf{W}_1'Y_1 \tag{5.137}$$

and

$$\text{var--cov}(\mathbf{d}_{1,IV}) = s_1^2(\mathbf{W}_1{}'\mathbf{Z}_1)^{-1}(\mathbf{W}_1{}'\mathbf{W}_1)(\mathbf{Z}_1{}'\mathbf{W}_1)^{-1} \tag{5.138}$$

where

$$s_1^2 = (\mathbf{Y}_1 - \mathbf{Z}_1\mathbf{d}_{1,IV})'(\mathbf{Y}_1 - \mathbf{Z}_1\mathbf{d}_{1,IV})/(n - g - k + 1) \tag{5.139}$$

However, in the case of an exactly identified equation the number excluded from the equation predetermined variables, $K - k$, is equal to the number included in the right hand side endogenous variables, $g - 1$. Therefore, these predetermined variables could be used as instruments in the place of $\mathbf{Y}_1$. In this case it can be proved that the ILS estimator is the same with IV estimator.

Example 5.7 A wages–prices link model for an EU member

Consider the wages–prices link model of Example 5.6. In this model the endogenous variables are p_t and w_t, and the exogenous variables are Y_t and U_t. The $\mathbf{Z}$ matrix in the prices equation is $\mathbf{Z}_1 = [1\ w\ Y]$ in which w may be correlated with the error term. The corresponding instrumental variables $\mathbf{W}$ matrix is $\mathbf{W}_1 = [1\ U\ Y]$. The $\mathbf{Z}$ matrix in the wages equation is $\mathbf{Z}_2 = [1\ p\ U]$ in which p may be correlated with the error term. The corresponding instrumental variables $\mathbf{W}$ matrix is $\mathbf{W}_2 = [1\ Y\ U]$.

Using the sample data of Table 5.2, and Eviews, the IV estimates of the structural equations of the model are

$$\hat{p}_t = -7.43590 + 0.61399w_t + 0.02483Y_t \qquad \bar{R}^2 = 0.7649$$

s.e.   (2.0600)   (0.2010)   (0.0066)

t   [3.6097]   [3.0547]   [3.7801]

$$\hat{w}_t = 10.5915 + 0.72569p_t - 0.79019U_t \qquad \bar{R}^2 = 0.6540$$

s.e.   (2.0076)   (0.1066)   (0.2879)

t   [5.2758]   [6.8051]   [2.7444]

$$\tag{5.140}$$

Because both equations are exactly identified, we see that the IV estimates in (5.140) are exactly the same with estimates found with the ILS method in (5.129). However, the estimates in (5.140) include also the standard errors estimated according to (5.138). The corresponding ILS estimates in (5.129) do not include their standard errors because due to non linear relationships between the structural and the reduced form coefficients the computation of the standard errors of the structural coefficients through standard errors of the reduced form coefficients is awkward.

6.1.4 Two stage least squares (2SLS) method

In an exactly identified equation the number of the excluded predetermined variables is equal to the number of the included explanatory endogenous variables. Therefore, in using the IV approach we can use for the endogenous variables, as many instruments as the number of the excluded predetermined variables.

For the overidentified equation the number of the excluded exogenous variables is greater than the number of the included explanatory endogenous variables. This means that in using the IV approach we can use as instruments any combination of excluded exogenous variables. However, by not using all the exogenous variables as instruments we lose information, a fact that will possibly yield inefficient estimates. Two stage least squares is an approach that uses as instruments all the exogenous variables to get consistent and efficient estimates.

Assume that the following equation, which has g endogenous and k predetermined variables, is an overidentified equation in a structural model of G endogenous and K predetermined variables.

$$Y_1 = \mathbf{Y}_1\boldsymbol{\gamma}_1 + \mathbf{X}_1\boldsymbol{\beta}_1 + \boldsymbol{\epsilon}_1 = \mathbf{Z}_1\boldsymbol{\delta}_1 + \boldsymbol{\epsilon}_1 \tag{5.141}$$

where

$$\mathbf{Z}_1 = [\mathbf{Y}_1 \ \mathbf{X}_1] \text{ and } \boldsymbol{\delta}_1' = [\boldsymbol{\gamma}_1 \ \boldsymbol{\beta}_1]$$

with

$$\mathbf{Y}_1 = [Y_2 \ Y_3 \dots \ Y_g], \ \mathbf{X}_1 = [X_1 \ X_2 \dots \ X_k] \text{ and } \mathbf{X} = [X_1 \ X_{k+1} \ X_{k+2} \dots \ X_K]$$

If $\mathbf{d}_1$ denotes an estimator of $\boldsymbol{\delta}_1$, the two stage least squares approach has the following two steps:

Step 1

Apply the OLS method to the following unrestricted reduced form equations

$$Y_i = \mathbf{X}\boldsymbol{\pi}_i + \upsilon_i \text{ for } i = 2, 3, \dots, g \tag{5.142}$$

to obtain the estimates of the reduced form coefficients $\mathbf{p}_i = (\mathbf{X}'\mathbf{X})^{-1}\mathbf{X}'Y_i$, where $\mathbf{p}_i$ estimate of $\boldsymbol{\pi}_i$, and use these estimates to obtain the predicted sample values of Y_i, i.e. $\hat{Y}_i = \mathbf{X}\mathbf{p}_i = \mathbf{X}(\mathbf{X}'\mathbf{X})^{-1}\mathbf{X}'Y_i$.

Step 2

Use the predicted sample values of Y_i, to construct matrix $\hat{\mathbf{Z}}_1 = [\hat{\mathbf{Y}}_1 \ \mathbf{X}_1]$ where $\hat{\mathbf{Y}}_1 = [\hat{Y}_2 \ \hat{Y}_3 \dots \ \hat{Y}_g]$ and apply the OLS method to the equation

$$Y_1 = \hat{\mathbf{Y}}_1\boldsymbol{\gamma}_1 + \mathbf{X}_1\boldsymbol{\beta}_1 + \boldsymbol{\eta}_1 = \hat{\mathbf{Z}}_1\boldsymbol{\delta}_1 + \boldsymbol{\eta}_1 \tag{5.143}$$

where $\boldsymbol{\eta}_1$ = error term, to get to 2SLS estimator

$$\mathbf{d}_{1,2SLS} = (\hat{\mathbf{Z}}_1'\hat{\mathbf{Z}}_1)^{-1}\hat{\mathbf{Z}}_1'Y_1 \tag{5.144}$$

The 2SLS estimator (5.144) in terms of the values of the original variables and for the *i*th equation is written as

$$\mathbf{d}_{i,2SLS} = (\hat{\mathbf{Z}}_i' \hat{\mathbf{Z}}_i)^{-1} \hat{\mathbf{Z}}_i' \mathbf{Y}_i = [(\mathbf{Z}_i' \mathbf{X})(\mathbf{X}'\mathbf{X})^{-1}(\mathbf{X}'\mathbf{Z}_i)]^{-1}(\mathbf{Z}_i' \mathbf{X})(\mathbf{X}'\mathbf{X})^{-1}\mathbf{X}'\mathbf{Y}_i \qquad (5.145)$$

and

$$\text{var-cov}(\mathbf{d}_{i,2SLS}) = s_i^2 (\hat{\mathbf{Z}}_i' \hat{\mathbf{Z}}_i)^{-1} \qquad (5.146)$$

where

$$s_i^2 = (\mathbf{Y}_i - \mathbf{Z}_i \mathbf{d}_{i,2SLS})'(\mathbf{Y}_i - \mathbf{Z}_i \mathbf{d}_{i,2SLS})/(n - g - k + 1) \qquad (5.147)$$

For an exactly identified equation the 2SLS estimator can be shown that it is the same with the ILS estimator, and it can be interpreted as an IV estimator (see Johnston, 1984).

One of the basic assumptions in the analysis above is that the structural error terms were not autoregressive. In the case that the error terms are autoregressive then we use the two-stage least squares with autoregression (2SLS/AR). We can distinguish two cases:

(a) 2SLS/AR(1) when there are no lagged endogenous variables

This is the simple case where

$$\epsilon_{i,t} = \rho_i \epsilon_{i,t-1} + \upsilon_{i,t} \text{ for } i = 1, 2, \ldots, G \qquad (5.148)$$

For the *i*th equation and the tth observation the usual quasi-differenced equation is the following:

$$\mathbf{Y}_{i,t} - \rho_i \mathbf{Y}_{i,t-1} = (\mathbf{Y}_{i,t}' - \rho_i \mathbf{Y}_{i,t-1}')\boldsymbol{\gamma}_i + (\mathbf{X}_{i,t}' - \rho_i \mathbf{X}_{i,t-1}')\boldsymbol{\beta}_i + \upsilon_{i,t} \qquad (5.149)$$

Because the problem of autoregression is a problem of efficiency and not of consistency, any consistent estimate of ρ_i, let us say an estimate based on the residuals of the 2SLS method, can be used in (5.149). Therefore, the steps for the 2SLS/AR(1) method may be the following:

Step 1

The same as step 1 of the 2SLS method. In other words, regress all the endogenous variables on all the exogenous variables of the model and keep their predictions, $\hat{\mathbf{Y}}_{i,t}$.

Step 2

The same as step 2 of the 2SLS method. In words, substitute the predictions of the endogenous variables found in step 1 in place of the corresponding endogenous variables on the right hand side of each equation and apply OLS. Having estimated $\mathbf{d}_i$, compute the 2SLS residuals e_i, and then compute the estimate of ρ_i, i.e. compute

$$\hat{\rho}_i = \frac{\Sigma e_{i,t} e_{i,t-1}}{\Sigma e_{i,t}^2} \tag{5.150}$$

Step 3

The results of steps 1 and 2 formulate the following model:

$$Y_{i,t} - \hat{\rho}_i Y_{i,t-1} = (\hat{\mathbf{Y}}'_{i,t} - \hat{\rho}_i \hat{\mathbf{Y}}'_{i,t-1}) \boldsymbol{\gamma}_i + (\mathbf{X}'_{i,t} - \hat{\rho}_i \mathbf{X}'_{i,t-1}) \boldsymbol{\beta}_i + \upsilon_{i,t} \tag{5.151}$$

and estimate it with OLS.

Step 4

The Cochrane–Orcutt iterative procedure may be used in step 3.

(b) 2SLS/AR(1) when there are lagged endogenous variables

In the presence of lagged endogenous variables the estimate of ρ_i is not consistent. In this case the steps for the 2SLS/AR(1) method may be modified as follows:

Step 1

Treat all lagged endogenous variables as the current endogenous variables, regress all the current endogenous variables on all the (strictly) exogenous variables of the model and keep their predictions, $\hat{\mathbf{Y}}_{i,t}$.

Step 2

The same as step 2 of the 2SLS method. In other words, substitute the predictions of the current and the lagged endogenous variables found in step 1 in place of the corresponding endogenous variables on the right hand side of each equation and apply OLS. Having estimated $\mathbf{d}_i$, compute the 2SLS residuals e_i, and then compute the estimate of ρ_i, using (5.150).

Step 3

The results of steps 1 and 2 formulate the following model

$$Y_{i,t} - \hat{\rho}_i Y_{i,t-1} = (\hat{\mathbf{Y}}'_{i,t} - \hat{\rho}_i \hat{\mathbf{Y}}'_{i,t-1}) \boldsymbol{\gamma}_i + (\mathbf{X}'_{i,t} - \hat{\rho}_i \mathbf{X}'_{i,t-1}) \boldsymbol{\beta}_i + \upsilon_{i,t} \tag{5.152}$$

and estimate it with OLS.

Step 4

The Cochrane–Orcutt iterative procedure may be used in step 3.

Example 5.8 A simple macroeconomic model for an EU member (2SLS)

Consider the following macroeconomic model for an EU member:

Private consumption function:

$$C_t = \alpha_0 + \alpha_1 Y_t + \alpha_2 C_{t-1} + \epsilon_{1t}, \, \alpha_1 > 0, \, 0 < \alpha_2 < 1 \tag{5.153}$$

Private gross investment function:

$$I_t = \beta_0 + \beta_1 Y_t + \beta_2 r_t + \beta_3 I_{t-1} + \epsilon_{2t}, \, \beta_1 > 0, \, \beta_2 < 0, \, 0 < \beta_3 < 1 \tag{5.154}$$

A money demand function:

$$r_t = \gamma_0 + \gamma_1 Y_t + \gamma_2 M_{t-1} + \gamma_3 P_t + \gamma_4 r_{t-1} + \epsilon_{3t}, \, \gamma_1 > 0, \, \gamma_2 < 0, \, \gamma_3 > 0, \, 0 < \gamma_4 < 1 \tag{5.155}$$

Income identity:

$$Y_t = C_t + I_t + G_t \tag{5.156}$$

where, using data from Table 5.3,

C_t = private consumption at constant market prices of year 1970 (million drs),

I_t = gross private investment at constant market prices of year 1970 (million drs),

G_t = government spending at constant market prices of year 1970 (million drs),

Y_t = total financial resources at constant market prices of year 1970 (million drs),

M_t = money supply (M1) at current prices (billion drs),

r_t = long-term interest rate (%),

P_t = consumer price index (base year $1970 = 1$).

The structural model of equations (5.153)–(5.156) has four endogenous variables (C_t, I_t, r_t, Y_t) and six predetermined variables (C_{t-1}, I_{t-1}, M_{t-1}, P_t, r_{t-1}, G_t, plus the constant). According to the order and rank conditions all structural equations are overidentified. The estimated equations with the 2SLS method, using Eviews, are shown below:

Table 5.3 Basic macroeconomic data of an EU member.

Year	Private consumption	Gross private investment	Government spending	Total financial resources	Money supply (M1)	Long-term interest rate	Consumer price index
1960	107808	19264	24548	151620	13.8	8.00	0.791684
1961	115147	19703	33606	168456	16.1	8.00	0.783142
1962	120050	22216	29571	171837	18.5	8.00	0.801758
1963	126115	24557	40299	190971	21.3	8.25	0.828688
1964	137192	30826	41594	209612	25.4	9.00	0.847185
1965	147707	35072	49745	232524	28.4	9.00	0.885828
1966	157687	36610	45184	239481	32.1	9.00	0.916505
1967	167528	34315	51488	253331	39.5	9.00	0.934232
1968	179025	43863	50624	273512	40.5	8.75	0.941193
				302005	43.7	8.00	0.969630
				323925	48.7	8.00	1.000000
				345852	54.5	8.00	1.033727
				375734	67.2	8.00	1.068064
				413665	83.9	9.00	1.228156
				386753	100.2	11.83	1.517795
				406436	115.7	11.88	1.701147
				426469	140.2	11.50	1.929906
				446054	165.6	12.00	2.159872
				469876	202.9	13.46	2.436364
				488350	235.4	16.71	2.838453
				482320	267.9	21.25	3.459030
				491383	335.3	21.33	4.081844
				506337	389.7	20.50	5.114169
				509173	449.4	20.50	6.067835
				508418	542.5	20.50	7.130602
				537249	675.1	20.50	8.435285
				536514	738.8	20.50	10.30081
				538496	843.4	21.82	11.91950
				562171	973.2	22.89	13.61448
				597552	1265.0	23.26	15.59285
				613571	1583.5	27.62	18.59539
				635692	1742.9	29.45	22.09116
				639987	1968.2	28.71	25.40122
				640143	2223.7	28.56	28.88346
				647725	2793.5	27.44	32.00385
				646025	3149.0	23.05	34.98085

$$293C_{t-1}$$

706)

$$(5.157)$$

377]

92 [0.762] SARGAN $=6.668$ [0.155]

where the figures in brackets next to the diagnostic tests show level of significance.

Private gross investment function:

$$\hat{I}_t = 6232.39 + 0.07278Y_t - 1097.613r_t + 0.64325I_{t-1}$$

s.e. (2932.5) (0.0223)   (347.874)   (0.1019)

t   [2.1252] [3.2685]   [3.1552]   [6.3115]

(5.158)

$\bar{R}^2 = 0.895$ DW $= 1.642$ BG $= 1.502$ [0.220] SARGAN $= 2.486$ [0.478]

The money demand function:

$$r_t = -1.46373 + 9.12(E-06)Y_t - 0.00839M_{t-1} + 0.53795P_t + 0.88081r_{t-1}$$

s.e.   (0.9257) (3.73(E-06))   (0.0039)   (0.3058)   (0.1046)

t   [1.5812] [2.4472]   [2.1520]   [1.7591]   [8.4230]

(5.159)

$\bar{R}^2 = 0.967$ DW $= 1.322$ BG $= 5.206$ [0.023] SARGAN $= 7.1344$ [0.028]

Although the estimates of the three structural equations are generally acceptable in terms of the a priori restrictions (signs of coefficients) and the statistical and diagnostic tests, the money demand function shows some autocorrelation in the residuals (DW and BG tests) and that the instruments are not all valid (SARGAN test). Therefore, we present below the estimates of the money demand function with the method of 2SLS/AR(1).

$$r_t = -1.86465 + 1.38(E-05)Y_t - 0.00837M_{t-1} + 0.55770P_t + 0.76147r_{t-1}$$

s.e.   (1.8378) (1.06(E-05))   (0.0041)   (0.3450)   (0.2669)

t   [1.0146] [1.2998]   [2.0483]   [1.6165]   [2.8525]

(5.160)

$\bar{R}^2 = 0.968$ DW $= 1.799$ $e_t = 0.40808e_{t-1}$
                              (0.3326) [1.2271]

In (5.160) the estimated correlation coefficient is not significant as it can be seen from the low t ratio.

6.1.5 *Limited information maximum likelihood (LIML) method*

This method utilises the likelihood of the endogenous variables included in the structural equation to be estimated only, and not all the endogenous variables of the model, and thus the name 'limited'. Assume, for example, that we want to estimate the following first equation of the model,

$$Y_1 = Y_1\gamma_1 + X_1\beta_1 + \epsilon_1 \tag{5.161}$$

where

$$Y_1 = [Y_2 \, Y_3 \, \ldots \, Y_g] \text{ and } X_1 = [X_1 \, X_2 \, \ldots \, X_k]$$

or the equation

$$Y_{1g}\gamma_{1g} - X_1\beta_1 = \epsilon_1 \tag{5.162}$$

where

$$Y_{1g} = [Y_1 \, Y_1] \text{ and } \gamma'_{1g} = [1 \; -\gamma_1]$$

The unrestricted reduced form equations for the endogenous variables included in the first structural equation is given by

$$Y_i = X\pi_i + v_i \text{ for } i = 1, 2, \ldots, g \tag{5.163}$$

where

$$X = [X_1 \, X_{k+1} \, X_{k+2} \, \ldots \, X_K] \text{ and } \pi_i = [\pi_{i1} \, \pi_{i2} \, \ldots \, \pi_{iK}]$$

Under the assumption that the reduced form disturbances are normally distributed, the log of their joint distribution L_1 is

$$\ln L_1 = -\frac{n}{2}[g\ln(2\pi) + \ln|\Omega_1|] - \frac{1}{2}\sum_t v'_{1t}\Omega_1^{-1}v_{1t} \tag{5.164}$$

where

$$v_{1t} = \begin{bmatrix} v_{1t} \\ v_{2t} \\ \vdots \\ v_{gt} \end{bmatrix} \text{ and } \Omega_1 = E(v_1 v'_1) = \begin{bmatrix} \omega_{11} & \omega_{12} & \ldots & \omega_{1g} \\ \omega_{21} & \omega_{22} & \ldots & \omega_{2g} \\ \vdots & \vdots & \vdots & \vdots \\ \omega_{g1} & \omega_{g2} & \ldots & \omega_{gg} \end{bmatrix}$$

According to Anderson and Rubin (1949), the limited information maximum likelihood approach refers to the maximisation of (5.164) subject to the restriction that $\text{rank}(\Omega_{gk*}) = g - 1$, where Ω_{gk*} is given in (5.103). However, it can be proved (see Johnston (1984), Kmenta (1986)) that the maximisation of (5.164) is equivalent with the minimisation of the *least variance ratio*, given by

$$\ell = \frac{\gamma'_{1g}W^*_{1g}\gamma_{1g}}{\gamma'_{1g}W_{1g}\gamma_{1g}}$$

$$\text{where } W^*_{1g} = Y'_{1g}Y_{1g} - Y'_{1g}X_1(X'_1X_1)^{-1}X'_1Y_{1g} \tag{5.165}$$

$$W_{1g} = Y'_{1g}Y_{1g} - Y'_{1g}X(X'X)^{-1}X'Y_{1g}$$

By noting that $\mathbf{c}_{1g}$ is an estimate of $\boldsymbol{\gamma}_{1g}$, $\mathbf{b}_1$ is an estimate of $\boldsymbol{\beta}_1$, and λ is the smallest characteristic root of the determinantal equation

$$|\mathbf{W}_{1g}^* - \ell\mathbf{W}_{1g}| = 0, \text{ which is a polynomial in } \ell, \tag{5.166}$$

then the LIML estimator is given by

$$(\mathbf{W}_{1g}^* - \lambda\mathbf{W}_{1g})\mathbf{c}_{1g} = 0 \tag{5.167}$$

and

$$\mathbf{b}_1 = (\mathbf{X}_1'\mathbf{X}_1)^{-1}(\mathbf{X}_1'\mathbf{Y}_{1g})\mathbf{c}_{1g} \tag{5.168}$$

The LIML estimator given in (5.167) and (5.168) has the same asymptotic variance-covariance matrix with the estimator of 2SLS. Furthermore, in the case of an exactly identified equation it can be shown that $\lambda = 1$, and therefore the LIML estimator produces the ILS estimator.

Example 5.9 A simple macroeconomic model for an EU member (LIML)

Consider the macroeconomic model for an EU member presented in Example 5.8. The estimated equations with the LIML method, using TSP, are shown below:

Private consumption function:

$$\hat{C}_t = 8423.51 + 0.14142Y_t + 0.78107C_{t-1}$$

$$\text{s.e.} \quad (2304.6) \quad (0.0538) \quad (0.0794)$$

$$\text{t} \quad [3.6551] \quad [2.6283] \quad [9.8403] \tag{5.169}$$

$$\bar{R}^2 = 0.999 \; DW = 1.923 \; \lambda = 1.247$$

where λ is the variance ratio.

Private gross investment function:

$$\hat{I}_t = 6254.74 + 0.0719Y_t - 1089.52r_t + 0.64687I_{t-1}$$

$$\text{s.e.} \quad (2935.7) \quad (0.0225) \quad (352.45) \quad (0.10239)$$

$$\text{t} \quad [2.1306] \quad [3.1951] \quad [3.0912] \quad [6.3175] \tag{5.170}$$

$$\bar{R}^2 = 0.895 \; DW = 1.649 \; \lambda = 1.087$$

Money demand function:

$$\hat{r}_t = -1.48583 + 9.28(E-06)Y_t - 0.00841M_{t-1} + 0.53967P_t + 0.87751r_{t-1}$$

s.e. (0.9266)   (3.74(E − 06))   (0.0039)   (0.3059)   (0.1047)

t   [1.6036]   [2.4843]   [2.1566]   [1.7643]   [8.3785]

(5.171)

$\bar{R}^2 = 0.968$ DW $= 1.312$ $\lambda = 1.312$

6.2 System methods of estimation

The common factor in the single equation estimation methods proposed till now is that in all methods we tried to bypass the correlation between the explanatory endogenous variables in the structural equations under estimation and their error terms. These single equation estimation techniques did not consider the possible correlation of disturbances among equations. Thus, although the single equation methods estimators were consistent, they were not asymptotically efficient. In this section we will present two methods of estimation of the structural parameters of a model that treat all equations simultaneously, and thus they are increasing the efficiency of the corresponding estimators, because they contain all possible information among equations, as such, for example, of the correlation among their error terms.

6.2.1 Three-stage least squares (3SLS) method

Repeating equation (5.130), a structural model may be written for all its structural equations as

$$Y_1 = \mathbf{Y}_1 \boldsymbol{\gamma}_1 + \mathbf{X}_1 \boldsymbol{\beta}_1 + \boldsymbol{\epsilon}_1 = \mathbf{Z}_1 \boldsymbol{\delta}_1 + \boldsymbol{\epsilon}_1$$

$$Y_2 = \mathbf{Y}_2 \boldsymbol{\gamma}_2 + \mathbf{X}_2 \boldsymbol{\beta}_2 + \boldsymbol{\epsilon}_2 = \mathbf{Z}_2 \boldsymbol{\delta}_2 + \boldsymbol{\epsilon}_2$$

$$\cdots$$

$$Y_G = \mathbf{Y}_G \boldsymbol{\gamma}_G + \mathbf{X}_G \boldsymbol{\beta}_G + \boldsymbol{\epsilon}_G = \mathbf{Z}_G \boldsymbol{\delta}_G + \boldsymbol{\epsilon}_G$$

(5.172)

where $\mathbf{Z}_i = [\mathbf{Y}_i\ \mathbf{X}_i]$ and $\boldsymbol{\delta}'_i = [\boldsymbol{\gamma}_i\ \boldsymbol{\beta}_i]$
    System (5.172) can be also written as

$$\mathbf{Y} = \mathbf{Z}\boldsymbol{\delta} + \boldsymbol{\epsilon}$$

where

$$\mathbf{Y} = \begin{bmatrix} Y_1 \\ Y_2 \\ \vdots \\ Y_G \end{bmatrix}, \mathbf{Z} = \begin{bmatrix} \mathbf{Z}_1 & \mathbf{0} & \cdots & \mathbf{0} \\ \mathbf{0} & \mathbf{Z}_2 & \cdots & \mathbf{0} \\ \vdots & \vdots & & \vdots \\ \mathbf{0} & \mathbf{0} & \cdots & \mathbf{Z}_G \end{bmatrix}, \boldsymbol{\delta} = \begin{bmatrix} \boldsymbol{\delta}_1 \\ \boldsymbol{\delta}_2 \\ \vdots \\ \boldsymbol{\delta}_G \end{bmatrix}, \boldsymbol{\epsilon} = \begin{bmatrix} \boldsymbol{\epsilon}_1 \\ \boldsymbol{\epsilon}_2 \\ \vdots \\ \boldsymbol{\epsilon}_G \end{bmatrix}$$

(5.173)

To avoid the problem of correlation between the explanatory endogenous variables $\mathbf{Y}_i$ and $\boldsymbol{\epsilon}_i$, for $i = 1, 2, \ldots, G$, we saw in the previous sections that we can use the predicted values of $\mathbf{Y}_i$, from the regression of $\mathbf{Y}_i$ on all the predetermined variables of the model, in the place of the corresponding explanatory endogenous variables. Then, system (5.172) is written as

$$\mathbf{Y}_1 = \hat{\mathbf{Y}}_1\boldsymbol{\gamma}_1 + \mathbf{X}_1\boldsymbol{\beta}_1 + \boldsymbol{\eta}_1 = \hat{\mathbf{Z}}_1\boldsymbol{\delta}_1 + \boldsymbol{\eta}_1$$
$$\mathbf{Y}_2 = \hat{\mathbf{Y}}_2\boldsymbol{\gamma}_2 + \mathbf{X}_2\boldsymbol{\beta}_2 + \boldsymbol{\eta}_2 = \hat{\mathbf{Z}}_2\boldsymbol{\delta}_2 + \boldsymbol{\eta}_2$$

$$\cdots \tag{5.174}$$

$$\mathbf{Y}_G = \hat{\mathbf{Y}}_G\boldsymbol{\gamma}_G + \mathbf{X}_G\boldsymbol{\beta}_G + \boldsymbol{\eta}_G = \hat{\mathbf{Z}}_G\boldsymbol{\delta}_G + \boldsymbol{\eta}_G$$

where $\hat{\mathbf{Z}}_i = [\hat{\mathbf{Y}}_i \ \mathbf{X}_i]$ and $\boldsymbol{\delta}_i = \begin{bmatrix} \boldsymbol{\gamma}_i \\ \boldsymbol{\beta}_i \end{bmatrix}$

and respectively system (5.174), in which the various possible identities are not included, is written as

$$\mathbf{Y} = \hat{\mathbf{Z}}\boldsymbol{\delta} + \boldsymbol{\eta}$$

where (5.175)

$$\mathbf{Y} = \begin{bmatrix} \mathbf{Y}_1 \\ \mathbf{Y}_2 \\ \vdots \\ \mathbf{Y}_G \end{bmatrix}, \hat{\mathbf{Z}} = \begin{bmatrix} \hat{\mathbf{Z}}_1 & \mathbf{0} & \cdots & \mathbf{0} \\ \mathbf{0} & \hat{\mathbf{Z}}_2 & \cdots & \mathbf{0} \\ \vdots & \vdots & \vdots & \vdots \\ \mathbf{0} & \mathbf{0} & \cdots & \hat{\mathbf{Z}}_G \end{bmatrix}, \boldsymbol{\delta} = \begin{bmatrix} \boldsymbol{\delta}_1 \\ \boldsymbol{\delta}_2 \\ \vdots \\ \boldsymbol{\delta}_G \end{bmatrix}, \boldsymbol{\eta} = \begin{bmatrix} \boldsymbol{\eta}_1 \\ \boldsymbol{\eta}_2 \\ \vdots \\ \boldsymbol{\eta}_G \end{bmatrix}$$

System (5.175) leads to the consistent 2SLS estimator, which according to (5.144) is the following:

$$\mathbf{d}_{2SLS} = (\hat{\mathbf{Z}}'\hat{\mathbf{Z}})^{-1}\hat{\mathbf{Z}}'\mathbf{Y} \tag{5.176}$$

If we knew a consistent estimate of matrix $\boldsymbol{\Omega} = E(\boldsymbol{\eta}\,\boldsymbol{\eta}')$, i.e. an estimate of the variance–covariance matrix of the disturbances in (5.175), then we could apply Aitken's (1935) generalised least squares (GLS) estimator, which is

$$\mathbf{d}_{GLS} = (\hat{\mathbf{Z}}'\boldsymbol{\Omega}^{-1}\hat{\mathbf{Z}})^{-1}\hat{\mathbf{Z}}'\boldsymbol{\Omega}^{-1}\mathbf{Y} \tag{5.177}$$

However, knowing the 2SLS consistent estimator in (5.176) we use it in order to compute a consistent estimate $\mathbf{W}$ of $\boldsymbol{\Omega}$ as follows:

$$\mathbf{W} = \begin{bmatrix} w_{11}\mathbf{I}_n & w_{12}\mathbf{I}_n & \cdots & w_{1G}\mathbf{I}_n \\ w_{21}\mathbf{I}_n & w_{22}\mathbf{I}_n & \cdots & w_{2G}\mathbf{I}_n \\ \vdots & \vdots & \vdots & \vdots \\ w_{G1}\mathbf{I}_n & w_{G2}\mathbf{I}_n & \cdots & w_{GG}\mathbf{I}_n \end{bmatrix}$$

where (5.178)

$$w_{ij} = \frac{1}{n - g_i + 1 - k_i} (\mathbf{Y}_i - \mathbf{Z}_i\mathbf{d}_{i,2SLS})'(\mathbf{Y}_i - \mathbf{Z}_i\mathbf{d}_{i,2SLS})$$

for $i, j = 1, 2, \ldots, G$ and $g_i + k_i \geq g_j + k_j$

By substituting (5.178) into (5.177) we obtain the 3SLS estimator which is

$$\mathbf{d}_{3SLS} = (\hat{\mathbf{Z}}'\mathbf{W}^{-1}\hat{\mathbf{Z}})^{-1}\hat{\mathbf{Z}}'\mathbf{W}^{-1}\mathbf{Y} \qquad (5.179)$$

with

$$\text{var–cov}(\mathbf{d}_{3SLS}) = (\hat{\mathbf{Z}}'\mathbf{W}^{-1}\hat{\mathbf{Z}})^{-1} \qquad (5.180)$$

Summarising, the three steps in the 3SLS method are the following:

Step 1

The same as step 1 of the 2SLS method. In words, regress all the endogenous variables on all the exogenous variables of the model and keep their predictions, $\hat{\mathbf{Y}}_{i,t}$.

Step 2

The same as step 2 of the 2SLS method. In other words, substitute the predictions of the endogenous variables found in step 1 in place of the corresponding endogenous variables on the right hand side of each equation and apply OLS. Having estimated $\mathbf{d}_{i,2SLS}$, compute matrix $\mathbf{W}$ in (5.178).

Step 3

Using the results of steps 1 and 2, apply the GLS estimator, i.e. compute the 3SLS estimator according to (5.179) and (5.180).

   If in place of the estimates $\mathbf{d}_{i,2SLS}$ of step 2, the estimates $\mathbf{d}_{i,3SLS}$ of step 3 are used, then the whole procedure may be iterated, to produce the *iterative three-stage least squares*. However, it has been proved (Madansky, 1964) that this iterative procedure does not improve the asymptotic efficiency and does not provide the maximum likelihood estimator. Furthermore, it has been proved that (Zellner and Theil, 1962) the exactly identified equations do not add relevant information in the estimation of the overidentified equations.

Example 5.10 A simple macroeconomic model for an EU member (3SLS)

Consider the macroeconomic model for an EU member presented in Example 5.8. The estimated equations with the 3SLS method, using Eviews, are shown below:

Private consumption function:

$$\hat{C}_t = 7835.32 + 0.16152Y_t + 0.75162C_{t-1}$$

s.e. (1976.1) (0.0419) (0.0618)

t [3.9651] [3.8522] [12.1551] $\qquad (5.181)$

$\bar{R}^2 = 0.999 \; DW = 1.911$

Private gross investment function:

$$\hat{I}_t = 6442.75 + 0.07281 Y_t - 1080.57 r_t + 0.63380 I_{t-1}$$

s.e. (2745.7)   (0.0197)   (314.13)   (0.0893)

t   [2.3465]   [3.6876]   [3.4399]   [7.0979]

$\bar{R}^2 = 0.896 \; DW = 1.911$

(5.182)

A money demand function:

$$\hat{r}_t = -1.8292 + 1.02(E - 05) Y_t - 0.00775 M_{t-1} + 0.46344 P_t + 0.88864 r_{t-1}$$

s.e. (0.8089)   (3.25(E − 06))   (0.0031)   (0.2401)   (0.0884)

t   [2.2614]   [3.1498]   [2.5383]   [1.9301]   [10.0569]

$\bar{R}^2 = 0.966 \; DW = 1.304$

(5.183)

Finally, the corresponding variance–covariance matrix **W** of the disturbances used in the estimation is the following:

$$\mathbf{W} = \begin{bmatrix} w_{C,C} & w_{C,I} & w_{C,r} \\ w_{I,C} & w_{I,I} & w_{I,r} \\ w_{r,C} & w_{r,I} & w_{r,r} \end{bmatrix} = \begin{bmatrix} 10806587 & 3119375 & -1949.103 \\ 3119375 & 21008012 & 1403.266 \\ -1949.103 & 1403.266 & 1.641795 \end{bmatrix}$$

6.2.2 *Full information maximum likelihood (FIML) method*

Let us write the simultaneous equations model in the form of (5.10) and (5.12), which is

$$\mathbf{\Gamma Y}_t + \mathbf{BX}_t = \mathbf{\epsilon}_t, \; t = 1, 2, \ldots, n \tag{5.184}$$

and

$$\mathbf{\epsilon}_t \sim N(\mathbf{0}, \mathbf{\Sigma}) \; E(\mathbf{\epsilon}_t \mathbf{\epsilon}_s') = \mathbf{0} \; E(\mathbf{\epsilon}_t \mathbf{\epsilon}_t') = \mathbf{\Sigma} \tag{5.185}$$

The full information maximum likelihood estimator is based on the entire system of equations and can be found by maximising the following log of the likelihood function (Kmenta, 1986),

$$L = -\frac{Gn}{2} \ln(2\pi) - \frac{n}{2} \ln|\mathbf{\Sigma}| + n \ln|\mathbf{\Gamma}| - \frac{1}{2} \sum_{t=1}^{n} (\mathbf{\Gamma Y}_t + \mathbf{BX}_t)' \mathbf{\Sigma}^{-1} (\mathbf{\Gamma Y}_t + \mathbf{BX}_t) \tag{5.186}$$

with respect to $\boldsymbol{\Gamma}$, $\mathbf{B}$ and $\boldsymbol{\Sigma}$. This estimator is consistent and asymptotically efficient. In the case that the simultaneous equations model contains identities, the variance–covariance matrix of the structural disturbances $\boldsymbol{\Sigma}$ is singular and therefore its inverse $\boldsymbol{\Sigma}^{-1}$ does not exist. In this case, and before constructing the likelihood function, the identities of the system can be eliminated by substituting them into the other equations.

Example 5.11 A simple macroeconomic model for an EU member (FIML)

Consider the macroeconomic model for an EU member presented in Example 5.7. The estimated equations with the FIML method, using TSP, are shown below:

Private consumption function:

$$\hat{C}_t = 8261.57 + 0.14666Y_t + 0.77342C_{t-1}$$

s.e. (4670.5) (0.0860)   (0.1236)

t   [1.7689] [1.7050]   [6.2582]

$\bar{R}^2 = 0.999 \; DW = 1.921$

(5.187)

Private gross investment function:

$$\hat{I}_t = 6365.42 + 0.07923Y_t - 1184.43r_t + 0.61406I_{t-1}$$

s.e. (5512.6) (0.0740)   (1148.4)   (0.2976)

t   [1.1547] [1.0708]   [1.0314]   [2.0635]

$\bar{R}^2 = 0.905 \; DW = 1.584$

(5.188)

Money demand function:

$$\hat{r}_t = -1.88352 + 10.6(E-06)Y_t - 0.00772M_{t-1} + 0.46338P_t + 0.87237r_{t-1}$$

s.e. (1.8536)   (6.91(E-06))   (0.0095)   (0.6965)   (0.1684)

t   [1.0161]   [1.5796]   [0.8110]   [0.6653]   [5.1812]

$\bar{R}^2 = 0.970 \; DW = 1.2842$

(5.189)

7 SEEMINGLY UNRELATED EQUATIONS

We saw in section 2 that in the case where matrix $\boldsymbol{\Gamma}$, in the structural form of a system of equations, is diagonal, i.e. it has the form

$$\boldsymbol{\Gamma} = \begin{bmatrix} \gamma_{11} & 0 & \cdots & 0 \\ 0 & \gamma_{22} & \cdots & 0 \\ \vdots & \vdots & \vdots & \vdots \\ 0 & 0 & \cdots & \gamma_{GG} \end{bmatrix} \tag{5.190}$$

the system of the structural equations is not a system of simultaneous equations but instead it is a *set of equations*. In this case each equation contains one and only one endogenous variable, e.g. its dependent variable.

Generally speaking, this set of equations may be written analytically as

$$Y_{1t} = \beta_{11}X_{11,t} + \beta_{12}X_{12,t} + \ldots + \beta_{1K_1}X_{1K_1t} + \epsilon_{1t}$$

$$Y_{2t} = \beta_{21}X_{21,t} + \beta_{22}X_{22,t} + \ldots + \beta_{2K_2}X_{2K_2t} + \epsilon_{2t}$$

$$\ldots$$

$$Y_{Gt} = \beta_{G1}X_{G1,t} + \beta_{G2}X_{G2,t} + \ldots + \beta_{GK_G}X_{GK_Gt} + \epsilon_{Gt} \tag{5.191}$$

for $t = 1, 2, \ldots, n$, or in matrix form as

$$\mathbf{Y}_i = \mathbf{X}_i\boldsymbol{\beta}_i + \boldsymbol{\epsilon}_i \text{ for } i = 1, 2, \ldots, G \tag{5.192}$$

where for the ith equation, $\mathbf{Y}_i$ is a $n \times 1$ vector of the values of the dependent variable, $\mathbf{X}_i$ is a $n \times K_i$ matrix of the values of the explanatory variables, $\boldsymbol{\epsilon}_i$ is a $n \times 1$ vector of the values of the error variable, and $\boldsymbol{\beta}_i$ is a $K_i \times 1$ vector of the corresponding regression coefficients. The equations in (5.192) may be also written together as

$$\begin{bmatrix} \mathbf{Y}_1 \\ \mathbf{Y}_2 \\ \vdots \\ \mathbf{Y}_G \end{bmatrix} = \begin{bmatrix} \mathbf{X}_1 & \mathbf{0} & \cdots & \mathbf{0} \\ \mathbf{0} & \mathbf{X}_2 & \cdots & \mathbf{0} \\ \vdots & \vdots & \vdots & \vdots \\ \mathbf{0} & \mathbf{0} & \cdots & \mathbf{X}_G \end{bmatrix} \begin{bmatrix} \boldsymbol{\beta}_1 \\ \boldsymbol{\beta}_2 \\ \vdots \\ \boldsymbol{\beta}_G \end{bmatrix} + \begin{bmatrix} \boldsymbol{\epsilon}_1 \\ \boldsymbol{\epsilon}_2 \\ \vdots \\ \boldsymbol{\epsilon}_G \end{bmatrix} \tag{5.193}$$

$$\underbrace{}_{Gn \times 1} \quad \underbrace{}_{Gn \times K} \quad \underbrace{}_{K \times 1} \quad \underbrace{}_{Gn \times 1}$$

where $K = K_1 + K_2 + \ldots + K_G$, or finally as

$$\mathbf{Y} = \mathbf{X}\boldsymbol{\beta} + \boldsymbol{\epsilon} \tag{5.194}$$

where the substitution is obvious.

With respect to the error terms we employ the following assumptions:
(1) The error terms have zero mean:

$$E(\epsilon_{it}) = 0, t = 1, 2, \ldots, n, i = 1, 2, \ldots, G \tag{5.195}$$

(2) For each equation i (=1, 2,..., G), the error terms have constant variance over time:

$$\text{var}(\epsilon_{it}) = E(\epsilon_{it}^2) = \sigma_i^2 = \sigma_{ii}, t = 1, 2, \ldots, n \qquad (5.196)$$

These variances may be different for different equations.

(3) For each equation i (=1, 2,..., G) and for two different time periods $t \neq s$ (=1, 2,..., n), the error terms are not autocorrelated:

$$\text{cov}(\epsilon_{it}, \epsilon_{is}) = E(\epsilon_{it}\epsilon_{is}) = 0, t \neq s \qquad (5.197)$$

(4) For the same time period t (=1, 2,..., n), the error terms of two different equations $i \neq j$ (=1, 2,..., G) may be correlated (*contemporaneous correlation*):

$$\text{cov}(\epsilon_{it}, \epsilon_{jt}) = E(\epsilon_{it}\epsilon_{jt}) = \sigma_{ij}, i \neq j \qquad (5.198)$$

(5) For two different equations $i \neq j$ (=1, 2,..., G) and for two different time periods $t \neq s$ (=1, 2,..., n), the error terms are not correlated:

$$\text{cov}(\epsilon_{it}, \epsilon_{is}) = E(\epsilon_{it}\epsilon_{js}) = 0, t \neq s, i \neq j \qquad (5.199)$$

In matrix form the assumptions above may be written as

$$E(\epsilon_i) = \mathbf{0}, E(\epsilon_i \epsilon_i') = \sigma_{ii}\mathbf{I}, E(\epsilon_i \epsilon_j') = \sigma_{ij}\mathbf{I}, \qquad (5.200)$$

or

$$E(\epsilon) = \mathbf{0}, E(\epsilon\epsilon') = \Omega = \Sigma \otimes \mathbf{I} \qquad (5.201)$$

where $\mathbf{I}$ is an identity matrix of order $n \times n$, $\otimes$ is the Kronecker product, and

$$\Sigma = \begin{bmatrix} \sigma_{11} & \sigma_{12} & \cdots & \sigma_{1G} \\ \sigma_{21} & \sigma_{22} & \cdots & \sigma_{2G} \\ \vdots & \vdots & \vdots & \vdots \\ \sigma_{G1} & \sigma_{G2} & \cdots & \sigma_{GG} \end{bmatrix} \text{ with } \Omega = \Sigma \otimes \mathbf{I} = \begin{bmatrix} \sigma_{11}\mathbf{I} & \sigma_{12}\mathbf{I} & \cdots & \sigma_{1G}\mathbf{I} \\ \sigma_{21}\mathbf{I} & \sigma_{22}\mathbf{I} & \cdots & \sigma_{2G}\mathbf{I} \\ \vdots & \vdots & \vdots & \vdots \\ \sigma_{G1}\mathbf{I} & \sigma_{G2}\mathbf{I} & \cdots & \sigma_{GG}\mathbf{I} \end{bmatrix} \qquad (5.202)$$

From the discussion it is seen that the only link between the equations of the set of equations in (5.191), or in (5.192), is the contemporaneous correlation. In other words, the only link is through the covariance σ_{ij} of the error terms of the *i*th and the *j*th equations. For this reason Zellner (1962) gave the name 'seemingly unrelated regression equations (SURE)' to these equations.

The methods used in order to estimate the set of equations in (5.191), or in (5.192), depend on the assumptions about the error terms. These methods are the following:

7.1 Ordinary least squares (OLS)

If we assume that from the five assumptions (5.195) to (5.199), only the assumptions (5.195) and (5.196) hold, then each equation in the set of equations in (5.191) could be estimated individually by the classical ordinary least squares method. The least squares estimator in this case is the best linear unbiased estimator, and is given by

$$\mathbf{b}_{i,OLS} = (\mathbf{X}_i' \mathbf{X}_i)^{-1} \mathbf{X}_i' \mathbf{Y}_i \qquad (5.203)$$

with

$$\text{var--cov}(\mathbf{b}_{i,OLS}) = s_i^2 (\mathbf{X}_i' \mathbf{X}_i)^{-1} \qquad (5.204)$$

where

$$s_i^2 = \frac{1}{n - K_i} \sum_{j=1}^{n} e_j^2 = \frac{1}{n - K_i} \mathbf{e}_i' \mathbf{e}_i \qquad (5.205)$$

Example 5.12 Food demand equations for an EU member (OLS)

Consider the following set of three linear demand equations for specific food products:

meat: $Q_{m,t} = \beta_{10} + \beta_{11} P_{m,t} + \beta_{12} P_{f,t} + \beta_{13} Y_t + \epsilon_{1t}$

vegetables: $Q_{v,t} = \beta_{20} + \beta_{21} P_{v,t} + \beta_{22} P_{d,t} + \beta_{23} Y_t + \epsilon_{2t}$ (5.206)

oil: $Q_{o,t} = \beta_{30} + \beta_{31} P_{o,t} + \beta_{33} Y_t + \epsilon_{3t}$

where (actual data for estimation purposes in parentheses):

Q_m = quantity demanded (consumption) of meat

Q_v = quantity demanded (consumption) of fruit and vegetables

Q_o = quantity demanded (consumption) of oils and fats

P_m = price (index) of meat

P_f = price (index) of fish

P_v = price (index) of fruit and vegetables

P_d = price (index) of dairy goods

P_o = price (index) of oils and fats

Y = personal disposable income

The a priori restrictions for the coefficients in the three demand equations are the following:

meat: $\beta_{11} < 0$ (normal good), $\beta_{12} > 0$ (substitute good), $\beta_{13} > 0$ (superior good) vegetables: $\beta_{21} < 0$ (normal good), $\beta_{22} > 0$ (substitute good), $\beta_{23} > 0$ (superior good)

oils: $\beta_{31} < 0$ (normal good), $\beta_{33} > 0$ (superior good)

Assuming that there is no contemporaneous correlation between the error terms of the three demand equations we can apply the OLS method. The corresponding results, using the data in Table 5.4, are shown below.

Table 5.4 Consumption expenditure for specific food products (at constant 1970 market prices, million drs), price indices for specific food products (base year 1970), and personal disposable income (at constant 1970 market prices, million drs) of an EU member

Year	Q_m	Q_v	Q_o	P_m	P_f	P_v	P_d	P_o	Y
1960	6978	11313	4977	0.730	0.631	0.764	0.727	0.635	117179.2
1961	7901	13094	5300	0.687	0.643	0.766	0.740	0.638	127598.9
1962	8660	12369	5430	0.668	0.665	0.779	0.767	0.647	135007.1
1963	9220	13148	5528	0.706	0.686	0.825	0.778	0.729	142128.3
1964	9298	15564	6186	0.826	0.723	0.778	0.829	0.732	159648.7
1965	11718	15132	6457	0.844	0.794	0.908	0.890	0.772	172755.9
1966	12858	16536	6428	0.874	0.854	0.977	0.946	0.786	182365.5
1967	13686	16659	5559	0.887	0.878	0.996	0.955	0.938	195611.0
1968	14468	18357	6582	0.872	0.890	0.952	0.974	0.857	204470.4
1969	15041	18733	7108	0.888	0.913	1.072	0.989	0.922	222637.5
1970	16273	19692	6681	1.000	1.000	1.000	1.000	1.000	246819.0
1971	17538	18865	7043	1.089	1.109	1.129	1.009	1.003	269248.9
1972	18632	19850	7223	1.139	1.181	1.198	1.016	1.038	297266.0
1973	19789	21311	7439	1.421	1.398	1.531	1.186	1.272	335521.7
1974	20333	21695	7311	1.643	1.600	1.918	1.582	1.696	310231.1
1975	21508	22324	7326	1.810	1.993	2.162	1.788	1.931	327521.3
1976	22736	21906	7638	2.100	2.159	2.709	2.008	1.957	350427.4
1977	25379	20312	6979	2.294	2.390	3.191	2.122	2.140	366730.0
1978	26151	22454	6973	2.499	3.169	3.733	2.504	2.444	390188.5
1979	26143	23269	7287	3.243	3.795	4.241	2.985	2.775	406857.2
1980	26324	24164	8103	3.862	6.850	5.294	3.764	3.440	401942.8
1981	24324	26572	7335	5.270	9.000	6.382	4.898	4.072	419669.1
1982	26155	25051	7591	6.709	10.177	7.450	5.945	4.727	421715.6
1983	27098	25344	8492	7.874	11.637	8.625	6.973	5.537	417930.3
1984	27043	25583	8580	8.949	13.951	10.682	8.319	6.774	434695.7
1985	27047	23402	8303	10.546	18.230	11.960	10.134	8.725	456576.2
1986	27136	23666	7287	12.262	23.003	13.128	12.514	10.254	439654.1
1987	28616	24639	7258	13.439	27.152	15.561	14.254	10.200	438453.5
1988	29624	26857	7328	15.092	29.460	16.553	15.776	11.007	476344.7
1989	30020	26431	7230	18.444	33.972	19.307	18.873	13.585	492334.4
1990	29754	25808	7261	21.632	40.391	23.881	22.757	17.181	495939.2
1991	29332	26409	7596	23.916	45.400	29.734	26.049	22.076	513173.0
1992	30665	26358	7638	27.246	51.814	31.481	29.590	20.751	502520.1
1993	31278	28120	7815	30.083	57.155	32.248	35.182	22.355	523066.1
1994	35192	28405	8126	31.827	63.540	36.397	40.331	26.603	520727.5
1995	36505	28388	8195	33.601	72.491	39.122	42.793	31.090	518406.9

Sources: Epilogi (1998).

A demand equation for meat:

$$\hat{Q}_{m,t} = 627.6114 - 1242.417P_{m,t} + 594.4882P_{f,t} + 0.06553Y_t, \ \bar{R}^2 = 0.981$$

s.e.   (664.79)   (299.14)   (140.87)   (0.0027)   (5.207)

t   [0.9441]   [4.1532]   [4.2207]   [24.592]

A demand equation for fruit and vegetables:

$$\hat{Q}_{v,t} = 8364.896 - 397.1046P_{v,t} + 329.6762P_{d,t} + 0.04062Y_t, \ \bar{R}^2 = 0.948$$

s.e.   (691.59)   (177.43)   (156.88)   (0.0027)   (5.208)

t   [12.081]   [2.2381]   [2.1015]   [15.002]

A demand equation for oils and fats:

$$\hat{Q}_{o,t} = 4819.17 - 32.6999P_{o,t} + 0.00724Y_t, \ \bar{R}^2 = 0.721$$

s.e.   (269.15)   (14.939)   (0.00095)   (5.209)

t   [17.905]   [2.1890]   [7.6547]

The estimates of the three demand equations are acceptable in the light of the signs of the regression coefficients and the reported statistics.

7.2 Generalised least squares (GLS)

In the case that assumption (5.198) holds then the OLS estimator is not efficient because it does not use information on the contemporaneous correlation. To obtain a more efficient estimator than the OLS estimator the set of the equations may be written in the form of the 'stacked' model of (5.193), In other words, the G different equations may be viewed as the single equation of (5.193), and the generalised least squares estimator may be used. This estimator, which is best linear unbiased estimator, is given by (Aitken, 1935).

$$\mathbf{b}_{\mathrm{GLS}} = (\mathbf{X}'\mathbf{\Omega}^{-1}\mathbf{X})^{-1}\mathbf{X}'\mathbf{\Omega}^{-1}\mathbf{Y} = [\mathbf{X}'(\mathbf{\Sigma}^{-1}\otimes\mathbf{I})\mathbf{X}]^{-1}\mathbf{X}'(\mathbf{\Sigma}^{-1}\otimes\mathbf{I})\mathbf{Y} \tag{5.210}$$

with

$$\mathrm{var\text{-}cov}(\mathbf{b}_{\mathrm{GLS}}) = (\mathbf{X}'\mathbf{\Omega}^{-1}\mathbf{X})^{-1} = [\mathbf{X}'(\mathbf{\Sigma}^{-1}\otimes\mathbf{I})\mathbf{X}]^{-1} \tag{5.211}$$

7.3 Seemingly unrelated regression estimator (SURE)

For the GLS estimator in (5.210) and (5.211) to have practical meaning we have to know matrices $\mathbf{\Omega}$ or $\mathbf{\Sigma}$, or in other words we have to know the variances–covariances

σ_{ij} in matrix (5.202). In order to obtain consistent estimates of these variances–covariances the following steps may be used (Zellner, 1962):

Step 1

Apply the OLS method to each of the equations individually and obtain the estimates of the regression coefficients according to (5.203), or

$$\mathbf{b}_{i,OLS} = (\mathbf{X}_i'\mathbf{X}_i)^{-1}\mathbf{X}_i'\mathbf{Y}_i \qquad (5.212)$$

Step 2

Using these OLS estimates obtain the corresponding residuals according to

$$\mathbf{e}_{i,OLS} = \mathbf{Y}_i - \mathbf{X}_i\mathbf{b}_{i,OLS} \qquad (5.213)$$

Step 3

Using the residuals estimated in (5.213) obtain consistent estimates s_{ij} of the variances–covariances σ_{ij} according to

$$s_{ij} = \frac{1}{[(n - K_i)(n - K_j)]^{1/2}}\ \mathbf{e}_{i,OLS}'\mathbf{e}_{i,OLS} = \frac{1}{[(n - K_i)(n - K_j)]^{1/2}} \sum_{t=1}^{n} e_{it}e_{jt} \qquad (5.214)$$

Step 4

Using the estimates in (5.214) construct the following variance–covariance matrices $\mathbf{S}$ and $\mathbf{W}$, as being estimates of the matrices $\boldsymbol{\Sigma}$ and $\boldsymbol{\Omega}$ respectively:

$$\mathbf{S} = \begin{bmatrix} s_{11} & s_{12} & \cdots & s_{1G} \\ s_{21} & s_{22} & \cdots & s_{2G} \\ \vdots & \vdots & \vdots & \vdots \\ s_{G1} & s_{G2} & \cdots & s_{GG} \end{bmatrix} \text{ with } \mathbf{W} = \mathbf{S} \otimes \mathbf{I} = \begin{bmatrix} s_{11}\mathbf{I} & s_{12}\mathbf{I} & \cdots & s_{1G}\mathbf{I} \\ s_{21}\mathbf{I} & s_{22}\mathbf{I} & \cdots & s_{2G}\mathbf{I} \\ \vdots & \vdots & \vdots & \vdots \\ s_{G1}\mathbf{I} & s_{G2}\mathbf{I} & \cdots & s_{GG}\mathbf{I} \end{bmatrix} \qquad (5.215)$$

Step 5

Substituting the estimates of (5.215) into (5.210) and (5.211) we obtain the **SUR estimator**, as

$$\mathbf{b}_{SUR} = (\mathbf{X}'\mathbf{W}^{-1}\mathbf{X})^{-1}\mathbf{X}'\mathbf{W}^{-1}\mathbf{Y} = [\mathbf{X}'(\mathbf{S}^{-1} \otimes \mathbf{I})\mathbf{X}]^{-1}\mathbf{X}'(\mathbf{S}^{-1} \otimes \mathbf{I})\mathbf{Y} \qquad (5.216)$$

with

$$\text{var–cov}(\mathbf{b}_{SUR}) = (\mathbf{X}'\mathbf{W}^{-1}\mathbf{X})^{-1} = [\mathbf{X}'(\mathbf{S}^{-1} \otimes \mathbf{I})\mathbf{X}]^{-1} \qquad (5.217)$$

Step 6

Having estimated the regression coefficients in (5.216), the whole process could be iterated by substituting these estimates into step 2 and then continue to the next steps. This process will yield the *'iterated SUR estimates'*, which correspond to the 'maximum likelihood estimates of a SUR model' (Oberhofer and Kmenta, 1974).

Example 5.13 Food demand equations for an EU member (SURE)

Consider the set (5.206) of the three linear demand equations for specific food products of Example 5.12. Assuming that there exists contemporaneous correlation between the error terms of these equations the SUR estimates for these equations are derived as follows:

Step 1

The OLS estimates of the regression coefficients of the three equations individually have been obtained in Example 5.12.

Steps 2, 3, 4

Using these estimates the residuals for the three equations have been calculated and furthermore, using these residuals the variance–covariance matrix **S** have been estimated as follows:

$$\mathbf{S} = \begin{bmatrix} 1145077 & 22757.26 & 134263.2 \\ 22757.26 & 1108305 & 202239.7 \\ 134263.2 & 202239.7 & 207761.2 \end{bmatrix} \qquad (5.218)$$

Steps 5, 6

Using (5.216) and (5.217) the SUR estimates have been obtained to be the following (results from Eviews):

The demand for meat:

$$\hat{Q}_{m,t} = 680.5815 - 1192.477 P_{m,t} + 572.2006 P_{f,t} + 0.06516 Y_t, \ \bar{R}^2 = 0.981$$

s.e.	(625.189)	(269.793)	(4.5010)	(0.0025)	(5.219)
t	[1.0886]	[4.4200]	[4.5010]	[26.2964]	

The demand for fruit and vegetables:

$$\hat{Q}_{v,t} = 8481.553 - 331.7326 P_{v,t} + 273.6894 P_{d,t} + 0.03992 Y_t, \ \bar{R}^2 = 0.948$$

s.e.	(643.04)	(151.121)	(133.489)	(0.0025)	(5.220)
t	[13.190]	[2.195]	[2.050]	[16.112]	

The demand for oils and fats:

$$\hat{Q}_{o,t} = 4831.429 - 31.4861P_{o,t} + 0.00718Y_t, \quad \bar{R}^2 = 0.721$$

s.e.   (257.48)   (14.266)   (0.0009)

t   [18.764]   [2.207]   [7.942]

(5.221)

The variance–covariance matrix **S** used in the final iteration of the iterative SUR estimation (Eviews) is the following:

$$\mathbf{S} = \begin{bmatrix} 114687 & 51998.11 & 136474.4 \\ 51998.11 & 1113091 & 209501.3 \\ 136474.4 & 209501.3 & 207802.8 \end{bmatrix}$$

(5.222)

Comparing the results in Example 5.13 (SUR) with those in Example 5.12 (OLS) we see that the estimates in Example 5.13 are in general more efficient (greater t ratios) than the corresponding estimates in Example 5.12. Generally, the greater the contemporaneous correlation, the more efficient the SUR estimates are with respect to the OLS estimates.

7.4 Hypothesis testing for seemingly unrelated equations

We saw thus far that in the case where there is no contemporaneous correlation there is no need to apply the SURE methodology because by applying the OLS methodology in each equation individually we will obtain efficient estimates. Furthermore, having stacked together in model (5.193) the individual equations of a set of G equations, the obvious question that arises is: are there linear restrictions among the regression coefficients of this 'stacked single' equation? Two tests have been suggested to deal with these problems.

(a) Testing for contemporaneous correlation

This test is formed as follows:

H_0: all covariances are zero ($\sigma_{ij} = 0$ for $i \neq j$)

H_a: at least one covariance is nonzero

(5.223)

Breusch and Pagan (1980) suggested that under the null hypothesis the Lagrange multiplier statistic λ, which is given by

$$\lambda = n \sum_{i=2}^{G} \sum_{j=1}^{i-1} r_{ij}^2$$

(5.224)

where

$$r_{ij}^2 = \frac{s_{ij}^2}{s_{ii}s_{jj}} \qquad (5.225)$$

follows asymptotically the χ^2 distribution with $G(G-1)/2$ degrees of freedom.

Example 5.14 Food demand equations for an EU member (testing for contemporaneous correlation)

Consider the set (5.206) of the three linear demand equations for specific food products of Example 5.12. According to (5.225), the residual correlation matrix $\mathbf{r}_{\mathrm{OLS}}$ of the OLS residuals (Example 5.12) and the residual correlation matrix $\mathbf{r}_{\mathrm{SUR}}$ of the SUR residuals (Example 5.13) are the following:

$$\mathbf{r}_{\mathrm{OLS}} = \begin{bmatrix} 1.0000 & 0.0202 & 0.2753 \\ 0.0202 & 1.0000 & 0.4215 \\ 0.2753 & 0.4215 & 1.0000 \end{bmatrix}, \mathbf{r}_{\mathrm{SUR}} = \begin{bmatrix} 1.0000 & 0.0460 & 0.2796 \\ 0.0460 & 1.0000 & 0.4356 \\ 0.2796 & 0.4356 & 1.0000 \end{bmatrix} \qquad (5.226)$$

According now to (5.224) the Lagrange multiplier statistic is

$$\lambda_{\mathrm{OLS}} = n(r_{21}^2 + r_{31}^2 + r_{32}^2) = 9.1390, \lambda_{\mathrm{SUR}} = n(r_{21}^2 + r_{31}^2 + r_{32}^2) = 9.7214, n = 36 \qquad (5.227)$$

Taking into account that $G=3$, the degrees of freedom are $G(G-1)/2=6$, and for a significant level of 0.05 we get from the tables of the chi-squared distribution that $\chi^2(3) = 7.81473$. Thus, because the value(s) of the Lagrange multiplier statistic(s) is greater than the critical level of the $\chi^2(3)$ distribution, the null hypothesis is rejected in favour of the alternative hypothesis. In other words, at least one covariance is non-zero, suggesting that there exists contemporaneous correlation.

(b) Testing linear restrictions on the coefficients

This test is formed as follows:

$$H_0: \mathbf{R\beta} = \mathbf{r}$$

$$H_a: \mathbf{R\beta} \neq \mathbf{r} \qquad (5.228)$$

where $\mathbf{R}$ is a known matrix of dimensions $J \times K$, $\mathbf{\beta}$ is the coefficients vector of dimensions $K \times 1$, and $\mathbf{r}$ is a known vector of dimensions $J \times 1$.

Under the null hypothesis, the usual Wald (1943) statistic is written in this case as:

$$g = (\mathbf{Rb} - \mathbf{r})'(\mathbf{RCR'})^{-1}(\mathbf{Rb} - \mathbf{r}) \sim \chi^2(J) \qquad (5.229)$$

where:

$$C = [X'(\Sigma^{-1} \otimes I)X]^{-1} \tag{5.230}$$

or

$$\lambda = \frac{g/J}{(Y - Xb)'(\Sigma^{-1} \otimes I)(Y - Xb)/(Gn - K)} \sim F(J, Gn - K) \tag{5.231}$$

However, for these tests to be operational matrix Σ in the formulas above must be replaced with its estimate S (see Judge *et al.*, 1985).

Example 5.15 Food demand equations for an EU member (testing linear restrictions on the coefficients)

Consider the set (5.206) of the three linear demand equations for specific food products of Example 5.12. According to (5.193) this set of equations may be written as:

$$\begin{bmatrix} Q_m \\ Q_v \\ Q_o \end{bmatrix} = \begin{bmatrix} 1 & P_m & P_f & Y & 0 & 0 & 0 & 0 & 0 & 0 & 0 \\ 0 & 0 & 0 & 0 & 1 & P_v & P_d & Y & 0 & 0 & 0 \\ 0 & 0 & 0 & 0 & 0 & 0 & 0 & 0 & 1 & P_o & Y \end{bmatrix} \begin{bmatrix} \beta_{10} \\ \vdots \\ \beta_{13} \\ \vdots \\ \beta_{23} \\ \vdots \\ \beta_{33} \end{bmatrix} = \begin{bmatrix} \epsilon_1 \\ \epsilon_2 \\ \epsilon_3 \end{bmatrix} \tag{5.232}$$

Assume now that we want to test if the marginal propensity to consume (demand) with respect to income is the same for all food commodities. In other words we want to test if the restrictions $\beta_{13} = \beta_{23} = \beta_{33}$, or equivalently the restrictions $\beta_{13} - \beta_{33} = 0$ and $\beta_{23} - \beta_{33} = 0$, are correct. According to (5.228) these restrictions are written:

$$R\beta = \begin{bmatrix} 0 & 0 & 0 & 1 & 0 & 0 & 0 & 0 & 0 & 0 & -1 \\ 0 & 0 & 0 & 0 & 0 & 0 & 0 & 1 & 0 & 0 & -1 \end{bmatrix} \begin{bmatrix} \beta_{10} \\ \vdots \\ \beta_{13} \\ \vdots \\ \beta_{23} \\ \vdots \\ \beta_{33} \end{bmatrix} = \begin{bmatrix} 0 \\ 0 \end{bmatrix} = r \tag{5.233}$$

Taking into account (5.233) and using the SUR estimation in Example 5.13, statistic (5.229) is equal to (estimation with Eviews) $g = 834.2011$. Because this value is greater than the 0.05 significant level critical value of $\chi^2(2) = 5.99146$, the null hypothesis is rejected in favour of the alternative hypothesis. In other words, at least one of the marginal propensities to consume is not equal to the other propensities to consume.

7.5　Seemingly unrelated regression estimator with autocorrelation in the disturbances (SURE/AR)

Until now we assumed, according to (5.197), that there is no autocorrelation in the disturbances. However, in the case where there is autocorrelation in the disturbances assumption (5.200) is written as:

$$E(\epsilon_i\epsilon_i') = \begin{bmatrix} 1 & \rho_i & \cdots & \rho_i^{n-1} \\ \rho_i & 1 & \cdots & \rho_i^{n-2} \\ \vdots & \vdots & \vdots & \vdots \\ \rho_i^{n-1} & \rho_i^{n-2} & \cdots & 1 \end{bmatrix} \quad E(\epsilon_i\epsilon_j') = \begin{bmatrix} 1 & \rho_j & \cdots & \rho_j^{n-1} \\ \rho_i & 1 & \cdots & \rho_j^{n-2} \\ \vdots & \vdots & \vdots & \vdots \\ \rho_i^{n-2} & \rho_i^{n-2} & \cdots & 1 \end{bmatrix} \quad (5.234)$$

where ρ_i is the autocorrelation coefficient of the disturbances in the ith equation.

Park (1967) suggested the following steps in order to estimate a seemingly unrelated equations model with autocorrelation in the disturbances:

Step 1

Apply the OLS method to each of the equations individually and obtain the estimates of the regression coefficients.

Step 2

Using these OLS estimates obtain the corresponding residuals e_i and estimate the autocorrelation coefficients $\hat{\rho}_i$ for each equation individually, by regressing $e_{i,t}$ on $e_{i,t-1}$.

Step 3

Using the estimated autocorrelation coefficients transform the data according to the following transformations:

$$Y_{i,t}^* = Y_{i,t} - \hat{\rho}_i Y_{i,t-1}, \; X_{i,t}^* = X_{i,t} - \hat{\rho}_i X_{i,t-1} \quad \text{for } t = 2, 3, \ldots, n$$
$$Y_{i,t}^* = \sqrt{1-\hat{\rho}_i^2}\,Y_{i,t}, \; X_{i,t}^* = \sqrt{1-\hat{\rho}_i^2}\,X_{i,t} \quad \text{for } t = 1 \tag{5.235}$$

Step 4

Apply the SUR estimator to the transformed data of step 3, i.e. apply the steps 3, 4, 5 and 6 of the methodology in section 7.3.

Example 5.16　Food demand equations for an EU member (SURE/AR)

Consider the set (5.206) of the three linear demand equations for specific food products in Example 5.12. Having estimated in Example 5.12 each equation individually by OLS, Table 5.5 presents the Durbin–Watson statistics and the corresponding estimated autocorrelation coefficients according to step 2.

Table 5.5 Durbin–Watson statistics and autocorrelation coefficients

	Demand for meat	Demand for fruits and vegetables	Demand for oils and fats
Durbin–Watson	1.42144	1.13530	0.98065
Autocorrelation	0.27993	0.39407	0.47848
(s.e.)	(0.16393)	(0.15077)	(0.15000)
[t-ratio]	[1.70759]	[2.61379]	[3.18986]

Notes: Estimates with Eviews.

Having transformed all data according to the estimated autocorrelation coefficients in Table 5.5 and the transformations in (5.235), the application of the SURE methodology to these transformed data yields the following results:

The demand for meat:

$$\hat{Q}_{m,t}^* = 851.5305 - 1092.406P_{m,t}^* + 529.8507P_{f,t}^* + 0.06410Y_t^*$$

s.e   (797.29)   (284.02)   (132.23)   (0.0031)

t   [1.0680]   [3.8462]   [4.0071]   [20.781]

$\bar{R}^2 = 0.966$ DW $= 1.835$

(5.236)

The demand for fruits and vegetables:

$$\hat{Q}_{v,t}^* = 8436.796 - 318.4453P_{v,t}^* + 266.7486P_{d,t}^* + 0.03979Y_t^*$$

s.e.   (876.388)   (167.744)   (147.342)   (0.0033)

t   [9.6268]   [1.8984]   [1.8104]   [12.1737]

$\bar{R}^2 = 0.871$ DW $= 1.757$

(5.237)

The demand for oils and fats:

$$\hat{Q}_{o,t}^* = 4851.646 - 20.50658P_t^* + 0.00690Y_t^*$$

s.e.   (389.834)   (20.060)   (0.00135)

t   [12.4454]   [1.0222]   [5.1100]

$\bar{R}^2 = 0.350$ DW $= 1.717$

(5.238)

Finally, the residual covariance matrix **S** and the residual correlation matrix **r** used in the estimation are the following:

$$\mathbf{S} = \begin{bmatrix} 1047576 & -24074.4 & 109603.4 \\ -24074.4 & 916728 & 134899 \\ 109603.4 & 134899 & 158488.4 \end{bmatrix} \mathbf{r} = \begin{bmatrix} 1.00000 & -0.02457 & 0.26899 \\ -0.02457 & 1.00000 & 0.35391 \\ 0.26899 & 0.35391 & 1.00000 \end{bmatrix}$$

(5.239)

From (5.224), using the estimated residual correlation matrix **r** in (5.239), it can be calculated that $\lambda = 7.1356$. This value is less than the 0.05 significant level critical value of $\chi^2(3) = 7.81473$ and thus, it can be argued that it was not necessary to use the SURE methodology to the transformed data, but an OLS application to these transformed data would yield efficient results.

8 DIAGNOSTIC TESTS FOR SIMULTANEOUS EQUATION MODELS

The diagnostic tests in a simultaneous equation model may be classified into the following categories:

1   Tests investigating the omission of predetermined variables from specific equations.
2   Tests investigating the serial correlation in the disturbances of the equations of the model.
3   Tests investigating the simultaneity of the variables involved in a simultaneous equation model.
4   Tests investigating the exogeneity of the variables involved in a simultaneous equation model.

8.1 Tests for omitted variables

These tests, which generally test the restrictions that overidentify an equation, may be formed as follows:

$$\text{H}_0\text{: appropriately omitted exogenous variables}$$
$$\text{H}_a\text{: inappropriately omitted exogenous variables} \tag{5.240}$$

Two such tests considered here are:

(a) Likelihood ratio tests (LR)

Anderson and Rubin (1950) suggested a likelihood ratio statistic, which is given by

$$\text{LR} = n(\lambda_i - 1) \sim \chi^2(K - k_i - g_i) \tag{5.241}$$

where λ_i is the smallest characteristic root of the determinantal equation of the LIML estimation (see section 6.1.5). In the case that the value of the statistic LR is greater than the critical value of the χ^2 distribution for a given significance level, then this means that exogenous variables have inappropriately omitted from the equation under examination.

Example 5.17 A simple macroeconomic model for an EU member (testing overidentifying restrictions, LR)

Using the model estimated in Example 5.9 with LIML. This model has $G = 4$ endogenous variables and $K = 6$ (excluding the constant) predetermined variables. For each equation we have the following information:

A private consumption function: $k_1 = 1 \; g_1 = 2 \quad \lambda_1 = 1.247$

A private gross investment function: $k_2 = 2 \; g_2 = 2 \quad \lambda_2 = 1.087$

A money demand function: $k_3 = 3 \; g_3 = 2 \quad \lambda_3 = 1.312$

Taking into account that $n = 35$, we can get from (5.241) and from tables of the chi-squared distribution, for 0.05 significance level, yields

$LR_1 = 8.645 > \chi^2(3) = 7.82$, meaning that H_0 is rejected,

$LR_2 = 3.045 < \chi^2(2) = 5.99$, meaning that H_0 is accepted, and

$LR_3 = 10.92 > \chi^2(1) = 3.84$, meaning that H_0 is rejected.

In other words, the LR test suggests that from the private consumption function and from the money demand function exogenous variables have been inappropriately omitted. However, Basmann (1960) found that this test rejects the null hypothesis too often.

(b) Lagrange multiplier tests (LM)

Hausman (1983) suggested the following steps in order to test omitted variables from an equation (see also Wooldridge, 1990):

Step 1

Estimate the specific equation with one of the single equation methods of estimation and save the residuals.

Step 2

Regress the saved residuals in step 1 on all the predetermined variables of the model plus the constant and note R^2.

Step 3

Use the Lagrange multiplier statistic LM, which is given by

$$LM = nR^2 \sim \chi^2(K - k_i - g_i) \tag{5.242}$$

in order to test the hypotheses in (5.240).

Example 5.18   A simple macroeconomic model for an EU member (testing overidentifying restrictions, LM)

Using the model estimated in Example 5.8 with 2SLS. From the regression of the saved residuals for each equation on all the predetermined variables of the model we obtained the following determination coefficients:

A private consumption function: $R^2 = 0.208369$

A private gross investment function: $R^2 = 0.080182$

A money demand function: $R^2 = 0.117510$

Taking into account that n = 35, we can get from (5.242) and from tables of the chi-squared distribution, for 0.05 significance level, yields

$LM_1 = 7.293 < \chi^2(3) = 7.82$, meaning that H_0 is accepted,

$LM_2 = 2.806 < \chi^2(2) = 5.99$, meaning that H_0 is accepted, and

$LM_3 = 4.113 > \chi^2(1) = 3.84$, meaning that H_0 is rejected.

In other words, the LM test suggests that from the money demand function only exogenous variables have inappropriately omitted.

8.2   Tests for serial correlation

The *i*th equation of a structural model according to (5.130) can be written as

$$Y_{i,t} = \mathbf{Y}_{i,t}\boldsymbol{\gamma}_i + \mathbf{X}_{i,t}\boldsymbol{\beta}_i + \epsilon_{i,t} \tag{5.243}$$

with matrix $\mathbf{Z}_t = [X_{1t} \, X_{2t} \, \dots \, X_{Kt}]$ being the matrix of all the predetermined variables and matrix $\mathbf{Y}_t = [Y_{1t} \, Y_{2t} \, \dots \, Y_{Gt}]$ being the matrix of all the endogenous variables in the model.

Wooldridge (1991) suggested the following procedure in order to test the serial correlation of any order in the disturbances of the *i*th equation of the model:

Step 1

Estimate the *i*th equation (5.243) by 2SLS and save the corresponding residuals $e_{i,t}$.

Step 2

Estimate the reduced form equations of the simultaneous equations model, i.e. regress each endogenous variable of the model on $\mathbf{Z}_t$, and save the fitted values of all the endogenous variables $\hat{\mathbf{Y}}_t$.

Step 3

Regress $e_{i,t}$ against $\hat{\mathbf{Y}}_{i,t}$, $\mathbf{X}_{i,t}$ and $e_{i,t-1}$, $e_{i,t-2}, \ldots, e_{i,t-p}$ and note R^2.

Step 4

Use the Lagrange multiplier statistic LM, which is given by

$$LM = (n - p)R^2 \sim \chi^2(p) \tag{5.244}$$

in order to test the null hypothesis of no serial correlation.

Example 5.19 A simple macroeconomic model for an EU member (testing serial correlation, LM)

Using the model estimated in Example 5.8 with 2SLS. From the regression of the saved residuals for each equation against the fitted values of its explanatory endogenous variables, its predetermined variables and the lagged values of the saved residuals, we obtained the following results using also the LM statistic in (5.244):

Private consumption function:

$R^2 = 0.002963$ for $p = 1$ with $n = 34$ and $R^2 = 0.057579$ for $p = 2$ with $n = 33$,

$LM_1 = 0.098 < \chi^2(1) = 3.84$ and $LM_2 = 1.785 < \chi^2(2) = 5.99$.

Private gross investment function:

$R^2 = 0.046305$ for $p = 1$ with $n = 34$ and $R^2 = 0.047381$ for $p = 2$ with $n = 33$,

$LM_1 = 1.528 < \chi^2(1) = 3.84$ and $LM_2 = 1.469 < \chi^2(2) = 5.99$.

Money demand function:

$R^2 = 0.098102$ for $p = 1$ with $n = 34$ and $R^2 = 0.194614$ for $p = 2$ with $n = 33$,

$LM_1 = 3.237 < \chi^2(1) = 3.84$ and $LM_2 = 6.033 > \chi^2(2) = 5.99$.

In other words, the LM test suggests that the disturbances in the money demand function exhibit some serial correlation of the second order (compare with results in (5.159)).

8.3 Simultaneity tests

We saw in previous sections that because the endogenous variables in a simultaneous equations model are correlated with the disturbances, the application of OLS to equations in which endogenous variables appear as explanatory variables will yield biased and inconsistent estimates. This failure of OLS, which was called 'simultaneous equations bias', becomes the starting point for investigating methods (alternative to OLS) of estimation which could yield consistent and efficient estimates. However, in the case where an explanatory endogenous variable is not correlated with the error term there is no need to use alternative methods of estimation. In fact, if we use these methods, when we should not, the estimates may be still consistent but inefficient. Therefore, there is a need for checking if the explanatory endogenous variables were correlated with the error terms become apparent.

Accordingly the test for simultaneous equations bias, or simultaneity, or the test for investigating the correlation between the explanatory endogenous variables and the error term, may be formed as follows:

H_0: there is no simultaneity (there is no correlation)

H_a: there is simultaneity (there is correlation)

(5.245)

Considering equation (5.243), Hausman (1976) suggested the following steps for testing the hypotheses in (5.245):

Step 1

Estimate the reduced form equations of the simultaneous equations model, i.e. regress each endogenous variable of the model on $\mathbf{Z}_t$, and save the fitted values of all the endogenous variables $\hat{\mathbf{Y}}_t$ and save also the corresponding reduced form residuals $\mathbf{u}_t$.

Step 2

Because $\mathbf{Y}_t = \hat{\mathbf{Y}}_t + \mathbf{u}_t$, substitute this expression of the explanatory endogenous variables into equation (5.243) and estimate by OLS the following equation:

$$Y_{i,t} = \hat{\mathbf{Y}}_{i,t}\boldsymbol{\gamma}_i + \mathbf{u}_{i,t}\boldsymbol{\gamma}_i + \mathbf{X}_{i,t}\boldsymbol{\beta}_i + \epsilon_{i,t}$$

(5.246)

For efficient estimation, Pindyck and Rubinfeld (1991) suggest the following equation to be estimated, instead of equation (5.246):

$$Y_{i,t} = \mathbf{Y}_{i,t}\boldsymbol{\gamma}_i + \mathbf{u}_{i,t}\boldsymbol{\gamma}_i + \mathbf{X}_{i,t}\boldsymbol{\beta}_i + \epsilon_{i,t}$$

(5.247)

Step 3

Use the F test (or the t test for one regression coefficient) to test the significance of the regression coefficients of the $\mathbf{u}_{i,t}$ variables. If the test shows significant coefficients, accept the alternative hypothesis in (5.245), i.e. accept that there is simultaneity. If the test shows non-significant coefficients, accept the null hypothesis in (5.245), i.e. accept that there is no simultaneity.

Example 5.20 A simple macroeconomic model for an EU member (testing simultaneity)

Using the model estimated in Example 5.8.

Denoting with $u_{C,t}$, $u_{I,t}$, $u_{r,t}$ and $u_{Y,t}$ the reduced form residuals corresponding to the reduced form equations with respect to C, I, r and Y, and following the steps of the Hausman test, the estimated equations of the model in step 2 are the following, where sign { } denotes the significance level of the corresponding estimate:

The private consumption function:

$$\hat{C}_t = 7519.307 + 0.17415\hat{Y}_t + 0.44296u_{Y,t} + 0.73293C_{t-1}$$

sign. {0.0001}   {0.0000}   {0.0000}   {0.0000}   (5.248)

$\bar{R}^2 = 0.999$

The private gross investment function:

$$\hat{I}_t = 5579.631 + 0.06765\hat{Y}_t + 0.53513u_{Y,t} - 1025.758\hat{r}_t - 596.8422u_{r,t} + 0.67607I_{t-1}$$

sign. {0.0087} {0.0001}   {0.0000}   {0.0004}   {0.0404}   {0.0000}

$\bar{R}^2 = 0.962$ F(for redundant $u_{Y,t}$ and $u_{r,t}$) = 38.549 {0.0000}   (5.249)

The money demand function:

$$\hat{r}_t = -1.46373 + 9.12(E-06)\hat{Y}_t - 2.35(E-05)u_{Y,t} - 0.00839M_{t-1} +$$

sign. {0.1233} {0.0202}   {0.4299}   {0.0391}

$$0.53795P_t + 0.88081r_{t-1}$$

sign. {0.0879}   {0.0000}

$\bar{R}^2 = 0.967$ (5.250)

According to step 3, the significance level for the regression coefficient of $u_{Y,t}$ in equation (5.248) using the t test is {0.0000}, the significance level for the regression coefficients of $u_{Y,t}$ and $u_{r,t}$ in equation (5.249) using the F test is {0.000}, and the significance level for the regression coefficient of $u_{Y,t}$ in equation (5.250) using the t test is {0.4299}. In other words, the simultaneity problem is present in the private consumption function and in the private gross investment function, whilst there is no simultaneity problem in the money demand function.

8.4 Exogeneity tests

Although the categorisation of the variables of a simultaneous equations model into endogenous and exogenous variables depends mainly on the theory that the model is trying to represent, Hausman (1976) extended his simultaneity test to be able to test if the explanatory endogenous variables are really endogenous variables. Considering the *i*th equation (5.243), with respect to variables $\mathbf{Y}_{i,t}$ the test may be formed as follows:

H_0: variables $\mathbf{Y}_{i,t}$ are exogenous

H_a: variables $\mathbf{Y}_{i,t}$ are endogenous

(5.251)

The steps of the test are given below:

Step 1

Estimate the reduced form equations of the simultaneous equations model, i.e. regress each endogenous variable of the model on $\mathbf{Z}_t$, and save the fitted values of all the endogenous variables $\hat{\mathbf{Y}}_t$.

Step 2

Estimate by OLS the following equation:

$$\mathbf{Y}_{i,t} = \mathbf{Y}_{i,t}\boldsymbol{\gamma}_i + \hat{\mathbf{Y}}_{i,t}\boldsymbol{\lambda}_i + \mathbf{X}_{i,t}\boldsymbol{\beta}_i + \boldsymbol{\epsilon}_{i,t}$$

(5.252)

Step 3

Use the F test (or the t test for one regression coefficient) to test the significance of the regression coefficients of the $\hat{\mathbf{Y}}_t$ variables. If the test shows significant coefficients, accept the alternative hypothesis in (5.251), i.e. accept that the corresponding variables are endogenous. If the test shows not significant coefficients, accept the null hypothesis in (5.251), i.e. accept that the corresponding variables are exogenous.

Example 5.21 A simple macroeconomic model for an EU member (testing exogeneity)

Assume the model estimated in Example 5.8.

Following the steps of the Hausman test for exogeneity, the estimated equations of the model in step 2 are:

A private consumption function:

$$\hat{C}_t = 7519.307 + 0.44296Y_t - 0.26881\hat{Y}_t + 0.73293C_{t-1}$$

sign.$\{0.0001\}$ $\{0.0000\}$ $\{0.0003\}$ $\{0.0000\}$

(5.253)

$\bar{R}^2 = 0.999$

A private gross investment function:

$$\hat{I}_t = 5579.631 + 0.53513Y_t - 0.46747\hat{Y}_t - 596.8422r_t - 428.916\hat{r}_t + 0.67607I_{t-1}$$

sign. $[0.0087]$ $[0.0000]$ $[0.0000]$ $[0.0404]$ $[0.2596]$ $[0.0000]$

$\bar{R}^2 = 0.962$ F(for redundant $\hat{Y}_t$ and $\hat{r}_t$) = 27.803 $[0.0000]$ \hfill (5.254)

A money demand function:

$$r_t = -1.46374 - 2.35(E-05)Y_t + 3.26(E-05)\hat{Y}_t - 0.00839M_{t-1} +$$

sign. $[0.1233]$ $[0.4299]$ $[0.2793]$ $[0.0391]$

$\qquad 0.53795P_t + 0.88081r_{t-1}$ \hfill (5.255)

$\qquad [0.0879]$ $[0.0000]$

$\bar{R}^2 = 0.967$

According to step 3, the significance level for the regression coefficient of $\hat{Y}_t$ in equation (5.253) using the t test is $[0.0003]$, the significance level for the regression coefficients of $\hat{Y}_t$ and $\hat{r}_t$ in equation (5.254) using the F test is $[0.0000]$, and the significance level for the regression coefficient of $\hat{Y}_t$ in equation (5.255) using the t test is $[0.2793]$. In other words, in the private consumption function variable Y_t behaves as an endogenous variable, in the private gross investment function at least one of the variables Y_t and r_t behave as an endogenous variable, and finally in the money demand function variable Y_t behaves as an exogenous variable (this final result is parallel with the result referring to the problem of no simultaneity in equation (5.250)).

9 A COMPARISON OF METHODS OF ESTIMATION

Table 5.6 presents briefly the results obtained in the preceding sections of this chapter referring to the estimation of a simple macroeconomic model for an EU member. The methods of estimation involve OLS, from single equation methods include 2SLS and FIML, and from systems methods of estimation include 3SLS and FIML. The obvious question is which of the estimates in Table 5.6 are 'preferable'.

Although a quick answer could be that the systems methods of estimation are preferable, because these possibly involve more information than the single equation methods of estimation, the answer is usually not as quick or as simple. Starting with OLS, we saw that this method if applied to a simultaneous equations model produces biased and inconsistent estimates. Going to single equation methods of estimation we saw that 2SLS and LIML are preferable to OLS because they produce consistent estimates. Furthermore, the systems methods of estimation 3SLS and FIML are preferable to 2SLS and LIML, because they produce more efficient estimates than the single methods of estimation. However, the preceding discussion refers to the asymptotic properties of the estimates of the various methods and not to their finite-sample properties. But in most applications the sample sizes are not infinite but finite and generally small.

Although in theory the systems methods look to be asymptotically preferable than the single equations methods, in practice we have to take into account the following disadvantages of systems methods:

1   Even in these days of high-speed computers, the computational burden for a moderate or large econometric model is quite big.
2   Possible specification errors in one or more equations of the system are transmitted to the other equations of the system, although these equations were correctly specified. Therefore, the systems methods of estimation are very sensitive to specification errors, whilst the single equation methods of estimation are not as sensitive. For the latter methods any specification error in one equation is stacked with that equation and does not affect the estimates of the rest of the equations.

Returning to the three-equation small macroeconomic model of an EU member, with 36 annually aggregate observations per variable, we see in Table 5.6 that the estimates of the regression coefficients are not very different from the methods of

Table 5.6 Estimates of a simple macroeconomic model for an EU member

	OLS	2SLS	LIML	3SLS	FIML
Consumption					
Constant	5250.619	7519.307	8423.51	7835.32	8261.57
	[2.8302]	[3.554]	[3.6551]	[3.9651]	[1.7689]
Y_t	0.25626	0.17415	0.14142	0.16152	0.14666
	[6.8976]	[3.6384]	[2.6283]	[3.8522]	[1.7050]
C_{t-1}	0.61213	0.73293	0.78107	0.75162	0.77342
	[11.1495]	[10.377]	[9.8403]	[12.1551]	[6.2582]
Investment					
Constant	6014.662	6232.39	6254.74	6442.75	6365.42
	[2.0652]	[2.1252]	[2.1306]	[2.3465]	[1.1547]
Y_t	0.08032	0.07278	0.07190	0.07281	0.07923
	[4.0370]	[3.2685]	[3.1951]	[3.6876]	[1.0708]
r_t	−1165.609	−1097.613	−1089.52	−1080.57	−1184.43
	[3.8046]	[3.1552]	[3.0912]	[3.4399]	[1.0314]
I_{t-1}	0.60591	0.64325	0.64687	0.63380	0.61406
	[6.1879]	[6.3115]	[6.3175]	[7.0979]	[2.0635]
Money					
Constant	−1.39435	−1.46373	−1.48583	−1.8292	−1.88352
	[1.5103]	[1.5812]	[1.6036]	[2.2614]	[1.0161]
Y_t	8.6(E − 06)	9.12(E − 06)	9.28(E − 06)	1.02(E − 05)	10.6(E − 06)
	[2.3285]	[2.4472]	[2.4843]	[3.1498]	[1.5796]
M_{t-1}	−0.00833	−0.00839	−0.00841	−0.00775	−0.00772
	[2.1367]	[2.1520]	[2.1566]	[2.5383]	[0.8110]
P_t	0.53255	0.53795	0.53967	0.46344	0.46338
	[1.7422]	[1.7591]	[1.7643]	[1.9301]	[0.6653]
r_{t-1}	0.89118	0.88081	0.87751	0.88864	0.87237
	[8.5598]	[8.4230]	[8.3785]	[10.0569]	[5.1812]

Notes: t-ratios in brackets.

estimation. Furthermore, assuming that the t-ratio is a consistent statistic to be used to hypothesis testing (because the sample size is greater than 30), and taking into account all the diagnostic tests performed in the previous sections, we believe that the 3SLS estimates are preferable for this simple model. This is because the 3SLS method yielded estimates of the regression coefficients which are similar in value with the other methods of estimation, but at the same time the 3SLS estimates are more efficient than the estimates by other methods as can be seen from the higher t-ratios.

DISCUSSION QUESTIONS

1 Explain what you understand by the 'identification problem'. Why should identification problems be dealt with prior to estimation?
2 'An equation is identified if the "rank" condition is satisfied'. Discuss.
3 Discuss the advantages of (2SLS) method. Under what circumstances indirect least squares might be used?
4 Explain the full information maximum likelihood method. Illustrated your answer with reference to a model of your own choice.

Bibliography

Aitken, A. (1935) 'On least squares and linear combinations of observations', *Proceedings of the Royal Statistical Society*, 55, 42–48.
Anderson, T. and Rubin, H. (1949) 'Estimation of the parameters of a single equation in a complete system of stochastic equations', *Annals of Mathematical Statistics*, 20, 46–63.
Anderson, T. and Rubin, H. (1950) 'The asymptotic properties of estimators of the parameters of a single equation in a complete system of stochastic equations', *Annals of Mathematical Statistics*, 21, 570–582.
Basmann, R. (1960) 'On finite sample distributions of generalised classical linear identifiability test statistics', *Journal of the American Statistical Association*, 55, 650–659.
Breusch, T. and Pagan, A. (1980) 'The LM test and its applications to model specification in econometrics', *Review of Economic Studies*, 47, 239–254.
Hausman, J. A. (1976) 'Specification tests in Econometrics', *Econometrica*, 46, 1251–1271.
Hausman, J. A. (1983) 'Specification and estimation of simultaneous equations models', in Z. Griliches and M. Intriligator (eds), *Handbook of Econometrics*, Amsterdam: North Holland.
Johnston, J. (1984) *Econometric Methods*, 3rd ed., New York: McGraw-Hill.
Judge, G. G., Griffiths, W. E., Hill, R. C., Lutkepohl, H. and Lee, T. C. (1985) *The Theory and Practice of Econometrics*, 2nd ed., New York: Wiley.
Kmenta, J. (1986) *Elements of Econometrics*, New York: McGraw-Hill.
Madansky, A. (1964) 'On the efficiency of three-stage least squares estimation', *Econometrica*, 32, 55.
Oberhofer, W. and Kmenta, J. (1974) 'A general procedure for obtaining maximum likelihood estimates in generalised regression models', *Econometrica*, 42, 579–590.
Park, R. (1967) 'Efficient estimation of a system of regression equations when disturbances are both serially and contemporaneously correlated', *Journal of the American Statistical Association*, 62, 500–509.
Phillips, A. W. (1958) 'The relation between unemployment and the rate of change of money wages in the United Kingdom, 1861–1957', *Economica*, XXV, 283–299.

Pindyck, R. S. and Rubinfeld, D. L. (1991) *Econometric Models and Economic Forecasts,* 3rd ed., New York: McGraw-Hill.

Wald, A. (1943) 'A note on the consistency of the maximum likelihood estimator', *Annals of Mathematical Statistics,* 20, 595–601.

Wooldridge, J. (1990) 'A note on the Lagrange multiplier and F-statistics for two stage least squares regression', *Economic Letters,* 34, 151–155.

Wooldridge, J. (1991) 'On the application of robust regression-rooted diagnostics to models of conditional means and conditional variables', *Journal of Econometrics,* 47, 5–46.

Zellner, A. (1962) 'An efficient method of estimating seemingly unrelated regressions and tests for aggregation bias', *Journal of the American Statistical Association,* 57, 348–368.

Zellner, A. and Theil, H. (1962) 'Three stage least squares: Simultaneous estimation of simultaneous equations', *Econometrica,* 30, 63–68.

6 Methodological strategies for dynamic modelling

Key points

- Specific – general methodology
- General – specific methodology
- Error correction models/cointegration

The traditional methodology: specific to general econometric modelling

Hendry, D. (1979 and 1985) used the label 'specific to general' as synonymous with traditional modelling and the specific regression strategy used by economists in 1950s, 1960s and 1970s. This approach to econometric modelling begins with a 'properly specified' single equation model which has stable regressors, a zero mean, is not autocorrelated and possesses a homoscedastic error term. Texts which delivered the spirit of this approach and popular in their time were Wannacott *et al.* (1970, 1979) and Koutsoyiannis (1973 and 1979). In a world of perfect stability the OLS estimation technique yielded the 'best linear unbiased estimators' (BLUE). Deviations from the assumptions regarding the error term violate the property of 'BLUE' so that if the error terms did not have a zero mean then the unbiased characteristic of the estimators is undermined; if the error term shows either heteroscedasticity or autocorrelation then the efficiency of the estimators is compromised. Moreover, if the regressors appeared stochastic then the efficiency and unbiasedness of the estimators is questioned.

Traditional econometric modelling assumed that since the optimality of the estimation would be demonstrated using OLS, applied econometrics in practical applications strove to estimate models using this method. Consequently, the approach adopted in the 1950s to 1970s failed to discuss why say an original specification of the regression equation (including assumptions about the error term) might be false. Hence no attempt was made to falsify underlying economic relationships so in practice in applied work researchers doggedly focused on developing estimations techniques (such as clones of generalised least squares or instrumental variable techniques) to overcome estimation problems. Failure of postulations about the white noise features of the error terms were viewed as local estimation issues to be overcome. Consequently, in applied work in the 1960s in industrial economics and macroeconomics investigators were pleased to report the final 'successful' equation. In discussing the final 'successful' equation a large number of unsuccessful equations would have been cast aside by researchers in pursuit of the final 'successful' equation. In industrial economics, the regressions run in the 1960s and 1970s studying a multi-faceted aspect of firm and industry structures have been severely criticised by many authors for simply discovering 'stylised facts' about industrial performance with no attempt to question the validity of the underlying theoretical relationships producing the single equation (Tirole, 1989). Indeed, this crude approach based on a 'stylised' view of theory where driven by a desire to seek verification of underlying theories. Hence, applied econometric work, particularly in industrial economics throughout the 1960s and 1970s was driven by a desire to corroborate models rather than 'test' them (Dixon, H., 1996). By using econometric models as a mechanism of corroboration the 'truth' of the specified model was maintained throughout the entire estimation process. Hence only those equations which failed to refute the hypothesised model were judged to be 'successful'. These approaches which typified much applied work led to fairly widespread condemnation of such investigations as being simply 'fishing trip' practices. Thus data mining refers to the practise of running a large number of regressors, often with different explanatory variables to find the unique equation which best supported the theory under consideration. This pursuit of the successful equation led Leamer (1978) to describe the practice as a 'fishing expedition' since the search for a 'successful' equation without judgement and purpose yields results which are impossible to evaluate.

Darnell and Evans (1990) argue that it is possible to modify the 'traditional' 'specific to general approach' by refocusing empirical studies such that econometricians seek to falsify theoretical hypotheses rather then running regressions until the equation of verification is discovered. Moreover, Darnell and Evans (ibid) usefully demonstrate that it is possible to achieve this within the 'classical' methodology as long as the empirical data analysis is conducted with judgement. This can be achieved if the empirical study has a well-defined research strategy with various stages in the process being demonstrably linked together. This approach then could replace the data mining features evident in previously published applied work. With the foregoing caveats in mind we can summarise the main features of the 'specific to general' traditional approach to econometric modelling. These are, therefore, that the investigation:

a   begins with theories (or falsifiable hypotheses derived from theories) which are drastic abstractions of reality;
b   formulates highly simplistic relationships (linear) to represent the hypotheses which reflect the theories;
c   estimates the linear equations from the data set using techniques which are 'optimal' only on the assumption that the econometric model is properly specified;
d   tests a few of the 'explicit' assumptions or utilises tests for 'implicit' assumptions such as tests for autocorrelation;
e   revises the original specification in the light of a restricted number of diagnostic tests;
f   re-estimates the model, as sets of hypotheses accordingly.

However, the approach nonetheless has obvious drawbacks these are:

a   that since each and every test is conditional on arbitrary assumptions which are tested later, then if any of these are rejected at any given stage in the investigation all previous inferences are invalidated;

b   that given the restricted number of diagnostic tests conducted, it is accordingly not always known if the 'best' model has been achieved by using this iterative methodology;

c   that the significance levels for the unstructured sequence of diagnostic tests is actually not known.

General to specific econometric modelling: Hendry's impact on applied econometrics

In the 1990s it has become increasingly popular to conduct time series econometrics using methods of autoregressive distributed lag methods as popularised as Hendry and Mizon. In a range of papers (Hendry and Mizon 1978, Hendry 1979 and 1980), these authors consistently attacked 'traditional' methods of econometric investigations and argued instead for the widespread adoption of 'general to specific' modelling procedures. From the mid-1980s through the 1990s, many influential models and a large number of econometric relationships were developed using these procedures. It is currently viewed as 'best practice' (Granger, 1997). Hendry's approach lies within the LSE tradition and has roots reaching back to Sargan's work on prices and wages (1964). A dominant feature of the methodology focuses on the role of the observable data in specification of the statistical model and the finally selected equation. Following the Hendry-Mizon (1978) approach we assume that economic theory leads to the following long run equilibrium relationship:

$$p_t = \alpha q_t \tag{6.1}$$

where p_t and q_t are measured in logarithms. Given this modelling approach, economic theory is seen as yielding statements about long run equilibrium behaviour. To capture 'short run dynamic adjustment behaviour' equation (6.1) is supplemented with lags and is recast in dynamic autoregressive distributed (ADL) form as:

$$g_t = \Sigma(\beta_j q_{t-j} + \partial_j p_{t-1}) \tag{6.2}$$

using the lag operator L, (6.2) may be written also as:

$$\partial(L)p_t = \beta(L)q_t \tag{6.3}$$

To estimate this equation an error term is affixed and the maximum lag length, m, is selected. Equation (6.3) represents the most general model and is purposely over-parameterised; then by data based simplifications it is reduced to the most parsimonious equation. Following Hendry and Richard (1983) the simplification process needs to respect six criteria. Thus the proposed simplified equation should:

1   be data admissible. This means that it must be logically possible for the data generating process to produce the specification selected;

2   be consistent with theory;

3   possess regressors which are weakly exogenous. Variables which breach the weak exogeneity condition should be modelled jointly;

4   show parameter consistency, especially if the equation is to be used for policy simulation or, forecasting;

5   be data coherent, meaning that the residuals are consistent with a white noise error structure;

6   encompassing, meaning that chosen equation must be capable of explaining the results of other specifications.

The move to specific modelling procedure then comprises several elements. First, theoretical long-run equilibrium features are assumed to provide an unsatisfactory starting point for the empirical analysis since the assumption in the approach is that the data reflect potentially disequilibrium processes; second, it is presupposed that these processes can be satisfactorily proxied by using lags of the variables nominated for inclusion in the equilibrium relationship; third, the simplification process of the autoregressive distributed lag model proceeds according to a set of sequential tests and the resulting equation is judged satisfactory only when it passes the test mounted by the six criteria. Hendry (1979) and his followers have dubbed this procedure 'general to specific modelling' and argued that the procedures followed make modelling decisions transparent.

As with the traditional methodology, the LSE tradition and Hendry's development of this adopt a verifications position. Accordingly Darnell and Evans (ibid.) argue that this approach borders on the arcane abstract empiricist approach in so far as the modelling procedure proposed by Hendry and others in the LSE tradition possesses only tenuous links with economic theory. For example, Darnell and Evans (ibid) argue that prescription for modelling, namely, 'test and test again' (Hendry, 1980) implies that the data generating process involves disequilibrium and equilibrium processes. When these processes are proxied by lags model specifications would ideally need to configure an equilibrium and disequilibrium model. Thus, the testing procedure would directly require the analyst to specify disequilibrium as well as equilibrium behaviour. Disequilibrium behaviour, however could not be represented by adding lagged adjustments but would involve the development of a theoretical model of optimal behaviour of agents acting out of equilibrium. However, strong advocacy of this problem is dangerous, in so far as Darnell and Evans (ibid) require that the analyst must be prepared to develop extensions to theory as well as verifiable testing procedures. This could seriously hinder developments of applied work. However, long run dynamic equilibria are a view of economic reality which cannot be observed. Hence, modelling the long run presumes researchers can observe, at least, the conditions which lead to equilibrium. In practice, these are non-observable too. Hence, the basic research motivation of 'general to specific' modelling is technically flawed. This flaw becomes apparent if it is argued that economic theory rarely addresses long-run dynamic equilibrium relationships and so provides little in the way of refutable hypotheses. Finally, if theory is developed in such a way as to derive refutable hypotheses in say a growth case, then Darnell and Evans (ibid) argue that a methodologically sound way of proceeding would be to use theoretical predictions as hypotheses capable of being falsified. By contrast, the use of econometric evidence as verification of theory is an explicit criterion for the evaluation of models using the 'general to specific' methodology

(Hendry and Richardson, 1983). However, the use of empirical analysis to refute theoretical hypotheses requires careful selection of the general model. In so far a 'general to specific' modelling fails to encompass the predictions of theory it means that theory is never subject to falsification tests but rather that theories are observed to be consistent with empirically derived long-run relationships. However, Granger (1997) and others involved in the debate about modelling the long run rebut many of the objections to 'general to specific modelling' raised by Darnell and Evans (ibid). In particular, Granger, a leading figure in the development of Cointegration procedures for modelling the long run argues criticisms of the type mounted by Darnell and Evans if followed, would paralyse econometric modelling of long-run relationships. Generally, Granger (ibid) Pesaran (1997) and Harvey, A. (1997) argue that best practice 'general to specific' modelling can be achieved within the context of economic theory or that until such time as economic theory provides a detailed framework about the dynamic specification of models, applied researchers should continue to solve these specification procedures empirically, within the framework of 'general to specific' modelling strategies. This modelling procedure can be achieved with the enveloping corpus of theory. Hence, utilisation of the 'general to specific' approach has moved on somewhat since the late 1970s when Hendry could write:

> 'until the model adequately characterises the data generation process it seems ... pointless trying to test hypotheses of interest to economic theory'
>
> (Hendry, 1979)

This statement left Hendry open to the charge that the testing of economic hypotheses using 'general to specific' strategies was of peripheral interest compared with specification of the data generating process. As a consequence Darnell and Evans (ibid) could argue that Hendry's approach was an example of naive empiricism or theory after measurement.

Although Darnell and Evans (ibid) are highly critical of examples of the 'state of the art' of dynamic modelling in the 1990s, many of their critical comments about the utilisation of 'general to specific' modelling procedures were valid and they produced a powerful statement of the limitations inherent in dynamic modelling investigations using this approach. The celebrated debate in the *Economic Journal* (1997) covered issues raised by Darnell and Evans to a degree which meets most objections. Pesaran in particular achieves most of what Darnell and Evans object to within the framework of economic theory. Most applied econometricians doing practical work nowadays aim to:

1 Begin with the most general specification which is reasonable to maintain within the context of theory.
2 Conduct the simplification process, which is undertaken by way of a sequence of tests, which aims to avoid the charge of measurement without theory or empirical ad hocery. Moreover, within the context of the 'general to specific' approach it should be noted that:

   a The significance levels for the sequences of testing is known.
   b The sequential testing procedures are used to select a data coherent model of economic relationships.

Nevertheless there still remain drawbacks with the 'general to specific' approach. These can be generally started as being that:

1   The chosen general model might only actually comprise a 'special' case of the data generation process so that diagnostic testing remains vitally important
2   Any data limitations and inconsistencies weaken the approach
3   Following the Darnell and Evans critique, there still remains no universally agreed uniquely optimal sequence for simplifying the 'general' model so the charge of ad hocery in, for example, selecting lags remains valid.

Methodological postscript

Econometrics is a branch of economics which incorporates, deduction, induction and inference. Since World War II econometrics developed a methodology which it was thought could lead to a rigorous identification of 'good' and 'bad' theories. The hope was that econometrics would yield a rich body of empirical knowledge similar to the corpus of knowledge held by the natural sciences. However, throughout the 1960s and 1970s applied work focused on problems such as how to obtain the best estimation procedure rather than towards testing hypotheses. Since then econometrics can be viewed as a vehicle for 'proving' economic theories. To an extent this feature continues in the work of Granger and Hendry. However, to the extent that much of their work concerns forecasting and seeks confirmation via application of cointegration procedures this might be sufficient to stave off criticism which is largely destructive of attempts to model economic time series.

Generally methodological issues in econometrics hinge on the validity of particular statistics and their underlying assumptions. If pre-testing leads to a non-rejection of the required assumptions then a dominant hypothesis can be tested. Procedures that pre-test assumptions are liable to error and a non-rejection of a null hypothesis may be the subject of Type II error. Thus the validity of the ultimate test is depend-

---

Chapter 6: Key terms

Specific to general (S-G) methodology   Econometric methodology based on a specific theoretical model. The model will be 'generalised' through diagnostic testing.

General to specific (G-S) methodology   An econometric methodology based on a 'general' dynamic model derived from the theory. The model is 'reduced' through diagnostic testing.

ARIMA   Autoregressive integrated moving average.

ECM   Error correction model.

SEMTSA   Structural econometric time series approach.

---

ent on the fact that no Type II errors were made at any stage. Ignoring the possibility of an allowance for Type II errors results in pre-test bias. Generally, therefore the possibility of committing a Type II error needs to be recognised in the final stages of estimation such that conclusions can be tempered with the recognition of the possibility of incurring pre-test bias, avoiding the confidence evident in published work in the 1960s and 1970s. The challenge for econometrics lies in recognition that it is always certain that the maintained hypotheses are the best, and that the maintained hypotheses reflect current methodological norms.

Time series econometrics: current practice

From the late 1960s until the early 1980s practising econometricians analysed time series data in a way that was quite distinct from procedures used by statisticians. Thus in these times econometricians used to formulate a traditional regression model to capture the behaviour of time series data. The profession focused on issues relating to autocorrelation in errors and simultaneity ignoring the dynamic features of the data. Moreover, it was implicitly argued that economic time series data were stationary around a deterministic trend. Time series statisticians by contrast ignored the role of econometric explanatory variables and analysed time series with powerful extrapolation techniques. Stationarity issues were obviated by differencing the data till they became stationary. However, studies began to appear which:

1   questioned the performance of econometric time series models vis-à-vis the forecasting of statistical time series approaches;
2   argued that economic time series were not stationary, making serious difficulties for statistics such as DW, R^2 and the 't' statistics.

These two issues spurred more research activity into the analysis of time series data by econometricians which is ongoing.

ARIMA models

After the publication by Box and Jenkins (1970), the phase 'time series analysis' became synonymous with the techniques promulgated by these authors in the context of forecasting. These authors eschewed the econometric model building approach using explanatory variables opting to rely instead on the prior behaviour of the variable being forecasted. Box-Jenkins procedures for the analysis of time series amount to powerful extrapolation methods. Hence, if X is the variable to be forecasted or modelled Box-Jenkins analysis starts by manipulation of X until it is assumed to be stationary such that its stochastic features do not vary with respect to time. This means that the mean of X, its covariance and other X values X_{t-n}, do not depend on t. This is checked by inspection of the estimated correlogram, a graph that plots the estimated nth-order autocorrelation coefficient, p_n as a function of n. For stationary variables the correlogram shows autocorrelations that disappear quickly as n becomes large.

  Economic time series are not inherently stationary because they are trending since the mean changes with time. Box and Jenkins argued that economic time series could be made stationary by differencing and found that one or two different operations are required. This creates a new time series X* which becomes the input for Box-Jenkins analysis: The general model for X* is written as: . . .

$$X^*_t = p_1 X^*_{t-1} + p_2 X^*_{t-2} + \ldots p_n X^*_{t-n} + e_t + g_1 e_{t-1} \ldots + g_p e_{t-p}$$

where f and g are the unknown parameters and e the distributed normal errors with a zero mean. The model above show X^*_t in terms of its own past values together with corrected past errors. Here there are no explanatory variables as there would be in a typical econometric model. This general model is called an ARIMA model for X. The autoregressive (AR) dimension of the model is the number of lagged values of X* and the number of lagged values of the error capture the moving average (MA) dimension of the model. Thus ARIMA means the autoregressive integrated moving average. There are three steps in the development of an ARIMA model, these are:

1   *Model selection:* that is the values for the number of times X_t is differenced to produce X*, the number of lagged values for X^*_t, and number of lagged values of the error term must be established. Most stationary time series can be modelled with low values for the number of lagged errors and lagged values of X*.
2   *Estimation:* the p and g parameters must be estimated usually by least squares approximation to the maximum likelihood estimator.
3   *Diagnostic testing:* thus the estimated model must be checked for adequacy and revised. The process may involve many attempts till a satisfactory model is obtained.

The critical stage is 1, that is model selection and identification. This process requires experience and judgement in the interpretation of some selected statistics together with inspection of the correlogram to fix the most appropriate model.

The structural econometric time series approach (SEMTSA)

In the early 1970s Box-Jenkins economic forecasting equations were out-performing econometric forecasting models regularly. As Box-Jenkins analysis was generalised to multi-variate Box-Jenkins where an entire vector of variables were modelled as an ARIMA process many observers thought that the days of econometric forecasting were numbered.

As a response to multivariate Box-Jenkins developments econometricians responded with a synthetic approach to time series incorporating both econometric and Box-Jenkins methodologies. This methodology was based on the proposition that dynamic structural equation econometric models are special cases of multivariate Box-Jenkins process in which prior constraints suggested by economic theory have been imposed on the parameters of the model. Moreover, if the exogenous variables in the econometric model are viewed as being generated by a complex ARIMA process, then each exogenous variable in the econometric model can be expressed as univariate Box-Jenkins ARIMA processes.

In the SEMTSA approach a typical econometric structural model is developed incorporating the used inputs from economic theory. The implied properties of the corresponding ARIMA equations are derived. Time series methods are then checked for consistency with restriction imposed on the theoretical model. Any inconsistencies would usually prompt a reappraisal of the econometric model: SEMTSA is, therefore, a procedure for finding and reappraising defects in the proposed structural theoretical models. Box-Jenkins ARIMA modelling is basically measurement without theory. Econometricians recognise this and argue that ARIMA models are good estimates of time series dependencies in the data. Hence ARIMA models are efficient benchmarks for forecasting but are not satisfactory as regards understanding economic relationships. ARIMA modelling discovers factors and issues that assist the formulation of econometric models. Hence ARIMA models should be viewed as complements to rather than rivals for traditional theoretical models.

Error correction models

ARIMA models of economic time series were generally more successful as forecasting mechanisms than traditional theoretical models because they were flexible with regards to dynamic specification. By contrast econometric structure models were static reflecting the dominant features of economic theory. The lack of attention to the dynamics of models was a normal consequence of the fact that economic theory can identify the long-run relationship between variables in equilibrium but lacked ability to specify time lags and dynamic adjustments in the process of reaching long-run equilibrium. Dynamic economic theories are of a limited number. Hence, ARIMA models have two key features which assist the modelling of time series data. These are that they are very flexible in the specification of the dynamic features of the time series and they spurn completely any views that economic theory might provide regarding long-run equilibrium.

Econometric models which incorporate knowledge from economic theory about long-run equilibrium relationships and simultaneously include a flexible lag structure allowing the actual data to play a key role in the specification of the models dynamic structure should be superior to the ARIMA methodology assuming economic theory is true. This type of approach does not yield a new version of the typical simultaneous equation models but instead the variables to be forecasted and modelled are encompassed by a single equation. In this approach, the only concession to the possibility of simultaneity is that if the variable fails to pass an exogeneity test then estimation is conducted using an instrumental variables procedure.

The role of economic theory in developing single equation models is twofold. First, economic theory suggests the explanatory variables to be included in the single equation. Second, economic theory isolates long-run equilibrium relationships among the economic variables. The single equation model is developed as follows: initially a traditional econometric model is devised and specified with a copious lag structure which is later whittled down by testing procedures on all explanatory variables, and lagged values of the dependent variable. Thereafter this equation is manipulated and formulated in a way that is more easily understood. This reformulating process involves providing a term showing the degree to which equilibrium is not reached. This last term, indeed the distinguishing feature of the approach is the error correction term since it reveals the 'error' in attaining equilibrium. This type of model has thus become known as the ECM error correction model. A simple example will suffice, consider the following long-run relationship suggested by theory:

$$X_t = \beta + \beta_1 Y_t + \beta_2 Y_{t-1} + \beta_3 Y_{t-2} + e_t$$

where X and Y are cast in logarithms, where theory suggests that X_t and Y_t grow at an equal rate, so that in equilibrium (X-Y) will appear as a constant value for the error. This relationship can be manipulated to yield:

$$\Delta X_t = \beta + \beta_1 Y_t + \beta_2 Y_{t-1} + (\beta_3 - 1)(Y_{t-1} - Y_{t-2}) + e_t$$

This is the error correction representation of the original specification; the last term is the error correction term which can be interpreted as representing states of disequilibrium. The concept can be explained as follows: if in error X grows too quickly, the error term gets bigger, since its coefficient is negative ($\beta_3 < 1$ for stationarity), ΔX_t falls, correcting the error.

The error correction model is stated in terms of differenced variables, with the error correction component measured in terms of the levels of variables. This feature gives this approach an edge over ARIMA models since in the latter all variables are differenced with no use made of the long run information provided by levels data.

Unit roots

Many studies have shown that most macroeconomic time series are not stationary, rather stationary with a deterministic trend. This creates a problem for econometricians since in the conditions of non-stationary data the normal properties of t and DW statistics and measure such as R^2 break down. Running regression with such data produces questionable, invalid or spurious results. A consequence of these discoveries is that it has become very important when analysing economic time series data to conduct initial tests for non-stationarity before proceeding with estimations. Tests for non-stationarity are not easy to perform. Box-Jenkins analysis uses inspection of the correlogram to estimate if a series is stationary or not. A key feature of this methodology which is also used by econometricians without any links to economic theory is that differencing creates stationarity. The meaning of integration in the ARIMA term means that a variance needs to be integrated d times, written I(d) to attain stationarity. Hence, a stationary variable is integrated of order zero, written I(0), a variable which must be differenced once to become stationary is to be I(1), or integrated of order one, etc. Economic variables are usually integrated at orders not larger than two and if non-stationary are normally I(2).

Let $X_t = X_{t-1} + e_t$, where e is a stationary error term, so e is I(0). Here X can be seen to be I(1) because $\Delta X_t = e_t$, which is I(0). Expressing these ideas more generally yields: $X_t = \alpha X_{t-1} + e_t$. So if the absolute value of α is smaller than 1, then X is I(0) and stationary, but if $\alpha = 1$, then X is I(1) and non-stationary. Hence, formal tests for stationarity are tests for $\alpha = 1$ and are dubbed tests for a 'unit root', where if the absolute value of α is greater than 1, a time series would be explosive and subject to chaotic trends.

In an ECM level variables enter the estimating equation in a specific way, hence they are mixed into a single entity. This encompasses the degree to which the system is in disequilibrium. If these levels variables are individually I(1) with a specific condition I(0) then their inclusion in the estimating equation will create spurious regressions.

With a case of non-stationarity, variables drift apart endlessly in an economic time series. By contrast stationary variables may drift too but due to economic forces they should be kept largely together. Examples are short- or long-run commodity prices, spot and forward exchange rates and exports and imports. Some variables are said to be cointegrated, if individually they are I(1) though a special linear combination is I(0). This cointegration mix is understood as an equilibrium relationship which implies that the contemplated variables must have an ECM representation. Hence in applied work cointegration possesses a formal framework for activating long-run equilibrium relationships. When a set of I(1) variables are cointegrated then regressing one on the others should produce residuals that are I(0); most tests for cointegration therefore take a form of a unit root test applied to the residuals resulting from estimation of the cointegrated relationship.

Checklist for practical economic time series studies

First, only use unit roots to establish the order of integration of the time series. Second, run the cointegrating regression suggested by theory. Third, apply a unit root test on the residuals from the regression to test for cointegration. Fourth, estimate the full ECM equation such that it is better to estimate the long run equilibrium relationship jointly with short run dynamics than estimating separately.

Long run modelling: the debate

With regards to time series analysis and long run modelling, the serial dependence inherent in economic time series has given rise to the continual problem of 'spurious regressions'. Much scepticism regarding the use of regression analysis in economics stems from this worry.

Since 1945 efforts to find a solution for the problem of 'spurious regression' have been mounted. Thus the Cochrane–Orcutt method for the calculation of regression coefficients following the assumption that the errors follow a first-order autoregressive pattern and tests for serial correlation. By the 1970s Box-Jenkins analysis of time series became a feature of much applied work, however, links with economic theory and economic models appeared remote. However, the 'success' of Box-Jenkins techniques as forecasting tools which continued to out-perform large-scale econometric models created a methodological debate which still rages today (*Economic Journal*, 1997). The core of this dispute concerns the issue that the investigation of time series involves two components, namely analysis and modelling. Stochastic trends in variables however imply that standard regression analysis may lead to ambiguous inferential findings. Proponents of cointegration argue that a clear advantage appears in the analysis of variables subject to stochastic trends since rules of thumb can be used to assess regression results. Moreover, proponents argue that using cointegration techniques to investigate an economic time series is the 'best' statistical way of approximating long-run equilibrium relationships.

The idea of cointegration (Granger, 1981) results from the analysis of the statistical features of linear combinations of two integrated variables which if stationary implies that the time series is cointegrated. So if two variables p_t and Y_t each contain a stochastic trend they can be called integrated variables; if moreover, p_t and Y_t are shown to be stationary then the two variables are cointegrated.

A typical modern cointegration study of an economic time series sets out the integration characteristics of time series variables at the beginning.

In time series analysis it is usually assumed that the error term is serially uncorrelated and is not correlated with the regression; moreover, it is further assumed that all regressors are stationary random variables. For large samples the estimated coefficients will be consistent. If these assumptions are breached then the null distribution of regression 't' and 'f' cannot be approximated by normal and f distributions. As regards the assumption that all regressors are stationary variables then for inferential purposes we are in a grey area of judgement. Hence, it is a matter of judgement to label p_t as a variable with a stochastic trend. However, it could be that more painstaking analysis of p_t might yield evidence to suggest that p_t is a 'non-stochastic' process. Again more detailed investigations of p_t might show that rather than an integrated process p_t might have non-stochastic determinants.

A key issue for cointegration analysis is this: if the dependent variable and at least one regressor are integrated, but there is no cointegrating relationship, the error term in the regression is integrated. If, in a regression of two variables the error term is itself cointegrated then there is no cointegration relationship in the variables under investigation, given that the parameter is not consistent but converges to a random variable.

Long-run equilibrium

The debate in the *Economic Journal* regarding modelling the long run focuses on the cointegration properties of time series variables since these are believed to provide a framework for testing hypotheses and assisting investigations at the specification stage. This approach arises from the belief that the economic forces of supply and demand through time operate on variables such that some combinations of the time series will not drift apart significantly, in the long run. Example of such variables are: interest rates, commodity prices, government income and spending and the value of sales, market shares and costs in an industry. Hence, Granger (1986) argues that the beliefs about long-term associations are clearly empirical questions. This is in direct contrast to Darnell and Evans (ibid).

The cointegration of an economic time series in Granger's view corresponds to the theoretical notion of a long-run dynamic equilibrium. This produces a way of testing propositions from theory. Suppose there are two series p_t and q_t both of which are integrated of order one, I(1) so that each has an infinite variance. If theory proposes a long-run equilibrium relationship:

$$q_t - \alpha p_t = 0$$

then a linear combination of the two series is stationary. These two series are cointegrated and their cointegration is a necessary condition for them to have a long-run stable relationship, otherwise p_t and q_t will diverge indefinitely without limit. The nature of this relationship requires consideration however, thus if,

$$q_t = \alpha p_t + \beta_t z_t + \text{error}$$

where in the sample values of z_t and the error are stationary, then for the period under study the stable relationships between p_t and q_t implies cointegration of these variables. However, stability in this context of the relationship would seem to imply cointegration. The stability only refers to the sample period and we cannot conclude that the long run is stable because environmental factors may destabilise the 'relationship' or z_t becomes non-stationary if different sample observations were used on the series for p_t and q_t. To an extent then the relationship of the variables p_t and q_t drawn from the cointegration equation may normally be insufficient. For example, the investigation may not have sufficient knowledge about how the long-run relationship works or may breakdown. Or more importantly, a cointegrating relationship may not yield circumstances, in which the hypothesised relationship could be fulfilled. Thus if cointegration is not found it may be due to the lack of an analytical treatment of subsidiary hypotheses. Generally lack of cointegration may not be sufficient evidence to scrap the hypothesised relationship.

In the example given of a cointegration analysis of the Beer Orders in UK beer supply and price indices it is legitimate for the investigation when evaluating the cointegration equation to consider its meaning: thus though construed in the analysis as a 'long run' supply relationship it could be argued that it might just as easily represent a 'demand' relationship, despite the fact that variables in the original specification of the model do not include normal demand variables such as substitutes or, levels of disposable income. Generally, therefore, cointegration procedures cannot lead to watertight conclusions about an hypothesised long-run relationship because they normally fail to model subsidiary hypotheses.

The debate in the *Economic Journal*, Granger (1997) touches on these points and argues that generally investigations should test for cointegration of variables and use derived information for subsequent model specifications. Nonetheless Granger argues (1986) that testing for cointegration can be seen as a pre-test to avoid spurious regressions. Moreover, Engle and Granger (1987) prove that if two variables p_t and q_t are both I(1) and are cointegrated then there always exists a generating process which has an error correction form:

$$\Delta q_t = -p_1 z_{t-1} + \text{lagged}(\Delta p_t, \Delta q_t) + \alpha(\beta)e_{1t}$$
$$\Delta p_t = -p_2 z_{t-1} + \text{lagged}(\Delta p_t, \Delta q_t) + \alpha(\beta)e_{2t}$$

Obviously the data in the error correction model must be cointegrated. The resulting equilibrium error in the model allows for the impact of long-run equilibrium theory into the models to be introduced. Such error-correction models should yield better short-run forecasts and yield long-run forecasts which are economically meaningful. From a modelling point of view a significant limitation of time series analysis is that there is no indication as to how the integrated processes occur. Thus cointegration studies do not always employ economic concepts in the initial stages of the study. In later stages economic theory only enters the analysis as a maintained hypothesis throughout. Cointegration then represents an inductive study strategy where the final equation can only prove and not falsify economic theories.

Darnell and Evans (ibid) argue that this is wholly open to the change of measurement without theory. They argue (pp. 142–143) that the practitioner of cointegration adopts an unsound methodological procedure of subsuming auxiliary hypotheses under the umbrella of 'integrated' variables and stationary white noise errors whilst the new proposition though derived from theory is translated into the cointegration concept. Because falsification of theories cannot be undertaken the use of cointegration cannot further our knowledge of economic theory because this requires a deeper philosophical approach which can be encapsulated as: if p_t and q_t can be described as a process I(1) what does this mean for theory? Failure to recognise this results in abstract empiricism or inductivism. This does seem a touch unfair and falls once again into the trap of potentially paralysing econometric modelling. Or, that econometricians using cointegration procedures within the general to specific modelling approach should start with a rigorous philosophical discussion about what a time series for prices and quantities for two variables might mean with reference to theory. This appears to be asking too much in our area where we cannot model actual situations under laboratory conditions or fix in some way environmental conditions. In a sense Granger's checklist for undertaking a rigorous cointegration study, using the general to specific approach seems a reasonable counter-argument to the view that all econometric models should be able to falsify theory. One is reminded here of the methodological debates according to the tests of the Leontief Paradox. Recall the 'paradox' being that Leontief input–output analysis found that contrary to all common knowledge regarding US capital endowments, the test results showed that USA exports were relatively labour intensive. Economists, however did not reject the Heckscher–Ohlin theory rather they modified the 'tests'. Most 'tests' turned out too consistent with the 'paradox' and inconsistent with the theory. Ohlin, however, could merely remark that the 'tests' were not proper tests of the theory and that data sets were inconsistent. Nonetheless despite the clear rejection by empirical methods the theory still holds centre ground today despite the development of rival theories. In a deeper sense therefore it is difficult to imagine a case in economic analysis where cointegration procedures would lead to the rejection of well established economic hypotheses and theories, in this sense the 'verificationist' approach seems appropriate.

DISCUSSION QUESTIONS

1   Distinguish the (S-G) approach to econometric modelling from the (G-S) approach. Explain the role of diagnostic testing in each approach.
2   Discuss the nature of the practical problems concerning the (G-S) approach to econometric modelling.
3   What are the implications of an 'ECM' framework for the behaviour of the economic agents? What is the meaning of the coefficient of an error correction term?
4   Under what circumstances can two or more 'ECM' type frameworks be combined?
5   Discuss the advantages/disadvantages of an ARIMA type framework for econometric modelling.

ADDITIONAL READING

Greene W. H. (2000) 'Econometric Analysis', 7th ed. New York–Prentice Hall.
Darnell A. (1990) 'The Limits of Econometrics', 4th ed. Cheltenham–Edward Elgar Evans. L.
Kennedy P. (1994) 'Econometrics' 4th ed. Macmillan.
Granger C. et al (Autumn 1996) 'Controversy: Modeling the Long Run', Economic Journal.
Article by Pesaran H., Harvey A. and Dixon et al (please see Chapter 2).

7 Time-series analysis:

Non-stationarity and cointegration

Key points

- Unit root tests
- Cointegration two variable case
- Cointegration multivariate case

1 INTRODUCTION

Economic analysis suggests that there is a long-run, or equilibrium, relationship between the variables involved in the economic theory under investigation. Applied econometric analysis in trying to estimate these long-run relationships implicitly considers the 'constancy doctrine' of the variables involved, in terms of means and variances being constant while not dependent on time. However, empirical research has shown that in most cases the 'constancy doctrine' is not satisfied by the time-series variables. Therefore, the classical t and F tests, etc. based on estimation methods which consider the 'constancy doctrine', or, in other words assume, without verification that the variables involved in the estimation are stationary, are no longer valid giving misleading inferences. This is known as the 'spurious' regression problem. As a result researchers in the field, in trying to correct imperfect estimations, proposed revisions to traditional estimation approaches. These revisions included the systematic examination of the *non-stationarity* of the variables and *cointegration*, among others.

Cointegration analysis has been regarded as perhaps the most revolutionary development in econometrics since the mid 1980s. In simple words, cointegration analysis refers to groups of variables that drift together, although which individually non-stationary in the sense that they tend upwards or downwards over time. This common drifting of variables makes linear relationships between these variables exist over long periods of time, thereby giving thus insight into equilibrium relationships of economic variables. However, if these linear relationships do not hold over long periods of time then the corresponding variables are said to be 'not co-integrated'.

Generally, cointegration analysis is a technique used in the estimation of the long-run or, equilibrium parameters in a relationship with non-stationary variables. It is a new method for specifying, estimating and testing dynamic models and therefore it can be used for testing the validity of underlying economic theories.

Furthermore, the usefulness of cointegration analysis is also seen in the estimation of the short-run or disequilibrium parameters in a relationship, because the latter estimation can utilise the estimated long-run parameters through cointegration methods. There are known as 'error correction models'.

Before we continue our analysis we present in the next section basic definitions of the concepts involved.

2 THE DEFINITIONS

The most common definitions in time-series analysis are the following:

Stochastic process

A family of real valued random variables $X_1, X_2, X_3, \ldots$, where the subscripts refer to successive time periods, is called 'stochastic process' and is denoted with $\{X_t\}$. Each of the random variables in the stochastic process has generally its own probability distribution and are not independent.

Time series

Consider that for each time period we get a sample of size one (one observation) on each of the random variables of a stochastic process. Therefore, we get a series of observations corresponding to each time period and to each different random variable. This series of observations is called a 'time series' and is denoted with X_t.

Realisation

The time series X_t is a 'single realisation' of the stochastic process $\{X_t\}$ because of lack of replication in sampling, or in other words because it is impossible to get another observation from the same random variable. We could say that the stochastic process is regarded as the 'population' and the time series is regarded as its realisation, i.e. as a 'sample' in cross-sectional data (one observation from each section, each random variable). For convenience, we will use the terms stochastic process and time series synonymously, and we will use the common notation of X_t to represent them.

Strict stationarity

Because each of the random variables X_t in a stochastic process has a probability distribution, a stochastic process can be specified by the joint probability distribution of its random variables. A stochastic process, and correspondingly a time series, is said to be 'strictly stationary' if its joint distribution is unchanged if displaced in time, or in other words if its joint distribution of any set of n variables $X_1, X_2, \ldots, X_n$ is the same as the joint distribution of $X_{1+k}, X_{2+k}, \ldots, X_{n+k}$ for all n and k.

Weak stationarity

Because it is generally difficult in practice to determine the joint distribution of a stochastic process, we can use instead the means, the variances, and the covariances

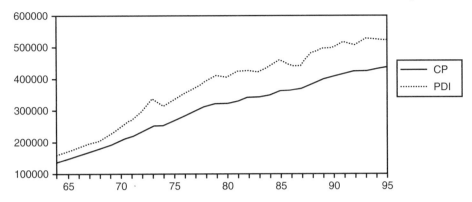

Figure 7.1 Private consumption and personal disposable income for an EU member, 1960–1995, annual data, millions of 1970 drs

of its random variables X_t, for $t = 1, 2, \ldots$ A stochastic process, and correspondingly a time series, is said to be 'weakly stationary' if:

1   The mean $E(X_t) = \mu$ is constant for all t.   (7.1)

2   The variance $\text{var}(X_t) = E(X_t - \mu)^2 = \sigma^2$ is constant for all t.   (7.2)

3   The covariance $\text{cov}(X_t, X_{t+k}) = E[(X_t - \mu)(X_{t+k} - \mu)] = \gamma_k$ (7.3)
    is constant for all t and $k \neq 0$.

Stationarity

In this chapter when we use the term 'stationary' we will refer to weak stationarity. Generally speaking, a stochastic process, and correspondingly a time series, is stationary if the means and variances are constant over time and the (auto)covariances between two time periods t and $t + k$, depend only on the distance (gap or lag) k between these two time periods and not on the actual time period t at which these covariances are considered.

Non-stationarity

If one or more of the three conditions for stationarity are not fulfilled, the stochastic process, and correspondingly the time series, is called 'non-stationary'. In fact, most time series in economics are non-stationary. In Figure 7.1, for example, the time series of private consumption (CP) and personal disposable income (PDI) for an EU member, since they are all trending consistently upwards are almost certain not to satisfy condition (7.1) for stationarity, and therefore they are non-stationary (actual data for these series can be found in Table 7.1).

Having defined stationarity, in what it follows we present some useful time series models:

White noise

This is a purely random process $\{\epsilon_t\}$, for t taking values from $-\infty$ to $+\infty$, where the ϵ_t are all identically and independently distributed (IID) with

1   mean $E(\epsilon_t) = 0$ for all t, (7.4)

2   variance $var(\epsilon_t) = \sigma^2$ for all t, and (7.5)

3   covariance $cov(\epsilon_t, \epsilon_{t+k}) = 0$ for all t and $k \neq 0$. (7.6)

The white noise time series, which is also written as

$$\epsilon_t \sim IID(0, \sigma^2) \tag{7.7}$$

is *stationary by definition*, since its means are zero, its variances are σ^2, and its covariances are zero, being therefore constant over time.

Random walk

This is a simple stochastic process $\{X_t\}$ with X_t being determined by

$$X_t = X_{t-1} + \epsilon_t \tag{7.8}$$

where ϵ_t is white noise.

The mean of X_t in (7.8) is given by

$$E(X_t) = E(X_{t-1} + \epsilon_t) = E(X_{t-1}) + E(\epsilon_t) = E(X_{t-1}), \text{ due to (7.4)}, \tag{7.9}$$

implying that the mean of X_t is constant over time. In order to find the variance of X_t, we use (7.8) which after successive substitutions is written as

$$X_t = X_0 + \epsilon_1 + \epsilon_2 + \ldots + \epsilon_t = X_0 + \sum_{t=1}^{t} \epsilon_t \tag{7.10}$$

where X_0 is the initial value of X_t, which is assumed to be any constant or could be also taken to be equal to zero. The variance of (7.10), taking into account (7.7), is given by

$$var(X_t) = \sum_{t=1}^{t} var(\epsilon_t) = t\sigma^2 \tag{7.11}$$

(7.11) shows that the variance of X_t is not constant over time but instead it increases with time. Therefore, because condition (7.2) for stationarity is not fulfilled, X_t, or the random walk time series is a *non-stationary* time series. However, if (7.8) is written in first differences, i.e. is written as

$$\Delta X_t = \epsilon_t \tag{7.12}$$

this first differenced new variable is stationary, because it is equal to ϵ_t which is stationary by definition.

Random walk with drift

This is the case of the stochastic process $\{X_t\}$ with X_t being determined by

$$X_t = \mu + X_{t-1} + \epsilon_t \tag{7.13}$$

where $\mu \neq 0$ is a constant and ϵ_t is white noise. The term 'drift' has been given to this process because if we write (7.13) as the first difference

$$\Delta X_t = X_t - X_{t-1} = \mu + \epsilon_t \tag{7.14}$$

this shows that the time series X_t 'drifts' upwards or downwards, depending on the sign of μ being positive or negative. The random walk with drift time series is also a non-stationary time series. The processes in (7.8) and in (7.13) are no longer 'random walks' if the assumption of white noise is relaxed to allow for autocorrelation in ϵ_t. However, even in cases of autocorrelation in ϵ_t, time series X_t will still be non-stationary.

Time trends

We call 'time trend' the tendency of a non-stationary series to move in one direction. Let us consider the following model:

$$X_t = \alpha + \beta t + \varphi X_{t-1} + \epsilon_t, \, \alpha \neq 0 \tag{7.15}$$

where ϵ_t is white noise and t is a time trend.

Using model (7.15) we can distinguish the following cases (Thomas, 1997):

Case 1: Stochastic trend

This is the case where $\beta = 0$ and $\varphi = 1$. Model (7.15) is written in this case as

$$X_t = \alpha + X_{t-1} + \epsilon_t \tag{7.16}$$

or

$$\Delta X_t = \alpha + \epsilon_t \tag{7.17}$$

From (7.17) it is seen that X_t trends upwards or downwards according to the sign of α being positive or negative respectively. This type of trend is known as 'stochastic trend'. Model (7.16) is called 'difference-stationary process (DSP)' because the non-stationarity in X_t can be eliminated by taking first differences of the time series (Nelson and Plosser, 1982).

Case 2: Deterministic trend

This is the case where $\beta \neq 0$ and $\varphi = 0$. Model (7.15) is written in this case as

$$X_t = \alpha + \beta t + \epsilon_t \tag{7.18}$$

From (7.18) it is seen that X_t trends upwards or downwards according to the sign of β being positive or negative respectively. This type of trend is known as a 'deterministic trend'. Model (7.18) is called 'trend-stationary process (TSP)' because the non-stationarity in X_t can be eliminated by subtracting the trend $(\alpha + \beta t)$ from the time series.

Case 3: Combined stochastic and deterministic trend

This is the case where $\beta \neq 0$ and $\varphi = 1$. Model (7.15) is written in this case as

$$X_t = \alpha + \beta t + X_{t-1} + \epsilon_t \tag{7.19}$$

From (7.19) it is seen that X_t trends upwards or downwards according to the combined effect of the parameters α and β. This type of trend is known as 'combined stochastic and deterministic trend'. To test the hypothesis that a time series is of a DSP type against being of a TSP type specific tests have to be employed, as those developed by Dickey and Fuller (1979, 1981).

Generalisations

We saw in (7.8) that the random walk process is the simplest non-stationary process. However, this process is a special case of

$$X_t = \varphi X_{t-1} + \epsilon_t \tag{7.20}$$

which is called a 'first-order autoregressive process (AR_1)'. This process is such stationary if for the parameter φ holds, that $-1 < \varphi < 1$. If it is either $\varphi < -1$ or $\varphi > 1$ then the process will be non-stationary.

Generalising, equation (7.20) is a special case of

$$X_t = \varphi_1 X_{t-1} + \varphi_2 X_{t-2} + \varphi_3 X_{t-3} + \ldots + \varphi_q X_{t-q} + \epsilon_t \tag{7.21}$$

which is called 'qth-order autoregressive process (AR_q)'. It can be proved (Greene, 1999) that this process is stationary if the roots of the 'characteristic equation'

$$1 - \phi_1 L - \phi_2 L^2 - \phi_3 L^3 - \ldots - \phi_q L^q = 0 \tag{7.22}$$

where L is the lag operator, are all greater than unity in absolute values. Otherwise, the process in (7.21) is non-stationary.

Integrated series

We saw in (7.12) that if the random walk series is differenced once the differenced series $\Delta X_t = X_t - X_{t-1}$ is stationary. In this case we say that the original non-stationary series X_t is 'integrated of order 1', and is denoted with I(1). Similarly, if a non-stationary series has to be differenced twice ($\Delta^2 X_t = \Delta X_t - \Delta X_{t-1}$) before it becomes stationary, the original series X_t is 'integrated of order 2', and is denoted with I(2). Generally, if a non-stationary series has to be differenced d times before it becomes

stationary, the original series X_t is 'integrated of order d', and is denoted with I(d). By definition I(0) denotes a stationary series, which means that this series is stationary without any differencing. However, if a series does not become stationary irrespectively of how many times it is differenced, it is called 'non-integrated'.

In the special case where the time series X_t is subject to 's' seasonality (say quarterly (s = 4), monthly (s = 12), etc), the definition of an integrated series is modified respectively (Engle *et al.* (1989), Charemza and Deadman (1997)). Thus, a non-stationary series is said to be 's-seasonally integrated of order (d, D)', and it is denoted with $SI_s(d, D)$, if in order to be transformed to a stationary series it has to be s-differenced D times and then the resulting series it has to be first differenced d times.

The properties of integrated series

If a series X_t is integrated of order d, then this is written as

$$X_t \sim I(d) \tag{7.23}$$

Generally, we distinguish the following properties of integrated series:

1   If $X_t \sim I(0)$ and $Y_t \sim I(1)$ then $Z_t = (X_t + Y_t) \sim I(1)$.
2   If $X_t \sim I(d)$ then $Z_t = (a + bX_t) \sim I(d)$, where a and b are constants.
3   If $X_t \sim I(0)$ then $Z_t = (a + bX_t) \sim I(0)$.
4   If $X_t \sim I(d_1)$ and $Y_t \sim I(d_2)$, then $Z_t = (aX_t + bY_t) \sim I(d_2)$, where $d_1 < d_2$.
5   If $X_t \sim I(d)$ and $Y_t \sim I(d)$, then $Z_t = (aX_t + bY_t) \sim I(d^*)$, where d* in most cases is equal to d. However, there are special cases, which we will see later in this chapter, where d* is less than d.

Chapter 7: Key terms

Spurious regression   Misleading results due to non-stationarity of the regressors.

Weak stationarity   The mean, variance and the covariances of a time series are time invariant (constant over time).

Unit root test   A statistical test designed to detect non-stationarity.

IDW   Integration Durbin-Watson statistic used to detect non-stationarity.

Cointegration   An econometric methodology employed to confirm the existence of a long run relationship among economic variables as postulated by the economic theory.

VAR (model)   Vector autoregression model

3 TESTING STATIONARITY: THE AUTOCORRELATION FUNCTION

Tests for stationarity may be divided into two groups; the traditional and the modern. The first group that uses the autocorrelation function is presented in this section, and the second group that uses the unit roots we present in the next section.

For a stochastic process we know that it is:

mean: $\quad E(X_t) = \mu$ (7.24)

variance: $\quad var(X_t) = E(X_t - \mu)^2 = \gamma_0$ (7.25)

covariance: $\quad cov(X_t, X_{t+k}) = E[(X_t - \mu)(X_{t+k} - \mu)] = \gamma_k$ (7.26)

hence the 'population autocorrelation function (P_ACF)' is given by

$$\rho_k = \frac{\gamma_k}{\gamma_0} = \frac{cov(X_t, X_{t+k})}{var(X_t)}$$ (7.27)

where ρ_k is the autocorrelation coefficient (AC) between X_t and X_{t-k}. The autocorrelation coefficient ρ_k takes values between -1 and $+1$ as verifiable from (7.27). The plot of ρ_k against k is called the 'population correlogram'. One basic property of the autocorrelation function is that it is an even function of lag k, i.e. it is $\rho_k = \rho_{-k}$. For other properties see Jenkins and Watts (1968).

For a realisation of a stochastic process, i.e. for a time series X_t, we know that it is:

mean: $\quad \bar{X} = \frac{1}{n} \Sigma X_t$ (7.28)

variance: $\quad \hat{\gamma}_0 = \frac{1}{n} \Sigma (X_t - \bar{X})^2$ (7.29)

covariance: $\quad \hat{\gamma}_k = \frac{1}{n} \Sigma (X_t - \bar{X})(X_{t+k} - \bar{X})$ (7.30)

hence the 'sample autocorrelation function (S_ACF)' is given by

$$\hat{\rho}_k = \frac{\hat{\gamma}_k}{\hat{\gamma}_0} = \frac{\Sigma (X_t - \bar{X})(X_{t+k} - \bar{X})}{\Sigma (X_t - \bar{X})^2}$$ (7.31)

where $\hat{\rho}_k$ is the estimated autocorrelation coefficient between X_t and X_{t-k}. The estimated autocorrelation coefficient $\hat{\rho}_k$ takes values between -1 and $+1$ as verifiable from (7.31). The plot of $\hat{\rho}_k$ against k is called the 'sample correlogram'. In what it follows when we refer to autocorrelation functions or correlograms we mean the sample equivalent.

As a rule of thumb, the correlogram can be used for detecting non-stationarity in a time series. As an example, let as consider the time series of private consumption for an EU member presented in Table 7.1. We have seen already in Figure 7.1 that this time series may be non-stationary. In Table 7.2 the estimated autocorrelation coefficients (AC) for this time series are reported, and the corresponding correlogram is presented in Figure 7.2. Furthermore, in Table 7.2 the estimated autocorrelation coefficients for the same time series, but differenced once are reported, with the corresponding correlogram in Figure 7.3.

By examining the correlogram in Figure 7.2 we see that the autocorrelation coefficients start from very high values ($\hat{\rho}_k = 0.925$ at lag $k = 1$) and their values decrease very slowly towards zero as k increases, showing thus a very slow rate of decay. On the contrary, the correlogram in Figure 7.3 shows that all autocorrelation

Table 7.1 Private consumption and personal disposable income for an EU member, (money at millions of 1970 drachmas)

Year	Private consumption	Personal disposable income	Consumption price index
1960	107808.0	117179.2	0.783142
1961	115147.0	127598.9	0.791684
1962	120050.0	135007.1	0.801758
1963	126115.0	142128.3	0.828688
1964	137192.0	159648.7	0.847185
1965	147707.0	172755.9	0.885828
1966	157687.0	182365.5	0.916505
1967	167528.0	195611.0	0.934232
1968	179025.0	204470.4	0.941193
1969	190089.0	222637.5	0.969630
1970	206813.0	246819.0	1.000000
1971	217212.0	269248.9	1.033727
1972	232312.0	297266.0	1.068064
1973	250057.0	335521.7	1.228156
1974	251650.0	310231.1	1.517795
1975	266884.0	327521.3	1.701147
1976	281066.0	350427.4	1.929906
1977	293928.0	366730.0	2.159872
1978	310640.0	390188.5	2.436364
1979	318817.0	406857.2	2.838453
1980	319341.0	401942.8	3.459030
1981	325851.0	419669.1	4.081844
1982	338507.0	421715.6	5.114169
1983	339425.0	417930.3	6.067835
1984	345194.0	434695.7	7.130602
1985	358671.0	456576.2	8.435285
1986	361026.0	439654.1	10.30081
1987	365473.0	438453.5	11.91950
1988	378488.0	476344.7	13.61448
1989	394942.0	492334.4	15.59285
1990	403194.0	495939.2	18.59539
1991	412458.0	513173.0	22.09116
1992	420028.0	502520.1	25.40122
1993	420585.0	523066.1	28.88346
1994	426893.0	520727.5	32.00385
1995	433723.0	518406.9	34.98085

Sources: Epilogi (1998).

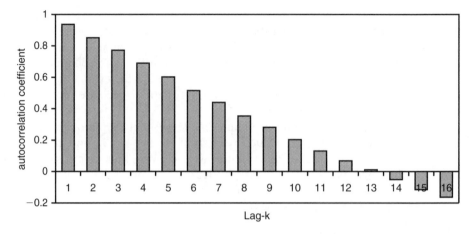

Figure 7.2 Correlogram for private consumption of an EU member

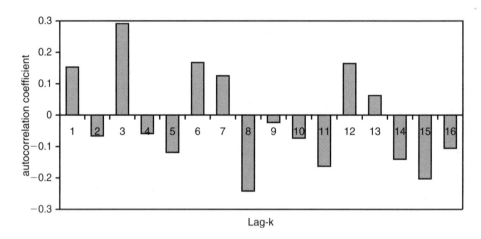

Figure 7.3 Correlogram for differenced private consumption of an EU member

coefficients are close to zero. The correlogram in Figure 7.2 is a typical correlogram for a non-stationary time series, whilst the correlogram in Figure 7.3 is a typical correlogram for a stationary time series. Generally, as a rule of thumb,

'if, in the correlogram of a time series the estimated autocorrelation coefficient $\hat{\rho}_k$ does not fall quickly as the lag k increases, this is an indication that the time series is non-stationary. By contrast, if in the correlogram of a time series the estimated autocorrelation coefficient $\hat{\rho}_k$ does fall quickly as the lag k increases then, this is an indication that the time series is stationary.'

However, because we have argued that the correlogram is a rule of thumb in detecting non-stationarity of a time series, more objective measures should be used for detecting the non-stationarity of a time series. Some of these are:

(a) Testing autocorrelation coefficients individually

Bartlett (1946) has shown that the sample autocorrelation coefficients approximately follow the normal distribution with zero mean and variance equal to 1/n, where n is the sample size. Therefore, the hypotheses for testing autocorrelation coefficients individually may be formulated as:

$$
\left.
\begin{array}{l}
H_0: \rho_k = 0, \text{ if } |\hat{\rho}_k| < t_{\alpha/2}\dfrac{1}{\sqrt{n}}, \text{ i.e. stationary} \\[3ex]
H_a: \rho_k \neq 0, \text{ if } |\hat{\rho}_k| > t_{\alpha/2}\dfrac{1}{\sqrt{n}}, \text{ i.e. non-stationary}
\end{array}
\right\} \tag{7.32}
$$

where $t_{\alpha/2}$ is the critical value of the t distribution (normal distribution) for an α level of significance.

In our case, of the time series of the private consumption in Table 7.1, the sample size is $n = 36$ and therefore $1/\sqrt{36} = 0.167$. If we assume that the level of significance is $\alpha = 0.05$ then $t_{\alpha/2} = 1.96$, and thus $t_{\alpha/2}1/\sqrt{n} = 0.327$. Comparing this value of 0.327 with the estimated autocorrelation coefficients in Table 7.2 of the original private consumption series we see that all the AC up to lag $k = 8$ are greater than the critical value of 0.327. Therefore, we accept the alternative hypothesis in (7.32), i.e. we accept the hypothesis that private consumption is a non-stationary time series. Furthermore, comparing this value of 0.327 with the estimated autocorrelation coefficients in Table 7.2 of the once differenced private consumption series we see that the absolute values of all the AC are less than the critical

Table 7.2 Autocorrelation coefficients (AC), Ljung-Box Q-statistic (LB_Q), and probability level (prob) for the private consumption time series for an EU member

Lag	Original series of private consumption			Differenced series of private consumption		
k	AC	LB_Q	Prob	AC	LB_Q	Prob
1	0.925	33.440	0.000	0.148	0.8333	0.361
2	0.848	62.362	0.000	−0.068	1.0138	0.602
3	0.768	86.800	0.000	0.291	4.4489	0.217
4	0.682	106.70	0.000	−0.058	4.5876	0.332
5	0.597	122.44	0.000	−0.119	5.1938	0.393
6	0.514	134.47	0.000	0.166	6.4184	0.378
7	0.430	143.21	0.000	0.122	7.1072	0.418
8	0.351	149.24	0.000	−0.245	9.9972	0.265
9	0.276	153.11	0.000	−0.026	10.032	0.348
10	0.202	155.27	0.000	−0.078	10.347	0.411
11	0.131	156.21	0.000	−0.163	11.778	0.381
12	0.065	156.45	0.000	0.161	13.237	0.352
13	0.002	156.45	0.000	0.062	13.465	0.413
14	−0.058	156.66	0.000	−0.136	14.604	0.406
15	−0.117	157.54	0.000	−0.203	17.279	0.302
16	−0.170	159.52	0.000	−0.102	17.983	0.325

Notes: Estimates via Eviews.

value of 0.327. Therefore, we accept the null hypothesis in (7.32), i.e. we accept the hypothesis that the once differenced private consumption is a stationary time series.

(b) Testing autocorrelation coefficients jointly

Box and Pierce (1970) in testing the autocorrelation coefficients jointly, developed the so-called 'Q-statistic', which is given by

$$Q = n \sum_{k=1}^{m} \hat{\rho}_k^2 \sim \chi^2(m) \tag{7.33}$$

where n is the sample size and m is the lag length used. Because this statistic is not valid for small samples, Ljung and Box (1978) proposed a variation of the statistic in (7.33), the LB_Q-statistic, as follows:

$$LB\_Q = n(n+2) \sum_{k=1}^{m} \left(\frac{\hat{\rho}_k^2}{n-k} \right) \sim \chi^2(m) \tag{7.34}$$

The statistic in (7.34) is more powerful both in small and large samples than the statistic in (7.33) which may be used in large samples only. The hypotheses for testing autocorrelation coefficients jointly may be formulated as:

$$\left. \begin{array}{l} H_0: \text{all } \rho_k = 0, \text{ if } LB\_Q < \chi_\alpha^2(m), \text{ i.e. stationary} \\[2mm] H_a: \text{not all } \rho_k = 0, \text{ if } LB\_Q > \chi_\alpha^2(m), \text{ i.e. non-stationary} \end{array} \right\} \tag{7.35}$$

where: $\chi_\alpha^2(m)$, is the critical value of the χ^2 distribution for an α level of significance and m degrees of freedom.

In this time series of the private consumption in Table 7.1, the sample size is n = 36 and the estimated autocorrelation coefficients reported in Table 7.2 extend up to lag length k = 16. Furthermore, in the same Table 7.2, the Ljung-Box Q-statistics (LB_Q) are reported for the specific lag lengths autocorrelation coefficients and with the corresponding probability levels for significance (prob). We see in Table 7.2 that all the probs are 0.000 for the autocorrelation coefficients of the original private consumption series. Therefore, we accept the alternative hypothesis in (7.35), i.e. we accept the hypothesis that private consumption is a non-stationary time series. Finally, none of the probs in Table 7.2 are less than the 0.05 significance level for the autocorrelation coefficients of the once differenced private consumption series. Therefore, we accept the null hypothesis in (7.35), i.e. we accept the hypothesis that the once differenced private consumption is a stationary time series.

4 TESTING FOR STATIONARITY: THE UNIT ROOTS METHODOLOGY

Consider the first-order autoregressive process in (7.20), i.e.

$$X_t = \varphi X_{t-1} + \epsilon_t \tag{7.36}$$

where ϵ_t is white noise. This process may also be written as

$$X_t - \varphi X_{t-1} = \epsilon_t \text{ or as } (1 - \varphi L)X_t = \epsilon_t \tag{7.37}$$

For (7.37) to be stationary the root of the characteristic equation $1 - \varphi L = 0$ must be greater than unity in absolute values. This equation has one root only, which is $L = 1/\varphi$, and thus stationarity requires $-1 < \varphi < 1$. Therefore, the hypotheses for testing the stationarity of X_t may be written as

$$\left.\begin{array}{l} H_0: |\phi| \geq 1, \text{ for non-stationarity} \\ H_a: |\phi| < 1, \text{ for stationarity} \end{array}\right\} \tag{7.38}$$

In the case that $\varphi = 1$, i.e. if the null hypothesis is true, then (7.36) is nothing else but the random walk process (7.8), which we saw that was non-stationary. This unity of φ is known as the 'unit root' problem, i.e. the problem of the non-stationarity of the corresponding process. In other words, a unit root is another way to express non-stationarity.

By subtracting X_{t-1} from both sides of (7.36), we get that $X_t - X_{t-1} = \varphi X_{t-1} - X_{t-1} + \epsilon_t$ which is also written as

$$\Delta X_t = \delta X_{t-1} + \epsilon_t \tag{7.39}$$

where Δ is the difference operator and $\delta = \varphi - 1$. In other words, (7.39) is another way to write (7.36). Assuming that φ is positive (this is true for most economic time series) the hypotheses in (7.38) may be written equivalently as

$$\left.\begin{array}{l} H_0: \delta \geq 0, \text{ for non-stationarity} \\ H_a: \delta < 0, \text{ for stationarity} \end{array}\right\} \tag{7.40}$$

In the case that $\delta = 0$, or equivalently $\varphi = 1$, i.e. if the null hypothesis is true, the corresponding process is non-stationary. In other words, non-stationarity, or, the unit root problem, may be expressed either as $\varphi = 1$, or as $\delta = 0$. One could then suggest that the problem of testing for non-stationarity of time series X_t reduces to testing if parameter $\varphi = 1$ in the regression of equation (7.36), or, if parameter $\delta = 0$ in the regression of equation (7.39). Such testing could be performed by using the two t-tests respectively

$$t_\varphi = \frac{\hat{\phi} - 1}{s_{\hat{\phi}}} \text{ or } t_\delta = \frac{\hat{\delta}}{s_{\hat{\delta}}} \tag{7.41}$$

where $s_{\hat{\phi}}$ and $s_{\hat{\delta}}$ are the estimated standard errors of the estimated parameters $\hat{\phi}$ and $\hat{\delta}$ respectively. However, the situation is more complex. Under the null hypothesis of non-stationarity, i.e. under $\varphi = 1$ or $\delta = 0$, the t values computed in (7.41) do not follow the usual student's t distribution, but they follow a non-standard and even asymmetrical distribution. Therefore, other distribution tables should be employed.

4.1 The Dickey–Fuller (DF) test

Dickey and Fuller (1979) on the basis of Monte–Carlo simulations, and under the null hypothesis of the existence of a unit root in the process generating of time series, have tabulated critical values for the t_δ statistic in (7.41) which they called them as the 'τ (tau) statistics'. These critical values are presented in Table 7.3. More recently these critical values have been extended by MacKinnon (1991) through Monte–Carlo simulations. Comparing the critical values in Table 7.3 with those of the standard t distribution we see that the τ values are much greater (in absolute values) than the corresponding t values. The Dickey–Fuller test for a unit root may then be constructed in the following two steps:

Step 1

Run the OLS regression in (7.39), i.e.

$$\Delta X_t = \delta X_{t-1} + \epsilon_t \qquad (7.42)$$

and save the usual t_δ ratio in (7.41).

Step 2

Decide about the existence of a unit root in the process generating of time series X_t according to the following hypothesis:

H_0: $\delta = 0$, for non-stationarity, if $t_\delta > \tau$

H_a: $\delta < 0$, for stationarity, if $t_\delta < \tau$ ⟵ much negative   (7.43)

where τ is the critical value from Table 7.3 for a given significance level. In other words, for a time series to be stationary the t_δ value must be *much negative*. Otherwise the time series is non-stationary.

Dickey and Fuller noticed that the τ critical values depend on the type of the regression equation in (7.42). Therefore they tabulated τ critical values when the regression equation contains a constant also, i.e. when equation (7.42) becomes

$$\Delta X_t = \alpha + \delta X_{t-1} + \epsilon_t \qquad \tau_\mu \qquad (7.44)$$

and when the regression equation contains a constant and a linear trend, i.e. when equation (7.42) becomes

$$\Delta X_t = \alpha + \beta t + \delta X_{t-1} + \epsilon_t \qquad \tau_\tau \qquad (7.45)$$

For equation (7.44) the corresponding τ critical values are called τ_μ and for equation (7.45) the corresponding τ critical values are called τ_τ. These critical values are also presented in Table 7.3. However, the test about the stationarity of a time series depends always on the coefficient δ of the regressor X_{t-1}.

Table 7.3 Critical values for the Dickey–Fuller τ-statistics

Sample size	Probability of a smaller value							
	0.01	0.025	0.05	0.10	0.90	0.95	0.975	0.99
No constant No Time (Statistic-τ)								
25	−2.66	−2.26	−1.95	−1.60	0.92	1.33	1.70	2.16
50	−2.62	−2.25	−1.95	−1.61	0.91	1.31	1.66	2.08
100	−2.60	−2.24	−1.95	−1.61	0.90	1.29	1.64	2.03
250	−2.58	−2.23	−1.95	−1.62	0.89	1.29	1.63	2.01
500	−2.58	−2.23	−1.95	−1.62	0.89	1.28	1.62	2.00
∞	−2.58	−2.23	−1.95	−1.62	0.89	1.28	1.62	2.00
Constant No Time (Statistic-τ_μ)								
25	−3.75	−3.33	−3.00	−2.62	−0.37	0.00	0.34	0.72
50	−3.58	−3.22	−2.93	−2.60	−0.40	−0.03	0.29	0.66
100	−3.51	−3.17	−2.89	−2.58	−0.42	−0.05	0.26	0.63
250	−3.46	−3.14	−2.88	−2.57	−0.42	−0.06	0.24	0.62
500	−3.44	−3.13	−2.87	−2.57	−0.43	−0.07	0.24	0.61
∞	−3.43	−3.12	−2.86	−2.57	−0.44	−0.07	0.23	0.60
Constant Time (Statistic-τ_τ)								
25	−4.38	−3.95	−3.60	−3.24	−1.14	−0.80	−0.50	−0.15
50	−4.15	−3.80	−3.50	−3.18	−1.19	−0.87	−0.58	−0.24
100	−4.04	−3.73	−3.45	−3.15	−1.22	−0.90	−0.62	−0.28
250	−3.99	−3.69	−3.43	−3.13	−1.23	−0.92	−0.64	−0.31
500	−3.98	−3.68	−3.42	−3.13	−1.24	−0.93	−0.65	−0.32
∞	−3.96	−3.66	−3.41	−3.12	−1.25	−0.94	−0.66	−0.33

Source: Fuller, W. (1976) *Introduction to Statistical Time Series,* New York: John Wiley.

Example 7.1 Testing the stationarity of the private consumption time series of an EU member (DF)

In Table 7.1 the private consumption (C_t) time series for an EU member is presented. We saw in section 7.3 that using the autocorrelation function methodology this time series may be non-stationary. Let us apply now the DF test on the same data. The corresponding to equations (7.44) and (7.45) OLS estimates for C_t are respectively the following:

$$\Delta \hat{C}_t = 12330.48 - 0.01091 C_{t-1}$$

$$t \quad [5.138] \quad [-1.339] \tag{7.46}$$

$$R^2 = 0.052 \ DW = 1.765$$

$$\Delta \hat{C}_t = 15630.83 + 346.4522t - 0.04536C_{t-1}$$

t   [1.966]   [0.436]   [−0.571]   (7.47)

R² = 0.057 DW = 1.716

In Table 7.4 the MacKinnon critical values for the rejection of the hypothesis of a unit root, evaluated from Eviews for equations (7.46) and (7.47), are reported.

Table 7.4 MacKinnon critical values for equations (7.46) and (7.47) and the DF t_δ ratios

Significance level	Equation (7.46) Constant, No trend (τ_μ) [$t_\delta = -1.339$]	Equation (7.47) Constant and trend (τ_τ) [$t_\delta = -0.571$]
0.01	−3.6289	−4.2412
0.05	−2.9472	−3.5426
0.10	−2.6118	−3.2032

Considering the hypotheses testing in (7.43) we see that for equation (7.46) the $t_\delta = -1.339$ value is greater than all the τ_μ critical values in Table 7.4 and thus, the null hypothesis is not rejected. Therefore, the private consumption time series exhibits a unit root, or in other words is a non-stationary time series. Similarly, for equation (7.47) the $t_\delta = -0.571$ value is greater than all the τ_τ critical values in Table 7.4 and thus, the null hypothesis is again not rejected. Therefore, the private consumption (C_t) time series exhibits a unit root, or in other words is a non-stationary time series.

Having found that the C_t time series is non-stationary, let us see how the first difference of this series (ΔC_t) behaves in terms of non-stationarity. By repeating the same exercise as we did with C_t we get for ΔC_t the following results:

$$\Delta^2 \hat{C}_t = 7972.671 - 0.85112\Delta C_{t-1}$$

t   [4.301]   [−4.862]   (7.48)

R² = 0.425 DW = 1.967

$$\Delta^2 \hat{C}_t = 10524.35 - 114.461t - 0.89738\Delta C_{t-1}$$

t   [3.908]   [−1.294]   [−5.073]   (7.49)

R² = 0.454 DW = 1.988

where $\Delta^2 X_t = \Delta X_t - \Delta X_{t-1}$. In Table 7.5 the MacKinnon critical values for the rejection of the hypothesis of a unit root, evaluated from Eviews for equations (7.48) and (7.49), are reported.

Considering the hypotheses testing in (7.43) we see that for equation (7.48) the $t_\delta = -4.862$ value is much less than all the τ_μ critical values in Table 7.5 and thus, the alternative hypothesis is accepted. Therefore, the private consumption differenced once time series is not exhibiting a unit root, or in other words is a stationary time series. Similarly, for equation (7.49) the $t_\delta = -5.073$ value is much less than all the τ_τ critical values in Table 7.5 and thus, the alternative hypothesis is again accepted. Therefore, the private consumption differenced once time series does not exhibit a unit root, or in other words is a stationary time series.

Summarising the results of this example we can say that since ΔC_t is stationary, i.e. using the terminology in section 7.2, it is an $I(0)$ stochastic process, and C_t is non-stationary, the private consumption time series is an $I(1)$ stochastic process. When the terminology of integration is used in examining the non-stationarity of a stochastic process, the Dickey–Fuller tests are also known as 'tests of integration' of a stochastic process.

Table 7.5 MacKinnon critical values for equations (7.48) and (7.49) and the DF t_δ ratios

Significance level	Equation (7.48) Constant, No trend (τ_μ) $[t_\delta = -4.862]$	Equation (7.49) Constant and trend (τ_τ) $[t_\delta = -5.073]$
0.01	−3.6353	−4.2505
0.05	−2.9499	−3.5468
0.10	−2.6133	−3.2056

4.2 The augmented Dickey–Fuller (ADF) test

We could generalise the DF test, if instead of equation (7.20), or (7.36), which is a first-order autoregressive process, we consider the general case of equation (7.21), i.e. equation

$$X_t = \varphi_1 X_{t-1} + \varphi_2 X_{t-2} + \varphi_3 X_{t-3} + \ldots + \varphi_q X_{t-q} + \epsilon_t \quad AP(8) \tag{7.50}$$

which is a qth-order autoregressive process. As a result of equation (7.50) we could consider even cases where the disturbances in equation (7.36) were not white noise, but instead they were serially correlated. In fact, if the disturbances in equation (7.36) were correlated then the DF test presented above is not valid.

Similarly to equation (7.39), in the general case of equation (7.50) we get that

$$\Delta X_t = \delta X_{t-1} + \delta_1 \Delta X_{t-1} + \delta_2 \Delta X_{t-2} + \ldots + \delta_{q-1} \Delta X_{t-q+1} + \epsilon_t \tag{7.51}$$

where $\delta = \varphi_1 + \varphi_2 + \varphi_3 + \ldots + \varphi_q - 1$ and the δ_js are general functions of the φs. The corresponding to equations (7.39), (7.44) and (745) are the following equations respectively:

$$\Delta X_t = \delta X_{t-1} + \sum_{j=2}^{q} \delta_j \Delta X_{t-j+1} + \epsilon_t \tag{7.52}$$

$$\Delta X_t = \alpha + \delta X_{t-1} + \sum_{j=2}^{q} \delta_j \Delta X_{t-j+1} + \epsilon_t \tag{7.53}$$

$$\Delta X_t = \alpha + \beta t + \delta X_{t-1} + \sum_{j=2}^{q} \delta_j \Delta X_{t-j+1} + \epsilon_t \tag{7.54}$$

Because the Dickey–Fuller equations (7.39), (7.44) and (7.45) have been 'augmented' with the lagged differenced terms to produce equations (7.52), (7.53) and (7.54) respectively, the usual DF test applied to the latter equations took the name of the 'augmented Dickey–Fuller (ADF) test'. In fact, both the critical values for the Dickey–Fuller τ-statistics in Table 7.3 still hold for the ADF test and the testing of hypotheses is still that in (7.43). In other words, if t_δ in the estimated with OLS equations (7.52), (7.53) or (7.54) is negative enough, then the corresponding times series will be stationary. Otherwise it will be non-stationary.

We said in the beginning of this section that one of the reasons for augmenting the initial Dickey–Fuller equations with extra lagged differenced terms was to eliminate possible autocorrelation in the disturbances. In order to see how many extra terms we have to include in the equations the usual Akaikes's information criterion (AIC) and Schwartz criterion (SC) could be employed. Furthermore, in order to test if the disturbances were not autocorrelated the usual Breusch–Godfrey, or Lagrange multiplier (LM), test could be used.

Example 7.2 Testing the stationarity of the consumer price index time series of an EU member (ADF)

In Table 7.1 the consumer price index (P_t) time series for an EU member is presented. Table 7.6 presents the AIC, SC, and LM statistics for the estimated for P_t augmented Dickey–Fuller equations (7.53) and (7.54).

From the results in Table 7.6 it is seen that the estimated equations with no extended differenced terms of P_t ($q = 0$) show that there exists autocorrelation in the residuals. There is no autocorrelation in the residuals when the equations are extended with one ($q = 1$) or two ($q = 2$) differenced terms of P_t. Furthermore, comparing the AIC and SC statistics between the various estimated versions of the equations it is seen that for equation (7.53) the minimum statistics are for $q = 1$, and for equation (7.54) the minimum statistics are also for $q = 1$. These 'best' estimated equations are the following.

$$\Delta \hat{P}_t = 0.07860 - 0.01692 P_{t-1} + 1.13822 \Delta P_{t-1}$$

$$t \quad [1.398] \quad [-0.809] \quad [7.099] \tag{7.55}$$

$$R^2 = 0.957 \; DW = 1.817$$

Table 7.6 AIC, SC and LM statistics for the P_t equations (7.53) and (7.54)

	$q=0$			$q=1$			$q=2$		
	AIC	SC	LM(1)	AIC	SC	LM(1)	AIC	SC	LM(1)
eq. (7.53)	−1.766	−1.677	21.329 (0.000)	−2.647	−2.513	0.527 (0.468)	−2.571	−2.389	0.133 (0.715)
eq. (7.54)	−2.101	−1.967	16.607 (0.000)	−2.717	−2.537	1.185 (0.276)	−2.670	−2.444	0.217 (0.641)

Notes: Estimates with Eviews, Probabilities in parentheses.

$$\Delta \hat{P}_t = -0.13369 + 0.01831t - 0.01171P_{t-1} + 0.96093\Delta P_{t-1}$$

t   [−1.134] [2.023]   [−0.582]   [5.453]   (7.56)

$R^2 = 0.962 \; DW = 1.746$

Using these 'best' estimated equations, in Table 7.7 the MacKinnon critical values for the rejection of the hypothesis of a unit root, evaluated from Eviews for equations (7.55) and (7.56), are reported. These critical values are nothing else but the same critical values as those in Table 7.5, showing thus that the MacKinnon critical values are not affected by the number of the extra differenced terms in the equations.

Considering the hypotheses testing in (7.43) we see that for equation (7.55) the $t_\delta = -0.809$ value is greater than all the τ_μ critical values in Table 7.7 and thus, the null hypothesis is not rejected. Therefore, the consumer price index time series exhibits a unit root, or in other words is a non-stationary time series. Similarly, for equation (7.56) the $t_\delta = -0.582$ value is greater than all the τ_τ critical values in Table 7.7 and thus, the null hypothesis is again not rejected. Therefore, the consumer price index (P_t) time series exhibits a unit root, or in other words is a non-stationary time series.

By repeating the same exercise as we did with P_t we observe that the differenced once time series, i.e. ΔP_t, is also a non-stationary time series. If we do the same for the differenced twice time series, i.e. $\Delta^2 P_t$, we find that it is a stationary time series. In other words, the consumer price index (P_t) is an I(2) stochastic process.

Table 7.7 MacKinnon critical values for equations (7.55) and (7.56) and the DF t_δ ratios

Significance level	Equation (7.55) Constant, No trend (τ_μ) [$t_\delta = -0.809$]	Equation (7.56) Constant and trend (τ_τ) [$t_\delta = -0.582$]
0.01	−3.6353	−4.2505
0.05	−2.9499	−3.5468
0.10	−2.6133	−3.2056

4.3 Testing joint hypotheses with the Dickey–Fuller tests

In the DF, or ADF, tests we have seen till now that the null hypothesis was with respect to parameter δ. Nothing was said for the other two deterministic parameters of the Dickey–Fuller regression equations, i.e. of the parameters α, referring to the constant, or drift, and β, referring to the linear deterministic trend, or time trend. Dickey and Fuller (1981) provided tests for testing jointly the parameters α, β and δ. The F-test, given by

$$F = \frac{(SSR_r - SSR_u)/r}{SSR_u/(n-k)} \tag{7.57}$$

where SSR_r = sum of squared residuals from the restricted equation

SSR_u = sum of squared residuals from the unrestricted equation

n = number of usable observations

k = number of the coefficients in the unrestricted equation

r = number of restrictions,

could be used in order to test joint hypotheses, following the usual Wald methodology for testing restrictions. However, because the F distribution is not standard for these tests Dickey and Fuller provided three additional F-statistics, called Φ_1, Φ_2 and Φ_3, according to the joint hypotheses to be tested. These three statistics are reported in Table 7.8.

The joint hypotheses are the following:

(1) When Dickey–Fuller regression equations of the form (7.54) are used:

$$H_0: \alpha = \beta = \delta = 0, \text{ if } F < \Phi_2$$
$$H_a: \text{not all } \alpha, \beta, \delta = 0, \text{ if } F > \Phi_2 \qquad (7.58)$$

(2) When Dickey–Fuller regression equations of the form (7.54) are used:

$$H_0: \beta = \delta = 0, \text{ if } F < \Phi_3$$
$$H_a: \text{not both } \beta \text{ and } \delta = 0, \text{ if } F > \Phi_3 \qquad (7.59)$$

Table 7.8 Critical values for the Dickey–Fuller Φ-statistics

Sample size	Probability of a smaller value							
	0.01	0.025	0.05	0.10	0.90	0.95	0.975	0.99
Statistic Φ_1								
25	0.29	0.38	0.49	0.65	4.12	5.18	6.30	7.88
50	0.29	0.39	0.50	0.66	3.94	4.86	5.80	7.06
100	0.29	0.39	0.50	0.67	3.86	4.71	5.57	6.70
250	0.30	0.39	0.51	0.67	3.81	4.63	5.45	6.52
500	0.30	0.39	0.51	0.67	3.79	4.61	5.41	6.47
∞	0.30	0.40	0.51	0.67	3.78	4.59	5.38	6.43
Statistic Φ_2								
25	0.61	0.75	0.89	1.10	4.67	5.68	6.75	8.21
50	0.62	0.77	0.91	1.12	4.31	5.13	5.94	7.02
100	0.63	0.77	0.92	1.12	4.16	4.88	5.59	6.50
250	0.63	0.77	0.92	1.13	4.07	4.75	5.40	6.22
500	0.63	0.77	0.92	1.13	4.05	4.71	5.35	6.15
∞	0.63	0.77	0.92	1.13	4.03	4.68	5.31	6.09
Statistic Φ_3								
25	0.74	0.90	1.08	1.33	5.91	7.24	8.65	10.61
50	0.76	0.93	1.11	1.37	5.61	6.73	7.81	9.31
100	0.76	0.94	1.12	1.38	5.47	6.49	7.44	8.73
250	0.76	0.94	1.13	1.39	5.39	6.34	7.25	8.43
500	0.76	0.94	1.13	1.39	5.36	6.30	7.20	8.34
∞	0.77	0.94	1.13	1.39	5.34	6.25	7.16	8.27

Source: Dickey, D. A. and Fuller, W. A. (1981) 'Likelihood ratio statistics for autoregressive time series with a unit root', *Econometrica*, 49, 4, 1057–1072.

(3) When Dickey-Fuller regression equations of the form (7.53) are used:

$$H_0: \alpha = \delta = 0, \text{ if } F < \Phi_1$$
$$H_a: \text{not both } \alpha \text{ and } \delta = 0, \text{ if } F > \Phi_1 \qquad (7.60)$$

Example 7.3 Testing joint hypotheses for the non-stationarity of the consumer price index time series of an EU member (ADF)

We found in Example 7.2 that the consumer price index (P_t) time series for an EU member is non-stationary. In this example we will illustrate the Dickey-Fuller joint hypotheses tests presented just above.

Therefore, using the estimated equation (7.56) we can test the joint hypotheses in (7.58) and in (7.59). In other words, for (7.58) we test that

$$H_0: \text{data are generated by } \Delta P_t = \delta_1 \Delta P_{t-1} + \epsilon_t \text{ if } F < \Phi_2$$
$$H_a: \text{data are not generated by } \Delta P_t = \delta_1 \Delta P_{t-1} + \epsilon_t \text{ if } F > \Phi_2 \qquad (7.61)$$

and for (7.59) we test that

$$H_0: \text{data are generated by } \Delta P_t = \alpha + \delta_1 \Delta P_{t-1} + \epsilon_t \text{ if } F < \Phi_3$$
$$H_a: \text{data are not generated by } \Delta P_t = \alpha + \delta_1 \Delta P_{t-1} + \epsilon_t \text{ if } F > \Phi_3 \qquad (7.62)$$

From the estimated equation (7.56), using the usual Wald statistic for redundant variables, for $n = 34$ usable observations, $k = 4$ estimated coefficients and $r = 3$ restrictions, we get that $F_2(\alpha = 0, \beta = 0, \delta = 0) = 2.277$. The value of this statistic $F_2 = 2.277$ is less than the critical value $\Phi_2(n = 25) = 5.68$ and $\Phi_2(n = 50) = 5.13$, for 0.05 significance level, reported in Table 7.8. Therefore, we accept the null hypothesis, i.e. we accept that P_t is random walk, since $\alpha = 0$ means the absence of a stochastic trend, $\beta = 0$ means the absence of deterministic trend also and $\delta = 0$ means non-stationarity.

From the estimated equation (7.56), using the usual Wald statistic for redundant variables, for $n = 34$ usable observations, $k = 4$ estimated coefficients and $r = 2$ restrictions, we get that $F_3(\beta = 0, \delta = 0) = 2.406$. The value of this statistic $F_3 = 2.406$ is less than the critical value $\Phi_3(n = 25) = 7.24$ and $\Phi_3(n = 50) = 6.73$, for 0.05 significance level, reported in Table 7.8. Therefore, we accept the null hypothesis, i.e. we accept that P_t is subject to a stochastic trend only, since $\beta = 0$ means the absence of deterministic trend and $\delta = 0$ means non-stationarity.

Using the estimated equation (7.55) we can test the joint hypotheses in (7.60). In other words, for (7.60) we test that

$$H_0: \text{data are generated by } \Delta P_t = \delta_1 \Delta P_{t-1} + \epsilon_t \text{ if } F < \Phi_1$$
$$H_a: \text{data are not generated by } \Delta P_t = \delta_1 \Delta P_{t-1} + \epsilon_t \text{ if } F > \Phi_1 \qquad (7.63)$$

From the estimated equation (7.55), using the usual Wald statistic for redundant variables, for n = 34 usable observations, k = 3 estimated coefficients and r = 2 restrictions, we get that $F_1(\alpha = 0, \delta = 0) = 1.246$. The value of this statistic $F_1 = 1.246$ is less than the critical value $\Phi_1(n = 25) = 5.18$ and $\Phi_1(n = 50) = 4.86$, for 0.05 significance level, reported in Table 7.8. Therefore, we accept the null hypothesis, i.e. we accept that P_t is random walk, since $\alpha = 0$ means the absence of stochastic trend and $\delta = 0$ means non-stationarity.

4.4 Testing conditional hypotheses with the Dickey–Fuller tests

Dickey and Fuller (1981) provided us with three symmetric critical τ_{ij} values, called $\tau_{\alpha\mu}$, $\tau_{\alpha\tau}$, and $\tau_{\beta\mu}$, reported in Table 7.9, for testing the drift parameter α and the linear time trend parameter β, conditionally upon $\delta = 0$.

These conditional hypotheses are the following:
(1) When Dickey–Fuller regression equations of the form (7.54) are used:

$$\left.\begin{array}{l} H_0: \alpha = 0 \text{ given that } \delta = 0, \text{ if } |t| < |\tau_{\alpha\tau}| \\ H_a: \alpha \neq 0 \text{ given that } \delta = 0, \text{ if } |t| > |\tau_{\alpha\tau}| \end{array}\right\} \tag{7.64}$$

Table 7.9 Critical values for the Dickey–Fuller τ_{ij}-statistics (symmetric distributions)

Sample size	*Probability of a smaller value*			
	0.90	*0.95*	*0.975*	*0.99*
Statistic – $\tau_{\alpha\mu}$				
25	2.20	2.61	2.97	3.41
50	2.18	2.56	2.89	3.28
100	2.17	2.54	2.86	3.22
250	2.16	2.53	2.84	3.19
500	2.16	2.52	2.83	3.18
∞	2.16	2.52	2.83	3.18
Statistic – $\tau_{\alpha\tau}$				
25	2.77	3.20	3.59	4.05
50	2.75	3.14	3.47	3.87
100	2.73	3.11	3.42	3.78
250	2.73	3.09	3.39	3.74
500	2.72	3.08	3.38	3.72
∞	2.72	3.08	3.38	3.71
Statistic – $\tau_{\beta\tau}$				
25	2.39	2.85	3.25	3.74
50	2.38	2.81	3.18	3.60
100	2.38	2.79	3.14	3.53
250	2.38	2.79	3.12	3.49
500	2.38	2.78	3.11	3.48
∞	2.38	2.78	3.11	3.46

Source: Dickey, D. A. and Fuller, W. A. (1981) 'Likelihood ratio statistics for autoregressive time series with a unit root', *Econometrica*, 49, 4, 1057–1072.

(2) When Dickey–Fuller regression equations of the form (7.54) are used:

$$H_0: \beta = 0 \text{ given that } \delta = 0, \text{ if } |t| < |\tau_{\beta\tau}|$$
$$H_a: \beta \neq 0 \text{ given that } \delta = 0, \text{ if } |t| > |\tau_{\beta\tau}|$$

(7.65)

(3) When Dickey–Fuller regression equations of the form (7.53) are used:

$$H_0: \alpha = 0 \text{ given that } \delta = 0, \text{ if } |t| < |\tau_{\alpha\mu}|$$
$$H_a: \alpha \neq 0 \text{ given that } \delta = 0, \text{ if } |t| > |\tau_{\alpha\mu}|$$

(7.66)

Example 7.4 Testing conditional hypotheses for the non-stationarity of the consumer price index time series of an EU member (ADF)

We found in Examples 7.2 and 7.3 that the consumer price index (P_t) time series for an EU member is non-stationary. In this example we will illustrate the Dickey–Fuller conditional hypotheses tests presented just above.

From the estimated equation (7.56) with $n = 34$ usable observations we get that $t_\alpha = -1.134$ and $t_\beta = 2.023$. For testing hypotheses (7.64) and (7.65) we have to compare these t values with the corresponding τ_{ij} critical values in Table 7.9. From Table 7.9 we have that $\tau_{\alpha\tau}(n = 25) = 3.20$ and $\tau_{\alpha\tau}(n = 50) = 3.14$, and $\tau_{\beta\tau}(25) = 2.85$ and $\tau_{\beta\tau}(50) = 2.81$. Therefore, since $|t_\alpha| < |\tau_{\alpha\tau}|$ and $|t_\beta| < |\tau_{\beta\tau}|$ we accept the null hypotheses in (7.64) and (7.65) respectively, i.e. we accept that given that P_t is non-stationary, there is neither a stochastic trend nor a deterministic trend generating the series.

From the estimated equation (7.55) with $n = 34$ usable observations we get that $t_\alpha = 1.398$. For testing hypotheses (7.66) we have to compare this t value with the corresponding τ_{ij} critical value in Table 7.9. From Table 7.9 we have that $\tau_{\alpha\mu}(n = 25) = 2.61$ and $\tau_{\alpha\mu}(n = 50) = 2.56$. Therefore, since $|t_\alpha| < |\tau_{\alpha\mu}|$ we accept the null hypotheses in (7.66), i.e. we accept that given that P_t is non-stationary, there is no stochastic trend generating the time series.

Summarising, we see that the results in Example 7.4 do not contradict the results obtained in Example 7.3. In other words, we could say that in the process of generating the non-stationary time series of the price consumer index of an EU member both stochastic trends and deterministic trends are absent.

4.5 A sequential procedure in the Dickey–Fuller tests when the data-generating process is unknown

The tests for unit roots of a time-series are not as simple as the tests presented in the previous sections suggest. Unfortunately, the tests for the presence of unit roots in a time series are conditional on the presence of deterministic drifts and trends and the tests for the presence of deterministic drifts and trends in a time series are conditional on the presence of unit roots (Campbell and Perron, 1991; Enders, 1995). Doldado *et al.* (1990), Holden and Perman (1994) and Enders (1995) presented sequential procedures for testing the unit roots in a time series if the data generating process is not known. These procedures are based on the following points:

1   The 'top to bottom' philosophy, meaning that the top (starting) point of the procedure should be the most general case and then, step by step, to move towards the lowest (finishing) point which should be the most specific case.
2   If it is known that the time series under investigation contains a drift or trend, then the null hypothesis of a unit root can be tested using the standard normal distribution.
3   Because the unit roots tests have low power in rejecting the null hypothesis of a unit root, if at any step of the sequential procedure of testing the null hypothesis is rejected, then the whole procedure ends, concluding that the time series under investigation is stationary.

A modified mixture of these sequential procedures could be the following:

Step 1

Estimate the following equation with OLS

$$\Delta X_t = \alpha + \beta t + \delta X_{t-1} + \sum_{j=2}^{q} \delta_j \Delta X_{t-j+1} + \epsilon_t \tag{7.67}$$

- Use statistics AIC and SC to find the proper number of differenced terms to be included in the equation.
- Use statistic LM to test for autocorrelation in the residuals.

Step 2

Use statistic τ_τ to test the null hypothesis $\delta = 0$ in equation (7.67).

- If the null hypothesis is rejected then time series X_t does not contain a unit root. You may stop the whole process concluding that the time series is stationary.
- If the null hypothesis is not rejected you must continue in order to test the drift and trend terms.

Step 3

Use statistic $\tau_{\beta\tau}$ to test the conditional null hypothesis $\beta = 0$ given $\delta = 0$, i.e. to test the significance of the trend term given that the time series contains a unit root. You may verify this test by using statistic Φ_3 to test the joint null hypothesis $\beta = \delta = 0$.

- If the null hypothesis is not rejected, i.e. if β is not significant, you may continue.
- If the null hypothesis is rejected, i.e. if β is significant, you have to perform the following test:
  Use the standard normal distribution to retest the null hypothesis $\delta = 0$.

- If the null hypothesis is rejected, i.e. if time series X_t does not contain a unit root, you may stop the process concluding that the time series is stationary.
- If the null hypothesis is not rejected, i.e. if time series X_t contains a unit root you conclude that $\beta \neq 0$ and $\delta = 0$.

Step 4

Use statistic $\tau_{a\tau}$ to test the conditional null hypothesis $\alpha = 0$ given $\delta = 0$, i.e. to test the significance of the drift term given that the time series contains a unit root. You may verify this test by using statistic Φ_2 to test the joint null hypothesis $\alpha = \beta = \delta = 0$.

- If the null hypothesis is not rejected, i.e. if α is not significant, you may continue.
- If the null hypothesis is rejected, i.e. if α is significant, you have to perform the following test:
  Use the standard normal distribution to retest the null hypothesis $\delta = 0$.
  - If the null hypothesis is rejected, i.e. if time series X_t does not contain a unit root, you may stop the process concluding that the time series is stationary.
  - If the null hypothesis is not rejected, i.e. if time series X_t contains a unit root you conclude that $\alpha \neq 0$ and $\delta = 0$.

Step 5

Estimate the following equation with OLS

$$\Delta X_t = \alpha + \delta X_{t-1} + \sum_{j=2}^{q} \delta_j \Delta X_{t-j+1} + \epsilon_t \tag{7.68}$$

Step 6

Use statistic τ_μ to test the null hypothesis $\delta = 0$ in equation (7.68).

- If the null hypothesis is rejected then time series X_t does not contain a unit root. You may stop the whole process concluding that the time series is stationary.
- If the null hypothesis is not rejected you must continue in order to test the drift term.

Step 7

Use statistic $\tau_{\alpha\mu}$ to test the conditional null hypothesis $\alpha = 0$ given $\delta = 0$, i.e. to test the significance of the drift term given that the time series contains a unit root. You may verify this test by using statistic Φ_1 to test the joint null hypothesis $\alpha = \delta = 0$.

- If the null hypothesis is not rejected, i.e. if α is not significant, you may continue.
- If the null hypothesis is rejected, i.e. if α is significant, you have to perform the following test:
  Use the standard normal distribution to retest the null hypothesis $\delta = 0$.
  - If the null hypothesis is rejected, i.e. if time series X_t does not contain a unit root, you may stop the process concluding that the time series is stationary.

- If the null hypothesis is not rejected, i.e. if time series X_t contains a unit root you conclude that $\alpha \neq 0$ and $\delta = 0$.

Step 8

Estimate the following equation with OLS

$$\Delta X_t = \delta X_{t-1} + \sum_{j=2}^{q} \delta_j \Delta X_{t-j+1} + \epsilon_t \qquad (7.69)$$

Step 9

Use statistic τ to test the null hypothesis $\delta = 0$ in equation (7.69).

- If the null hypothesis is rejected you conclude that time series X_t does not contain a unit root, i.e. it is stationary.
- If the null hypothesis is not rejected you conclude that time series X_t contains a unit root, i.e. it is non-stationary.

Example 7.5 Using a sequential procedure for testing the non-stationarity of the consumer price index time series of an EU member (ADF)

This example is in fact a summary of the tests performed in Examples 7.2, 7.3 and 7.4. Table 7.10 summarises relevant results from these examples.

According to the steps presented above a brief explanation of these results is as follows:

Step 1

The results for equation (7.67) are shown. There is no autocorrelation in the residuals.

Table 7.10 Results of a sequential procedure for testing unit roots of the consumer price index time series (P) of an EU member

Equation	Intercept	Trend	P_{t-1}	ΔP_{t-1}	LM(1)	$F_3\,(\Phi_3)$	$F_2\,(\Phi_2)$	$F_1\,(\Phi_1)$
(7.67)	−0.1337	0.0183	−0.0117	0.9609	1.185	2.406	2.277	
	[−1.134]	[2.023]	[−0.582]	[5.453]	(0.276)			
(7.68)	0.0786		−0.0169	1.1382	0.527			1.246
	[1.398]		[−0.809]	[7.099]	(0.468)			
(7.69)			−0.0153	1.1597	0.919			
			[−0.723]	[7.161]	(0.345)			

Notes: t-ratios in brackets; probabilities in parentheses.

Step 2

Because $t_\delta = -0.582 > \tau_\tau = -3.5468$ the time series is containing a unit root.

Step 3

Because $|t_\beta| = 2.023 < |\tau_{\beta\tau}| = 2.81$ it may be $\beta = 0$. This result is also verified by $F_3 = 2.406 < \Phi_3 = 6.73$.

Step 4

Because $|t_\alpha| = 1.134 < |\tau_{\alpha\tau}| = 3.14$ it may be $\alpha = 0$. This result is also verified by $F_2 = 2.277 < \Phi_2 = 5.13$.

Step 5

The results for equation (7.68) are shown. There is no autocorrelation in the residuals.

Step 6

Because $t_\delta = -0.809 > \tau_\mu = -2.9499$ the time series is containing a unit root.

Step 7

Because $|t_\alpha| = 1.398 < |\tau_{\alpha\mu}| = 2.56$ it may be $\alpha = 0$. This result is also verified by $F_1 = 1.246 < \Phi_1 = 4.86$.

Step 8

The results for equation (7.69) are shown. There is no autocorrelation in the residuals.

Step 9

Because $t_\delta = -0.723 > \tau = -1.95$ the time series is containing a unit root.

FINAL CONCLUSION

The consumer price index (P_t) time series for an EU member is containing a unit root without drift and without trend.

4.6 Modifications, extensions, and problems of the unit root tests

Since the original Dickey–Fuller, and the augmented Dickey–Fuller test was developed, various modifications and/or extensions of this test were proposed. We have seen already some of these tests in the previous sections. However, we will refer very briefly to some of these and other tests.

(a) The multiple roots tests

Dickey and Pantula (1987) in order to test if a time series has more than one unit roots, proposed the successive testing of the time series differenced accordingly. In other words, if a time series has possibly one unit root, a simple version of the estimated equation we saw is the following:

$$\Delta X_t = \alpha + \delta X_{t-1} + \epsilon_t \tag{7.70}$$

If the time series has possibly two roots, the estimated equation is the following:

$$\Delta^2 X_t = \alpha + \delta \Delta X_{t-1} + \epsilon_t \tag{7.71}$$

If the time series has possibly three roots, the estimated equation is the following:

$$\Delta^3 X_t = \alpha + \delta \Delta^2 X_{t-1} + \epsilon_t \tag{7.72}$$

and so on. For each equation the usual DF, or ADF, procedure should be applied (see Examples 7.1 and 7.2).

(b) The seasonal unit roots tests

When a time series X_t is measured s times per year, Dickey *et al.* (1984) proposed the following regression equation to be used:

$$\Delta_s Z_t = \delta Z_{t-s} + \sum_{j=1}^{q} \delta_j \Delta_s X_{t-j} + \epsilon_t \tag{7.73}$$

where $\Delta_s Z_t = Z_t - Z_{t-s}$, and

$$Z_t = X_t - \sum_{j=1}^{h} \hat{\lambda}_j X_{t-j} \tag{7.74}$$

with the $\hat{\lambda}_j$s being the estimates of the λ_js obtained from the following regression:

$$\Delta_s X_t = \sum_{j=1}^{h} \lambda_j \Delta_s X_{t-j} + \eta_t \tag{7.75}$$

The test on unit roots could be based on the student's t statistic of the δ coefficient in the regression (7.73). Osborn *et al.* (1988) instead of using $\Delta_s Z_t$ as the dependent variable in the regression equation (7.73) they proposed variable $\Delta_s X_t$. Furthermore, Hylleberg *et al.* (1990) proposed a more general test to deal with cyclical movements at different frequencies, and therefore to test the corresponding unit roots (for more see Charemza and Deadman, 1997).

In the simple case where the seasonal pattern of a time series X_t measured s times per time period is purely deterministic, the following regression equation could be used:

$$\Delta\hat{\eta}_t = \delta\hat{\eta}_{t-1} + \sum_{j=1}^{q} \delta_j\Delta\hat{\eta}_{t-j} + \epsilon_t \tag{7.76}$$

where $\hat{\eta}_t$ are the estimated residuals of η_t derived from the following regression equation:

$$X_t = \alpha_s + \sum_{j=1}^{s-1} \alpha_j D_{jt} + \eta_t \tag{7.77}$$

where D_{jt} are $s-1$ dummy variables. In other words, $\hat{\eta}_t$ could be considered as a deseasonalised time series in the place of X_t. For the testing of unit roots the usual DF, or ADF, procedure could be used for coefficient δ of the regression equation (7.76) (Dickey *et al.* (1986), Enders (1995)).

(c) The Phillips-Perron tests

One of the basic assumptions of the DF tests is that the disturbances are independently and identically distributed (IID). Phillips and Perron (1988) proposed statistics, named the Z statistics, where the assumption of IID is relaxed. However, because the modifications to the DF tests proposed by Phillips-Perron are non-parametric we believe that this test, although popular, is beyond the scope of this book.

(d) A quick test for non-stationarity

Sargan and Bhargava (1983) suggested a quick test, although it is not very popular, for testing the possible non-stationarity of a time series X_t. Their test uses the same formula as the known formula for computing the Durbin–Watson statistic, which is

$$IDW = \frac{\Sigma(X_t - X_{t-1})^2}{\Sigma(X_t - \bar{X})^2} \tag{7.78}$$

where $\bar{X}$ = arithmetic mean of X_t. Due to the formula (7.78) this statistic is called the 'integration Durbin–Watson (IDW) statistic'. However, if the time series X_t is regressed on a constant, the estimate of the constant is the arithmetic mean of the time series X_t, and the one corresponding to this regression DW statistic is the IDW statistic computed in (7.78). In this regression, if the time series X_t is non-stationary, so will the corresponding residuals because

$$X_t = a + e_t \tag{7.79}$$

where a = estimated intercept and e_t = residuals. If the value of IDW is low, say lower than 0.5, time series X_t is suspected to be non-stationary. If the value of IDW is close to 2, time series X_t is stationary.

As an example, the value of IDW for the private consumption (C_t) time series of an EU member is equal to 0.010, indicating that this time series is non-stationary, as we found in Example 7.1. Furthermore, for the same time series differenced once (ΔC_t) it is IDW = 1.692, indicating that time series ΔC_t is stationary, as we also found in Example 7.1.

(e) *Problems of unit roots tests and recommendations*

The basic problem that is encountered in the use of the Dickey–Fuller tests is the *lack of power* of the tests. The power of a test is its ability to detect a false null hypothesis and it is measured by the probability of rejecting the null hypothesis when it is false. It has been proved, using Monte Carlo simulations, that the power of the unit root tests is very low. In other words, although a time series may be stationary, the unit root tests may fail to detect this, and suggest that the time series is non stationary.

Because most macroeconomic time series X_t are trended upwards the unit roots tests often indicate that these series are non-stationary. A usual procedure to transform the trended upwards time series to roughly remaining constant over time is to consider their percentage growth, i.e. to consider the time series $(X_t - X_{t-1})/X_{t-1}$. When the values of a time series are positive, by getting the natural logarithms of the series, i.e. $x_t = \ln(X_t)$, we can use the 'lower case' series x_t in the unit roots tests. In trying to detect if the time series x_t has a unit root we use the difference $\Delta x_t = x_t - x_{t-1}$. However, because it is approximately true that

$$\Delta x_t = \ln X_t - \ln X_{t-1} = \ln\left(\frac{X_t}{X_{t-1}}\right) \approx \frac{X_t - X_{t-1}}{X_{t-1}} \tag{7.80}$$

in the unit roots tests the time series are often taken in logarithms instead of natural levels. To see this, assume the upwards trending time series of the consumer price index (P_t), for example, which is given by

$$P_t = P_0 e^{gt + \eta_t} \tag{7.81}$$

where g = constant rate of growth and η_t = error term. If we take the logarithms of both sides of (7.81) we obtain that

$$\ln P_t = \ln P_0 + gt + \eta_t \tag{7.82}$$

which can be written as

$$\ln P_t - \ln P_{t-1} = gt - g(t-1) + \eta_t - \eta_{t-1}$$

or

$$p_t - p_{t-1} = g + \epsilon_t \tag{7.83}$$

where $p_t = \ln P_t$ and $\epsilon_t = \eta_t - \eta_{t-1}$. In the case that ϵ_t is white noise equation (7.83) is nothing else but a random walk with drift, as shown earlier in (7.14).

5 COINTEGRATION: THE TWO VARIABLE CASE

Let us consider Friedman's permanent income hypothesis in the simple version that total private consumption (C_t) is the sum of permanent private consumption $(C_{p,t})$ and transitory private consumption (ϵ_t). Assuming that permanent private consumption is proportional to permanent personal disposable income (Y_t) we could write that the private consumption function given by

$$C_t = C_{p,t} + \epsilon_t = \beta_1 Y_t + \epsilon_t \qquad (7.84)$$

where $1 \geq \beta_1$ (=proportionality parameter) > 0. In estimating this function with OLS, using the data for an EU member presented in Table 7.1 and under the assumption that permanent personal disposable income is equal to personal disposable income, we will get

$$\hat{C}_t = 0.80969 Y_t$$

t   [75.5662]

$$R^2 = 0.9924 \; DW = 0.8667 \qquad (7.85)$$

The estimates in (7.85), apart from the low Durbin–Watson statistic, are very good. The t statistic is very high, showing that the regression coefficient is significant, and the R^2 is also very high, indicating a very good fit. However, these estimates may be misleading because the two time series involved in the equation are trended or non-stationary random processes, as can be seen in Figure 7.1. As a consequence, the OLS estimator is not consistent and the corresponding usual inference procedures are not valid. These regressions where the results look very good in terms of R^2 and t statistics but the variables involved are trended time series have been called 'spurious regressions' by Granger and Newbold (1974). In fact Granger and Newbold suggested that if in a regression with trended time series variables the DW statistic is low and the determination coefficient R^2 is high we should suspect that the estimated equation suffers possibly from spurious situations. As a rule of thumb, if $R^2 > DW$ spurious regression should be suspected, as in equation (7.85), for example where $R^2 = 0.9924 > DW = 0.8667$.

Taking into account that most time series in economics are non-stationary (Nelson and Plosser, 1982) then the situation of getting spurious regression results is very often. A usual methodology to avoid the problem of non-stationarity of the time series is to use in regressions the first differences of the time series. However, using relationships where the variables are expressed in differences is like referring to the short-run or disequilibrium state of the phenomenon under investigation and not to its long-run or equilibrium state, where the variables are expressed in their original levels, as most economic theories suggest.

The obvious question that arises from this discussion is that if the problem of spurious regression is always present when we use non-stationary time series or if there are cases where although we are using non-stationary time series the problem is not there. The answer to this question is that if the trending time series involved in a regression are 'trending together', or in other words are on the 'same wavelength', then the regression results may not be spurious and thus the usual inferences depending on the t and F statistics may be valid. This 'synchrony' of the non-stationary time series is the idea behind the concept of 'cointegration' (Gujarati, 1995).

5.1 Cointegration: the concept

In the case of equation (7.84), the permanent income hypothesis, where the transitory private consumption (ϵ_t) is by definition a stationary time series, requires that the linear combination of the two time series C_t and Y_t, i.e. the time series $C_t - \beta_1 Y_t$, must be stationary because this series is equal to the stationary series ϵ_t. However, it is more than certain that both time series C_t and Y_t are non-stationary. In fact these series are integrated of order one, i.e. it is $C_t \sim I(1)$ and $Y_t \sim I(1)$. In other words, although $C_t \sim I(1)$ and $Y_t \sim I(1)$ the permanent income hypothesis requires that their linear combination $\epsilon_t = C_t - \beta_1 Y_t$ to be stationary, i.e. to be $\epsilon_t = C_t - \beta_1 Y_t \sim I(0)$. In such a case we say that the time series C_t and Y_t are cointegrated. But let us formulate this result more generally. Engle and Granger (1987) developed the following definition:

Cointegration of two variables

Two time series Y_t and X_t are said to be cointegrated of order (d, b), where $d \geq b \geq 0$, if both time series are integrated of order d, and there exists a linear combination of these two time series, say $a_1 Y_t + a_2 X_t$, which is integrated of order $(d - b)$. In mathematical terms, this definition is written:

$$\text{If } Y_t \sim I(d) \text{ and } X_t \sim I(d), \text{ then } Y_t, X_t \sim CI(d, b) \text{ if } a_1 Y_t + a_2 X_t, \sim I(d-b) \quad (7.86)$$

where CI is the symbol of cointegration. The vector of the coefficients that constitute the linear combination of the two series, i.e. $[a_1, a_2]$ in (7.86), is called 'cointegrating vector'.

We can distinguish the following two special cases which we will investigate in this chapter:

1 The case where d = b, resulting in $a_1 Y_t + a_2 X_t \sim I(0)$, which means that the linear combination of the two time series is stationary, and therefore, $Y_t, X_t \sim CI(d, d)$.
2 The case where d = b = 1, resulting in $a_1 Y_t + a_2 X_t \sim I(0)$, which means that the linear combination of the two time series is stationary, and therefore, $Y_t, X_t \sim CI(1, 1)$.

Let us consider the following relationship where $Y_t \sim I(1)$ and $X_t \sim I(1)$:

$$Y_t = \beta_0 + \beta_1 X_t \quad (7.87)$$

This relationship is in a *long-run equilibrium* when

$$0 = Y_t - \beta_0 - \beta_1 X_t \quad (7.88)$$

The deviation from the long-run equilibrium, called the 'equilibrium error', ϵ_t, is given by

$$\epsilon_t = Y_t - \beta_0 - \beta_1 X_t \qquad (7.89)$$

For the long-run equilibrium to have meaning, i.e. to exist, the equilibrium error in (7.89) should fluctuate around the equilibrating zero value, as shown in (7.88). In other words, the equilibrium error ϵ_t should be a stationary time series, i.e. it should be $\epsilon_t \sim I(0)$ with $E(\epsilon_t) = 0$. According to the definition in (7.86), because $Y_t \sim I(1)$ and $X_t \sim I(1)$, and the linear combination $\epsilon_t = Y_t - \beta_0 - \beta_1 X_t \sim I(0)$, we can say that Y_t and X_t are cointegrated of order $(1, 1)$, i.e. it is $Y_t, X_t \sim CI(1, 1)$. The cointegrating vector is $[1, -\beta_0, -\beta_1]$. It can be proved that in the two variable case and under the assumption that the coefficient of one of the variables is normalised to equal unity the cointegrating vector, i.e. the linear combination of the two time series, *is unique*.

Combining the results above we could say that the cointegration between two time series is another way to express the existence of a long-run equilibrium relationship between these two time series. Therefore, by considering that Y_t and X_t are cointegrated and that the equilibrium error ϵ_t is stationary with zero mean, we can write that

$$Y_t = \beta_0 + \beta_1 X_t + \epsilon_t \qquad (7.90)$$

and be sure that equation (7.90) will not produce spurious results. Stock (1987) proved that for large samples the OLS estimator for equation (7.90) is 'super-consistent', i.e. it is consistent and very efficient, because it converges faster to the true values of the regression coefficients than the OLS estimator involving stationary variables. However, Baneriee *et al.* (1986) showed that for small samples the OLS estimator is biased and the level of biasedness depends on the value of R^2; the higher the R^2 the lower the level of biasedness. Finally, according to Granger (1986) if we want to avoid spurious regression situations we should test before any regression if the variables involved are cointegrated, something that we present in the next section.

5.2 Cointegration: the tests

In this section we will present two simple methods for testing for cointegration between two variables.

(a) The Engle–Granger approach

Suppose that we want to test if the two variables Y_t and X_t are cointegrated. This approach, which is called the Engle–Granger (EG) test, or the augmented Engle–Granger (AEG) test, suggests the following steps:

Step 1

Find the order of integration of both variables using the unit roots methodology present in section 4. There are three cases: (1) If the order of integration of the two variables is the same, something that the concept of cointegration requires, continue to the next step. (2) If the order of integration of the two variables is different, you may conclude that the two variables are not cointegrated. (3) If the two variables are stationary the whole testing process stops because you can use the standard regression techniques for stationary variables.

Step 2

If the two variables are integrated of the same order, say I(1), estimate with OLS the long-run equilibrium equation

$$Y_t = \beta_0 + \beta_1 X_t + \epsilon_t \tag{7.91}$$

which in this case it is called the 'cointegrating regression' and save the residuals, e_t, as an estimate of the equilibrium error, ϵ_t. Although the estimated cointegrating vector $[1, -b_0, -b_1]$ is a consistent estimate of the true cointegrating vector $[1, -\beta_0, -\beta_1]$, this is not true for the estimated standard errors of these coefficients. For this reason the estimated standard errors are often not quoted with the cointegrating regression.

Step 3

For the two variables to be cointegrated the equilibrium errors must be stationary. To test this stationarity apply the unit roots methodology presented in the section 4 to the estimated equilibrium errors saved in the previous step. You could use, for example, the Dickey–Fuller test, or the augmented Dickey–Fuller test, to time series e_t, which involves the estimation of a version of the following equation with OLS

$$\Delta e_t = \delta e_{t-1} + \sum_{j=2}^{q} \delta_j \Delta e_{t-j+1} + v_t \tag{7.92}$$

Two things we should take into account in applying the DF, or ADF, tests:

1   Equation (7.92) does not include a constant term, because by construction the OLS residuals e_t are centred around zero.
2   Because the estimate of δ in (7.92) is downward biased, due to the fact that by construction the OLS methodology seeks to produce stationary residuals e_t, the usual Dickey–Fuller τ statistics are not appropriate for this test. Engle and Granger (1987), Engle and Yoo (1987), MacKinnon (1991), and Davinson and MacKinnon (1993) presented critical values for this test, which are even more negative than the usual Dickey–Fuller τ statistics. In Table 7.11 critical values for these cointegration tests are presented.

Step 4

Obtain conclusions about the cointegration of the two variables according to the following hypotheses:

$$\left. \begin{array}{l} H_0: \delta = 0, \text{ for non-stationarity of } e_t, \text{ i.e. for non-cointegration, if } t_\delta > \tau \\[6pt] H_a: \delta < 0, \text{ for stationarity of } e_t, \text{ i.e. for cointegration, if } t_\delta < \tau \end{array} \right\} \tag{7.93}$$

Table 7.11 Critical values for the EG, or AEG, cointegration tests

No. of variables	m = 2			m = 3			m = 4		
	Significance levels								
Sample size	*0.01*	*0.05*	*0.10*	*0.01*	*0.05*	*0.10*	*0.01*	*0.05*	*0.10*
25	−4.37	−3.59	−3.22	−4.92	−4.10	−3.71	−5.43	−4.56	−4.15
50	−4.12	−3.46	−3.13	−4.59	−3.92	−3.58	−5.02	−4.32	−3.98
100	−4.01	−3.39	−3.09	−4.44	−3.83	−3.51	−4.83	−4.21	−3.89
∞	−3.90	−3.33	−3.05	−4.30	−3.74	−3.45	−4.65	−4.10	−3.81

Source: Thomas, R. L. (1997), based on MacKinnon (1991).
m = number of variables in the cointegrating regression.

Example 7.6 Cointegration between private consumption and personal disposable income for an EU member (EG, or AEG)

In Table 7.1 the private consumption (C_t) and the personal disposable income (Y_t) for an EU member are presented. Following the steps above we will try to find if these two variables are cointegrated.

Step 1

Finding the order of integration of the two variables: Taking into account criteria AIC, SC and LM(1) we obtained:

For the private consumption variable (see Example 7.1):

$$\Delta \hat{C}_t = 12330.48 - 0.01091 C_{t-1}$$

$$t \qquad [5.138] \qquad [-1.339] \tag{7.94}$$

$$R^2 = 0.052 \; DW = 1.765$$

and

$$\Delta^2 \hat{C}_t = 7972.671 - 0.85112 \Delta C_{t-1}$$

$$t \qquad [4.301] \qquad [-4.862] \tag{7.95}$$

$$R^2 = 0.425 \; DW = 1.967$$

For the personal disposable income variable:

$$\Delta \hat{Y}_t = 19903.93 - 0.02479 Y_{t-1}$$

$$t \qquad [3.054] \qquad [-1.387] \tag{7.96}$$

$$R^2 = 0.055 \; DW = 2.270$$

and

$$\Delta^2 \hat{Y}_t = 12889.39 - 1.11754 \Delta Y_{t-1}$$

t   [3.983]   [−6.270]   (7.97)

$R^2 = 0.551$ DW $= 2.014$

Table 7.12 presents the MacKinnon critical values for the rejection of the hypothesis of a unit root, evaluated from Eviews for equations (7.94) to (7.97).

From the figures in Table 7.12 it is seen that both C_t and Y_t are non-stationary and that both ΔC_t and ΔY_t are stationary. In other words both C_t and Y_t are integrated of order one, i.e. it is $C_t \sim I(1)$ and $Y_t \sim I(1)$. Therefore we can proceed to step 2.

Step 2

The long-run estimated equilibrium relationship, or the cointegrating regression is the following:

$$\hat{C}_t = 11907.23 + 0.779585 Y_t$$

t   [3.123]   [75.566]   (7.98)

$R^2 = 0.994$ DW $= 1.021$

Using the estimated cointegrating vector $[1, -11907.23, -0.779585]$ we have estimated and saved the estimated equilibrium errors e_t.

Step 3

Testing the stationarity of e_t. Taking into account the criteria AIC, SC and LM(1) we obtained:

$$\Delta \hat{e}_t = -0.51739 e_{t-1}$$

t   [−3.150]   (7.99)

$R^2 = 0.224$ DW $= 1.948$

Table 7.12 MacKinnon critical values for equations (7.94) to (7.97) and the DF t_δ ratios

Significance level	Critical values for equations (7.94) and (7.96) [eq. (7.94) $t_\delta = -1.339$] [eq. (7.96) $t_\delta = -1.387$]	Critical values for equations (7.95) and (7.97) [eq. (7.95) $t_\delta = -4.862$] [eq. (7.97) $t_\delta = -6.272$]
0.01	−3.6289	−3.6353
0.05	−2.9472	−2.9499
0.10	−2.6118	−2.6133

Step 4

Comparing the $t_{\delta} = -3.150$ value from equation (7.99) with the critical values in Table 7.11 for m = 2, we see that this value is more or less equal to the critical values for a 0.10 level of significant. In other words, if we assume a significant level equal to 0.10, or 0.11, then we accept the alternative hypothesis in (7.93), i.e. we accept that e_t is stationary, meaning thus that variables C_t and Y_t are cointegrated, or that there exists a long-run equilibrium relationship between these two variables. However, if we work with a significant level less than 0.10, then these two variables are not cointegrated and we cannot say that there exists a long-run equilibrium relationship between private consumption and personal disposable income for an EU member.

(b) The Durbin–Watson approach

This approach is very simple and is based on the following two steps:

Step 1

Estimate the cointegrating regression (7.91), save the residuals e_t, and compute the Durbin–Watson statistic, which now is called the 'cointegrating regression Durbin-Watson (CRDW) statistic', as usually, i.e. as

$$ \mathrm{CRDW} = \frac{\Sigma(e_t - e_{t-1})^2}{\Sigma(e_t - \bar{e})^2}, \text{ where } \bar{e} = \text{arithmetic mean} \qquad (7.100) $$

Step 2

Obtain conclusions about the cointegration of the two variables according to the following hypotheses:

$$ \left. \begin{array}{l} H_0\text{: non-stationarity of } e_t, \text{ i.e. for non-cointegration, if CRDW} < d \\ H_a\text{: stationarity of } e_t, \text{ i.e. for cointegration, if CRDW} > d \end{array} \right\} \qquad (7.101) $$

In (7.101) the critical d values, with the null hypothesis being d = 0, have been computed by Sargan and Bhargava (1983) and by Engle and Granger (1987). These critical values are 0.511, 0.386 and 0.322 for significant levels 0.01, 0.05 and 0.10 respectively.

Example 7.7 Cointegration between private consumption and personal disposable income for an EU member (CRDW)

Here we repeat Example 7.6 using the methodology of the CRDW.

Step 1

From the estimated cointegrating regression (7.98) we see that CRDW = 1.021.

Step 2

Because the CRDW = 1.021 is greater than the critical values noted above, the alternative hypothesis in (7.101) is accepted, and thus we can conclude that private consumption and personal disposable income may be cointegrated.

(c) The autoregressive distributed lag (ADL) model approach

The Engle–Granger approach for testing cointegration between two variables Y_t and X_t depends crucially on the estimated equilibrium error e_t, or in other words on the estimation of the long-run equilibrium relationship. Phillips and Loretan (1991) suggested that instead of the cointegrating equation (7.91), which it is possibly misspecified due to the omission of lagged variables, we could use the autoregressive distributed lag (ADL) equation

$$Y_t = \alpha + \sum_{j=1}^{k} \alpha_j Y_{t-j} + \sum_{j=0}^{k} \beta_j X_{t-j} + \epsilon_t \qquad (7.102)$$

In the steady-state long-run equilibrium, the variables take the same values for all periods, i.e. it is $Y_t = Y_{t-1} = Y_{t-2} = \ldots = Y^*$ and $X_t = X_{t-1} = X_{t-2} = \ldots = X^*$, and therefore the steady-state long-run equilibrium relationship becomes

$$Y^* = \frac{\alpha}{1 - \sum_{j=1}^{k} \alpha_j} + \frac{\sum_{j=0}^{k} \beta_j}{1 - \sum_{j=1}^{k} \alpha_j} X^* = \alpha^* + \beta^* X^* \qquad (7.103)$$

where the cointegrating vector is $[1, -\alpha^*, -\beta^*]$. Using this cointegrating vector the equilibrium error is $\epsilon^* = Y^* - \alpha^* - \beta^* X^*$. If $[1, -a^*, -b^*]$ is the estimated cointegrating vector, obtained by estimating first, (7.102) with OLS and secondly, by substituting the estimates into (7.103), the estimated equilibrium error is given by $e_t = Y_t - a^* - b^* X_t$. This estimated equilibrium error could be used in testing for cointegration between the two variables Y_t and X_t following the steps of the Engle-Granger approach presented above.

Example 7.8 Cointegration between private consumption and personal disposable income for an EU member (ADL)

Here we repeat example 7.6 using the methodology of the ADL.

Step 1

It has been shown in Example 7.6 that both variables, private consumption and personal disposable income are integrated of order one.

Step 2

An estimate with OLS of a version of equation (7.102) is shown below:

$$\hat{C}_t = 9254.83 + 0.79945C_{t-1} + 0.27887Y_t - 0.12502Y_{t-1}$$

$$t \quad [4.725] \quad [9.832] \quad [6.193] \quad [-1.962] \quad\quad (7.104)$$

$$R^2 = 0.999 \; DW = 1.947$$

Substituting the estimates from (7.104) into the corresponding parameters of (7.103) we get that

$$a^* = \frac{a}{1 - a_1} = \frac{9254.83}{1 - 0.79945} = 46147.25,$$

$$b^* = \frac{b_0 + b_1}{1 - a_1} = \frac{0.27887 - 0.12502}{1 - 0.79945} = 0.76714$$

and therefore the estimated cointegrating vector is $[1, -46147.25, -0.76714]$. This cointegrating vector is used to construct the estimated equilibrium error $e_t = Y_t - 46147.25 - 0.76714X_t$.

Step 3

Testing the stationarity of e_t. Taking into account the criteria AIC, SC and LM(1) we obtained:

$$\Delta\hat{e}_t = -0.03202e_{t-1} - 0.41477\Delta e_{t-1} - 0.36552\Delta e_{t-2}$$

$$t \quad [-0.759] \quad [-2.362] \quad [-2.059] \quad\quad (7.105)$$

$$R^2 = 0.217 \; DW = 2.047$$

Step 4

Comparing the $t_8 = -0.759$ value from equation (7.105) with the critical values in Table 7.11 for $m = 2$, we see that this value is much greater than all the critical values noted in the table. This means that the equilibrium error is non-stationary indicating thus that private consumption and personal disposable income are not cointegrated.

Summarising the last three examples we could say that in Example 7.6 we found that variables C_t and Y_t were on the borderline of being cointegrated, in Example 7.7 we found that these variables are cointegrated, and finally in example 7.8 we found that the same two variables are not cointegrated. The results of these examples show that cointegration tests 'lack power' and fail to recover cointegration between two variables even when these variables are cointegrated. Therefore, we should use cointegration tests with great caution.

5.3 Cointegration: the estimation of error correction models (ECM)

The most important possibly result in cointegration analysis is the so-called 'Granger representation theorem' (Granger, 1986; Engle and Granger, 1987). According to this theorem if two variables Y_t and X_t are cointegrated then there is a long-run relationship between them. Of course, in the short run these variables may be in disequilibrium, with the disturbances being the equilibrating error ϵ_t. The dynamics of this short-run disequilibrium relationship between these two variables can always be described by an 'error correction model' (ECM) introduced by Sargan (1964). This error correction model which connects the short-run and the long-run behaviour of the two variables is given by

$$\Delta Y_t = \text{lagged}(\Delta Y_t, \Delta X_t) + \lambda \epsilon_{t-1} + \upsilon_t, \quad -1 < \lambda < 0 \tag{7.106}$$

where $Y_t \sim I(1)$, $X_t \sim (1)$, $Y_t, X_t \sim CI(1, 1)$, $\epsilon_t = Y_t - \beta_0 - \beta_1 X_t \sim I(0)$, $\upsilon_t =$ white noise disturbance term and $\lambda =$ short-run adjustment coefficient.

In (7.106) all variables are stationary because Y_t and X_t being integrated of order one, their differences ΔY_t and ΔX_t are integrated of order zero. Furthermore, the equilibrium error ϵ_t is integrated of order zero because variables Y_t and X_t are cointegrated. In other words, one could say that equation (7.106) could be estimated by OLS. However, this is not the case because the equilibrium error ϵ_t is not an observable variable. Therefore, before any estimation of equation (7.106) values of this error should be obtained.

Engle and Granger (1987) proposed the following two-step methodology in estimating equation (7.106):

Step 1

Estimate the cointegrating regression (7.91), get the consistent estimated cointegrating vector $[1, -b_0, -b_1]$ and use it in order to obtain the estimated equilibrium error $e_t = Y_t - b_0 - b_1 X_t$.

Step 2

Estimate the following equation by OLS:

$$\Delta Y_t = \text{lagged}(\Delta Y_t, \Delta X_t) + \lambda e_{t-1} + \upsilon_t \tag{7.107}$$

In estimating (7.107) we should take care about the following:

1 Use appropriate statistics, as AIC, SC and LM for example, in order to decide about the proper number of lags for the differenced variables to be used.
2 Use if appropriate the non-lagged differenced variable X_t.
3 Include in the equation other differenced 'exogenous' variables, as long as they are integrated of order one, in order to improve fit.

Example 7.9 Estimating an error correction model between private consumption and personal disposable income for an EU member (ECM)

Let us consider from Example 7.6 that the two variables C_t and Y_t are cointegrated.

Step 1

From the cointegrating regression estimated in (7.98), or

$$\hat{C}_t = 11907.23 + 0.779585Y_t$$

$$t \quad [3.123] \qquad [75.566] \tag{7.108}$$

$$R^2 = 0.994 \ DW = 1.021$$

we get the residuals, e_t.

Step 2

According to (7.107), using the criteria AIC, SC and t-ratios, a version of an estimated error correction model is the following:

$$\Delta\hat{C}_t = 5951.557 + 0.28432\Delta Y_t - 0.19996e_{t-1}$$

$$t \quad [7.822] \qquad [6.538] \qquad [-2.486] \tag{7.109}$$

$$R^2 = 0.572 \ DW = 1.941 \ LM(1) = 0.007\{p = 0.934\}$$

The results in (7.109) show that short-run changes in personal disposable income Y_t affect positively private consumption C_t. Furthermore, because the short-run adjustment coefficient is significant, it shows that 0.19996 of the deviation of the actual private consumption from its long-run equilibrium level is corrected each year.

6 COINTEGRATION: THE MULTIVARIATE CASE

Following the structure of the presentation in section 5 we try in this section to generalise from the two variables case to the more than two variables, or multivariate, case.

6.1 Cointegration: the concept

Engle and Granger (1987) developed the following general definition:

Cointegration of more than two variables

k time series X_{1t}, X_{2t}, ..., X_{kt}, are said to be cointegrated of order (d, b), where $d \geq b \geq 0$, if all time series are integrated of order d, and there exists a linear combination of these k time series, say $a_1X_{1t} + a_2X_{2t} + \ldots + a_kX_{kt}$, which is integrated of order (d − b). In mathematical terms, this definition is written:

$$\text{If } X_{1t} \sim I(d), X_{2t} \sim I(d), \ldots, X_{kt} \sim I(d) \text{ then } X_{1t}, X_{2t}, \ldots, X_{kt} \sim CI(d, b)$$
$$\text{if } a_1X_{1t} + a_2X_{2t} + \ldots + a_kX_{kt} \sim I(d-b) \tag{7.110}$$

The vector of the coefficients that constitute the linear combination of the k time series, i.e. $[a_1, a_2, \ldots, a_k]$ in (7.110), is the 'cointegrating vector'.

We can distinguish the following two special cases which we will investigate in this chapter:

1   The case where d = b, resulting in $a_1X_{1t} + a_2X_{2t} + \ldots + a_kX_{kt} \sim I(0)$, which means that the linear combination of the k time series is stationary, and therefore, X_{1t}, X_{2t}, ..., $X_{kt} \sim CI(d, d)$.

2   The case where d = b = 1, resulting in $a_1X_{1t} + a_2X_{2t} + \ldots + a_kX_{kt} \sim I(0)$, which means that the linear combination of the k time series is stationary, and therefore, X_{1t}, X_{2t}, ..., $X_{kt} \sim CI(1, 1)$.

Extending section 5 let us consider the following three variable relationship where $Y_t \sim I(1)$, $X_t \sim I(1)$ and $Z_t \sim I(1)$:

$$Y_t = \beta_0 + \beta_1X_t + \beta_2Z_t \tag{7.111}$$

This relationship is in a *long-run equilibrium* when

$$0 = Y_t - \beta_0 - \beta_1X_t - \beta_2Z_t \tag{7.112}$$

The deviation from the long-run equilibrium, i.e. the *equilibrium error*, ϵ_t, is given by

$$\epsilon_t = Y_t - \beta_0 - \beta_1X_t - \beta_2Z_t \tag{7.113}$$

As in the two variable case, for the long-run equilibrium to have meaning, i.e. to exist, the equilibrium error in (7.113) should fluctuate around the equilibrating zero value, as shown in (7.112). In other words, the equilibrium error ϵ_t should be a stationary time series, i.e. it should be $\epsilon_t \sim I(0)$ with $E(\epsilon_t) = 0$. According to the definition in (7.110), because $Y_t \sim I(1)$, $X_t \sim I(1)$ and $Z_t \sim I(1)$, and the linear combination $\epsilon_t = Y_t - \beta_0 - \beta_1 X_t - \beta_2 Z_t \sim I(0)$, we can say that Y_t, X_t and Z_t are cointegrated of order (1, 1), i.e. it is Y_t, X_t, $Z_t \sim CI(1, 1)$. The cointegrating vector in this case is $[1, -\beta_0, -\beta_1, -\beta_2]$.

In the two variables case and under the assumption that the coefficient of one of the variables is normalised to equal unity we said in section 5 that the cointegrating vector, is unique. However, in the multivariate case this is not true. It has been shown that if a long-run equilibrium relationship exists between k variables, then these variables are cointegrated, whilst if k variables are cointegrated, then there exists at least one long-run equilibrium relationship between these variables. In other words, in the multivariate case the cointegrating vector *is not unique*.

It can be proved (Greene, 1999) that in the case of k variables, there can only be up to $k - 1$ linearly independent cointegrating vectors. The number of these linearly independent cointegrating vectors is called 'cointegrating rank'. Therefore, the cointegrating rank in the case of k variables may range from 1 to $k - 1$. As a consequence, in the case where more than one cointegrating vectors exist, it may be impossible without out-of-sample information to identify the long-run equilibrium relationship (Enders, 1995). This is the case because cointegration is a purely statistical concept and it is in fact 'a-theoretical' in the sense that cointegrated relationships need not have any economic meaning (Maddala, 1992).

6.2 Cointegration: the tests

The tests presented in section 5.2 for cointegration of two variables may be generalised to include k variables. In this section we present the generalisation of the Engle-Granger approach only.

The Engle–Granger approach

Suppose we want to test if the $k + 1$ variables Y_t, X_{1t}, $X_{2t}, \ldots, X_{kt}$ are cointegrated. This approach, which is called the Engle–Granger (EG) test, or the augmented Engle–Granger (AEG) test, suggests the following steps:

Step 1

Find the order of integration of all variables using the unit roots methodology presented in section 4. If the order of integration of all variables is the same, something that the concept of cointegration requires, continue to the next step. However, it is possible to have a mixture of different order variables where subsets of the higher-order variables are cointegrated to the order of the lower-order variables (Cuthbertson *et al.* 1992). We will not consider these possibilities in this book.

Step 2

If all the variables are integrated of the same order, say $I(1)$, estimate with OLS the long-run equilibrium equation

$$Y_t = \beta_0 + \beta_1 X_{1t} + \beta_2 X_{2t} + \ldots + \beta_k X_{kt} + \epsilon_t \qquad (7.114)$$

which in this case is the *cointegrating regression* and save the residuals, e_t, as an estimate of the equilibrium error, ϵ_t. Although the estimated cointegrating vector $[1, -b_0, -b_1, -b_2, \ldots, -b_k]$ is a consistent estimate of the true cointegrating vector $[1, -\beta_0, -\beta_1, -\beta_2, \ldots, -\beta_k]$, this is not true for the estimated standard errors of these coefficients.

Step 3

For the variables to be cointegrated the equilibrium errors must be stationary. To test this stationarity apply the unit roots methodology presented in section 4 to the estimated equilibrium errors saved in the previous step. You could use, for example, the Dickey–Fuller test, or the augmented Dickey–Fuller test, to time series e_t, which involves the estimation of a version of the following equation with OLS

$$\Delta e_t = \delta e_{t-1} + \sum_{j=2}^{q} \delta_j \Delta e_{t-j+1} + \upsilon_t \qquad (7.115)$$

As in section 5.2, two things we should take into account in applying the DF, or ADF, tests:

1   Equation (7.115) is not including a constant term, because by construction the OLS residuals e_t are centred around zero.
2   Because the estimate of δ in (7.115) is downward biased, due to the fact that by construction the OLS methodology seeks to produce stationary residuals e_t, the usual Dickey–Fuller τ statistics are not appropriate for this test. Critical values, as those presented in Table 7.11 should be used. All these critical values depend on the number of variables included in the cointegrating regression.

Step 4

Obtain conclusions about the cointegration of the variables according to the following hypotheses:

$$\left. \begin{array}{l} H_0\colon \delta = 0, \text{ for non-stationarity of } e_t, \text{ i.e. for non-cointegration, if } t_\delta > \tau \\ H_a\colon \delta < 0, \text{ for stationarity of } e_t, \text{ i.e. for cointegration, if } t_\delta < \tau \end{array} \right\} \qquad (7.116)$$

Example 7.10 Cointegration between private consumption, personal disposable income, and inflation rate for an EU member (EG, or AEG)

In Table 7.1 the private consumption (C_t), the personal disposable income (Y_t) and the consumer price index (P_t) for an EU member are presented. We used formula (7.80) in order to compute the inflation rate (Z_t). Following these steps we can determine if these three variables are cointegrated.

Step 1

Finding the order of integration of the three variables: Taking into account criteria AIC, SC and LM(1) we obtained:

Variables C_t and Y_t are integrated of order one, i.e. it is $C_t \sim I(1)$ and $Y_t \sim I(1)$ (see Example 7.6).

For the inflation rate variable:

$$\Delta \hat{Z}_t = 0.02233 - 0.18357 Z_{t-1}$$

$$t \qquad [1.889] \quad [-1.991] \tag{7.117}$$

$$R^2 = 0.110 \ DW = 2.097$$

and

$$\Delta^2 Z_t = 0.00266 - 1.12517 \Delta Z_{t-1}$$

$$t \qquad [0.394] \quad [-6.296] \tag{7.118}$$

$$R^2 = 0.561 \ DW = 2.031$$

Table 7.13 presents the MacKinnon critical values for the rejection of the hypothesis of a unit root, evaluated from Eviews for equations (7.117) to (7.118).

From the figures in Table 7.13 it is seen that Z_t is non-stationary and that ΔZ_t is stationary, i.e. it is $Z_t \sim I(1)$. Summarising, all the three variables are integrated of order one and therefore we can proceed to step 2.

Step 2

The long-run estimated equilibrium relationship, or the cointegrating regression is the following:

$$\hat{C}_t = 8746.662 + 0.81087 Y_t - 73481.43 Z_t$$

$$t \qquad [2.220] \quad [50.831] \quad [-2.396] \tag{7.119}$$

$$R^2 = 0.995 \ DW = 1.518$$

Table 7.13 MacKinnon critical values for equations (7.117) to (7.118) and the DF t_δ ratios

Significance level	Critical values for equation (7.117) $[t_\delta = -1.991]$	Critical values for equation (7.118) $[t_\delta = -6.296]$
0.01	−3.6353	−3.6422
0.05	−2.9499	−2.9527
0.10	−2.6133	−2.6148

Using the estimated cointegrating vector $[1, -8746.662, -0.81087, +73481.43]$ we have estimated and saved the estimated equilibrium errors e_t.

Step 3

Testing the stationarity of e_t. Taking into account the criteria AIC, SC and LM(1) we obtained:

$$\Delta\hat{e}_t = -0.78108e_{t-1}$$

$$\text{t} \qquad [-4.455] \qquad\qquad\qquad\qquad\qquad (7.120)$$

$$R^2 = 0.375 \; DW = 1.900$$

Step 4

Comparing the $t_\delta = -4.455$ value from equation (7.120) with the critical values in Table 7.11 for $m = 3$, we see that this value is less than the critical values for a 0.05 level of significant. In other words, we accept the alternative hypothesis in (7.116), i.e. we accept that e_t is stationary, meaning thus that variables C_t, Y_t and Z_t are cointegrated.

If we go back to Example 7.6 we saw there that by using the Engle–Granger cointegration test we were not sure if variables C_t and Y_t were cointegrated. In this example, using the exactly same methodology, we found that variables C_t, Y_t and Z_t are cointegrated. This should not be surprising if we consider that in Example 7.6 we possibly made a specification error in assuming that in the long-run private consumption was dependent on personal disposable income only. In this example, having possibly corrected this specification error, by also including the level of inflation rate of the economy, we were able to reach more definite conclusions.

6.3 Cointegration: the estimation of error correction models (ECM)

In the case of more than two variables the Granger two step estimation methodology for the error correction equation, that we have seen in section 5.3, still holds under the assumption that one cointegrating vector exists. In this case, the error correction model which connects the short-run and the long-run behaviour of the $k + 1$ variables is given by

$$\Delta Y_t = lagged(\Delta Y_t, \Delta X_{1t}, \Delta X_{2t}, \ldots, \Delta X_{kt}) + \lambda\epsilon_{t-1} + \upsilon_t, \; -1 < \lambda < 0 \qquad (7.121)$$

where $Y_t \sim I(1), X_{1t} \sim I(1), \ldots, X_{kt} \sim I(1)$, $Y_t, X_{1t}, \ldots, X_{kt} \sim CI(1, 1)$, $\epsilon_t = Y_t - \beta_0 - \beta_1 X_{1t} - \ldots - \beta_k X_{kt} \sim I(0)$, $\upsilon_t =$ white noise disturbance term and $\lambda =$ short-run adjustment coefficient.

Under the assumption of the existence of only one cointegrating vector that connects the cointegrated variables, the OLS estimation applicable to the cointegrating equation will give consistent estimates. In the case where there are more than one cointegrating vectors the Engle-Granger methodology is no longer valid because it is not producing consistent estimates. In this case we have to use the methods presented in the next section.

The Engle and Granger (1987) two-step methodology in estimating equation (7.121), under the assumption of the existence of only one cointegrating vector, is the following:

Step 1

Estimate the cointegrating regression (7.114), get the consistent estimated cointegrating vector $[1, -b_0, -b_1, \ldots, -b_k]$ and use it in order to obtain the estimated equilibrium error $e_t = Y_t - b_0 - b_1 X_{1t} - \ldots - b_k X_{kt}$.

Step 2

Estimate the following equation by OLS:

$$\Delta Y_t = \text{lagged}(\Delta Y_t, \Delta X_{1t}, \Delta X_{2t}, \ldots, \Delta X_{kt}) + \lambda e_{t-1} + \upsilon_t \qquad (7.122)$$

In estimating (7.122) we should take care about the following:

1 Use appropriate statistics, as AIC, SC and LM for example, in order to decide about the proper number of lags for the differenced variables to be used.
2 Use if appropriate the non-lagged differenced variables X_t.
3 Include in the equation other differenced 'exogenous' variables, as long as they are integrated of order one, in order to improve fit.

Example 7.11 Estimating an error correction model between private consumption, personal disposable income and inflation rate for an EU member (ECM)

Let us consider from Example 7.10 that the three variables C_t, Y_t and Z_t are cointegrated. Furthermore, assume that there exists one cointegrating vector only.

Step 1

From the cointegrating regression estimated in (7.119), or

$$\hat{C}_t = 8746.662 + 0.81087 Y_t - 73481.43 Z_t$$

$$t \quad [2.220] \quad [50.831] \quad [-2.396] \qquad\qquad (7.123)$$

$$R^2 = 0.995 \ DW = 1.518$$

we get the residuals, e_t.

Step 2

According to (7.121), using the criteria AIC, SC and t-ratios, a version of an estimated error correction model is the following:

$$\Delta \hat{C}_t = 0.50572 \Delta C_{t-1} + 0.34563 \Delta Y_t - 76297.96 \Delta Z_t - 0.37675 e_{t-1}$$

t   [6.370]   [7.479]   [−3.107]   [−2.786]   (7.124)

$R^2 = 0.511 \; DW = 2.254 \; LM(1) = 0.264 \{p = 0.607\}$

The results in (7.124) show that short-run changes in personal disposable income Y_t and in the inflation rate affect, respectively, positively and negatively private consumption C_t. Furthermore, because the short-run adjustment coefficient is significant, it shows that 0.37675 of the deviation of the actual private consumption from its long-run equilibrium level is corrected each year.

7 VECTOR AUTOREGRESSIONS AND COINTEGRATION

We said in section 6 that the Engle–Granger methodology produces some problems when more than two variables are involved in the cointegrating exercise. In this case, the Johansen (1988) and Stock and Watson (1998) approaches to cointegration may be considered. Because these approaches make use of the vector autoregression models we will very briefly present them in the next section.

7.1 Vector autoregression (VAR) models

Assume the following simultaneous equation model in its structural form.

$$C_t = \alpha_0 + \alpha_1 Y_t + \alpha_2 C_{t-1} + \epsilon_{1t}$$
$$Y_t = \beta_0 + \beta_1 Y_{t-1} + \beta_2 C_{t-1} + \epsilon_{2t} \qquad (7.125)$$

where C = consumption and Y = income. The rationale of the first equation of this model could be that current consumption depends on current income and on lagged consumption due to habit persistence. The rationale of the second equation may be that current income depends on lagged income and on lagged consumption because higher consumption indicates higher demand which produces higher economic growth and therefore higher income.

Corresponding to model (7.125) a reduced form model can by found to be the following:

$$C_t = \pi_{10} + \pi_{11} C_{t-1} + \pi_{12} Y_{t-1} + \upsilon_{1t}$$
$$Y_t = \pi_{20} + \pi_{21} C_{t-1} + \pi_{22} Y_{t-1} + \upsilon_{2t} \qquad (7.126)$$

A distinctive property of the reduced form model (7.126) is that all its endogenous variables are expressed in terms of its lagged endogenous variables only. In this model there are no other 'exogenous' variables. This model constitutes of a 'vector autoregressive model of order 1', because the highest lag length of its variables is one, and is denoted with VAR(1).

Generally, a system of m variables of the form

$$Y_{1t} = \alpha_{10} + \alpha_{11,1}Y_{1,t-1} + \alpha_{12,1}Y_{1,t-2} + \ldots + \alpha_{1k,1}Y_{1,t-k} + \ldots + \alpha_{11,m}Y_{m,t-1} + \alpha_{12,m}Y_{m,t-2}$$

$$+ \ldots + \alpha_{1k,m}Y_{m,t-k} + \upsilon_{1t}$$

$$\ldots \tag{7.127}$$

$$Y_{mt} = \alpha_{m0} + \alpha_{m1,1}Y_{1,t-1} + \alpha_{m2,1}Y_{1,t-2} + \ldots + \alpha_{mk,1}Y_{1,t-k} + \ldots + \alpha_{m1,m}Y_{m,t-1}$$

$$+ \alpha_{m2,m}Y_{m,t-2} + \ldots + \alpha_{mk,m}Y_{m,t-k} + \upsilon_{mt}$$

is called a 'vector autoregressive model, or process, of order k', or VAR(k). In matrix form (7.127) is written

$$\mathbf{Y}_t = \boldsymbol{\delta} + \mathbf{A}_1\mathbf{Y}_{t-1} + \ldots + \mathbf{A}_k\mathbf{Y}_{t-k} + \boldsymbol{\upsilon}_t = \boldsymbol{\delta} + \sum_{j=1}^{k} \mathbf{A}_j\mathbf{Y}_{t-j} + \boldsymbol{\upsilon}_t \tag{7.128}$$

where

$$\mathbf{Y}_t = \begin{bmatrix} Y_{1t} \\ Y_{2t} \\ \vdots \\ Y_{mt} \end{bmatrix} \quad \boldsymbol{\delta} = \begin{bmatrix} \alpha_{10} \\ \alpha_{20} \\ \vdots \\ \alpha_{m0} \end{bmatrix} \quad \mathbf{A}_j = \begin{bmatrix} \alpha_{11,j} & \alpha_{12,j} & \cdots & \alpha_{1m,j} \\ \alpha_{21,j} & \alpha_{22,j} & \cdots & \alpha_{2m,j} \\ \vdots & \vdots & \vdots & \vdots \\ \alpha_{m1,j} & \alpha_{m2,j} & \cdots & \alpha_{mm,j} \end{bmatrix} \quad \boldsymbol{\upsilon}_t = \begin{bmatrix} \upsilon_{1t} \\ \upsilon_{2t} \\ \vdots \\ \upsilon_{mt} \end{bmatrix} \tag{7.129}$$

In the cases where in the system (7.127) the lag lengths are not the same in all the equations of the system, this model is called 'near vector autoregressive', or 'near VAR'.

The assumptions that usually follow a VAR model are the assumptions of a reduced form simultaneous equations model, i.e.:

$$\upsilon_{it} \sim N(0, \omega_{ii}), \text{ for all t, and } i = 1, 2, \ldots, m, \text{ where } \omega_{ii} = \text{var}(\upsilon_{it})$$

$$E(\upsilon_{it}\upsilon_{is}) = 0, \text{ for } t \neq s, \text{ and } i = 1, 2, \ldots, m$$

$$E(\upsilon_{it}\upsilon_{jt}) = \omega_{ij}, \text{ for all t, and } i, j = 1, 2, \ldots, m, \text{ where } \omega_{ij} = \text{cov}(\upsilon_{it}, \upsilon_{jt}) \tag{7.130}$$

or in matrix form

$$\boldsymbol{\upsilon}_t \sim N(\mathbf{0}, \boldsymbol{\Omega}), \text{ with } E(\boldsymbol{\upsilon}_t\boldsymbol{\upsilon}_s') = \mathbf{0} \text{ and } \boldsymbol{\Omega} = E(\boldsymbol{\upsilon}_t\boldsymbol{\upsilon}_t') = \begin{bmatrix} \omega_{11} & \omega_{12} & \cdots & \omega_{1m} \\ \omega_{21} & \omega_{22} & \cdots & \omega_{2m} \\ \vdots & \vdots & \vdots & \vdots \\ \omega_{m1} & \omega_{m2} & \cdots & \omega_{mm} \end{bmatrix} \tag{7.131}$$

A vector stochastic process $\{\mathbf{Y}_t\}$ is called stationary if (Judge *et al.* 1988):

$$\left.\begin{array}{l} E(\mathbf{Y}_t) = \mu \text{ for all t} \\ \text{var}(Y_{jt}) < \infty \text{ for } j = 1, 2, \ldots, m \text{ and all t} \\ \text{cov}(\mathbf{Y}_t, \mathbf{Y}_{t+k}) = E[(\mathbf{Y}_t - \mu)(\mathbf{Y}_{t+k} - \mu)'] = \mathbf{\Gamma}_k \text{ for all t} \end{array}\right\} \tag{7.132}$$

Furthermore, a VAR(k) process is stationary if its means and covariance matrices are bounded and the polynomial defined by the determinant

$$|\mathbf{I} - \mathbf{A}_1\lambda - \mathbf{A}_2\lambda^2 - \ldots - \mathbf{A}_k\lambda^k| = 0 \tag{7.133}$$

has all its roots outside the complex unit circle (Judge *et al.* 1988).

Under the assumptions written above, the parameters of a VAR(k) model can be consistently estimated with OLS. Therefore, for the *i*th equation the OLS estimator is given by

$$\mathbf{a}_i = (\mathbf{X'X})^{-1}\mathbf{X'Y}_i \sim N(\alpha_i, \omega_{ii}(\mathbf{X'X})^{-1}) \tag{7.134}$$

where

$$\mathbf{a}_i = [a_{i0}\ a_{i1,1} \ \ldots \ a_{ik,m}] \text{ is an estimate of } \alpha_i = [\alpha_{i0}\ \alpha_{i1,1} \ \ldots \ \alpha_{ik,m}]$$
$$\mathbf{Y}_i = [Y_{i1}\ Y_{i2} \ \ldots \ Y_{in}]' \text{ and } \mathbf{X}_i = [\mathbf{1}\ \mathbf{Y}_{1,-1} \ \ldots \ \mathbf{Y}_{m,-k}] \tag{7.135}$$

Consistent estimates w_{ij} of the parameters ω_{ij} in (7.131) are given by

$$w_{ij} = \frac{(\mathbf{Y}_i - \mathbf{Xa}_i)'(\mathbf{Y}_i - \mathbf{Xa}_i)}{n} \text{ or } \frac{(\mathbf{Y}_i - \mathbf{Xa}_i)'(\mathbf{Y}_i - \mathbf{Xa}_i)}{n - mk - 1} \tag{7.136}$$

We have to note here that the generalised least squares (GLS) estimator if applied to (7.127) will give the same exactly results as the OLS estimator because matrix **X** is the same for all equations. The seemingly unrelated regressions (SUR) estimator could be applied to the near VAR model to improve efficiency.

The preceding estimation method assumes that the lag length, i.e. the order of the VAR, is known. In cases where the VAR order is large we have a major problem in VAR analysis; the problem of *over-parameterisation*. However, in most cases the VAR order is not known and therefore it has to be selected. Common tests for selecting the VAR order are the following. In all these tests it is assumed that the number of observations is n, and thus k presample values for all the variables must be considered:

(a) The likelihood ratio (LR) test

This test depends on the usual likelihood ratio statistic given by

$$LR = 2[\ln \ell_u - \ln \ell_r] \sim \chi2(v) \tag{7.137}$$

where

$\ln \ell_u = $ log of likelihood of the complete in coefficients (unrestricted) equation

$\ln \ell_r = $ log of likelihood of the smaller in coefficients (restricted) equation

$v = $ number of restrictions imposed

Assuming that the coefficients of a VAR(k) model the corresponding to the lagged variables are given by the matrix $\mathbf{A} = [\mathbf{A}_1 \mathbf{A}_2 \ldots \mathbf{A}_k]$, the test works by testing in a sequence the following hypotheses, starting from a large assumed lag length k.

$H_0: \mathbf{A}_k = 0$ vs. $H_a: \mathbf{A}_k \neq 0$

$H_0: \mathbf{A}_{k-1} = 0$ vs. $H_a: \mathbf{A}_{k-1} \neq 0$ given that $\mathbf{A}_k = 0$

$H_0: \mathbf{A}_{k-2} = 0$ vs. $H_a: \mathbf{A}_{k-2} \neq 0$ given that $\mathbf{A}_k = \mathbf{A}_{k-1} = 0$

$\ldots$

$H_0: \mathbf{A}_1 = 0$ vs. $H_a: \mathbf{A}_1 \neq 0$ given that $\mathbf{A}_k = \mathbf{A}_{k-1} = \ldots = \mathbf{A}_2 = 0$ \hfill (7.138)

The test stops when a null hypothesis is rejected using the LR statistic and the VAR order q, for $k \geq q \geq 1$, is selected accordingly. However, since estimation methods require white noise errors a higher value of q might finally be used in the estimations (Holden and Perman, 1994).

(b) The Akaike information criterion (AIC) and Schwartz criterion (SC)

The usual Akaike's information criterion and Schwartz criterion in the VAR context are defined as follows:

$$AIC(q) = \ln|\mathbf{W}_q| + \frac{2m^2 q}{n} \tag{7.139}$$

and

$$SC(q) = \ln|\mathbf{W}_q| + \frac{m^2 q}{n} \ln(n) \tag{7.140}$$

where m = number of equations, n = common sample size, q = lag length, and $\mathbf{W}$ is the estimated residual covariance matrix $\mathbf{\Omega}$ evaluated for VAR(q). The VAR order q is selected for the corresponding minimum value of the criterion.

Example 7.12 Estimating a VAR model for the private consumption and the personal disposable income for an EU member (VAR)

Suppose that we want to estimate a VAR model for the variables of the private consumption (C_t) and personal disposable income (Y_t) of an EU member, presented in Table 7.1.

The first thing that we have to do is to specify the VAR order. Because the data are annual it seems unlikely that the lag length will be more than k = 3. Therefore, keeping the values of the variables for the first three years as pre-sample values, Table 7.14 presents the statistics from the estimation of the model for various lag lengths ranging from k = 1 to k = 3.

For the LR test it is $v = 4$, because going down from one lag length to an immediately lower lag length we exclude one lag on each of the two variables in each of the two equations. For a 0.05 significant level the critical value for the $\chi^2(4)$ distribution is 9.4877. Because going down from $q = 3$ to $q = 1$ none of the LR values in Table 7.14 is greater that $\chi^2(4) = 9.4877$, this means that none of the null hypotheses in (7.138) is rejected. Therefore, this test indicates that the proper VAR order of this model is $q = 1$. The same VAR order $q = 1$ is indicated by the other two statistics, AIC and SC, because these statistics take their minimum value for $q = 1$.

Estimates of the VAR(1) model are given below:

$$\hat{C}_t = 12581.43 + 0.96780C_{t-1} + 0.01657Y_{t-1}$$

$$t \quad [4.537] \quad [8.579] \quad [0.189] \tag{7.141}$$

$$R^2 = 0.9978$$

and

$$\hat{Y}_t = 11929.05 + 0.60371C_{t-1} + 0.50776Y_{t-1}$$

$$t \quad [1.613] \quad [2.007] \quad [2.174] \tag{7.142}$$

$$R^2 = 0.9903$$

Table 7.14 Statistics for the consumption–income VAR model for an EU member

Lag length (q)	Log likelihood (ln ℓ)	AIC(q)	SC(q)	LR
1	−619.1697	35.20985	35.34726	–
2	−617.4926	35.23003	35.45905	3.3542
3	−616.0448	35.26455	35.58518	2.8956

Notes: Estimates with Eviews.

7.2 VAR and cointegration

Let us consider the VAR model of (7.128), with m variables, or

$$\mathbf{Y}_t = \sum_{j=1}^{k} \mathbf{A}_j \mathbf{Y}_{t-j} + \upsilon_t \tag{7.143}$$

where for simplicity the intercept has been excluded. Assume also that all its m variables are either simultaneously integrated of order one, or of order zero.

Model (7.143) can be rewritten as

$$\Delta\mathbf{Y}_t = \mathbf{B}\mathbf{Y}_{t-1} + \sum_{j=1}^{k-1} \mathbf{B}_j \Delta\mathbf{Y}_{t-j} + \upsilon_t \tag{7.144}$$

where

$$\mathbf{B} = -(\mathbf{I} - \mathbf{A}_1 - \mathbf{A}_2 - \ldots - \mathbf{A}_k) \tag{7.145}$$

and

$$\mathbf{B}_j = -(\mathbf{A}_{j+1} - \mathbf{A}_{j+2} - \ldots - \mathbf{A}_{j+k}) \text{ for } j = 1, 2, \ldots, k-1 \tag{7.146}$$

Model (7.144) looks like an 'error correction model', and if all its m variables are integrated of order one then variables $\Delta \mathbf{Y}_{t-j}$ are stationary. This model can consistently be estimated under the assumption that its variables are cointegrated, so $\mathbf{BY}_{t-1}$ is also stationary.

It can be proved that (see Engle and Grander (1987) and Johansen (1989) for the original works, or Enders (1985) and Charemza and Deadman (1997) for good presentations):

1   If the rank of matrix **B** is zero, then all the elements in this matrix are zero. Therefore, in (7.144) the error correction mechanism $\mathbf{BY}_{t-1}$ does not exist, meaning that there is no long-run equilibrium relationship between the variables of the model. Thus, these variables are not cointegrated. The VAR model could be formulated in terms of the first differences of the variables.

2   If the rank of matrix **B** is equal to m, i.e. its rows are linearly independent, the vector process $\{\mathbf{Y}_t\}$ is stationary, meaning that all variables are integrated of order zero, and therefore the question of cointegration does not arise. The VAR model could be formulated in terms of the levels of all variables.

3   If the rank of matrix **B** is r, where $r < m$, i.e. its rows are not linearly independent, it can be shown that this matrix can be written as

$$\mathbf{B} = \mathbf{D} \cdot \mathbf{C}' \tag{7.147}$$

where **D** and **C** are matrices of $m \times r$ dimensions. Matrix **C** is called the 'cointegrating matrix', and matrix **D** is called the 'adjustment matrix'. In the case where $\mathbf{Y}_t \sim I(1)$ then $\mathbf{C}'\mathbf{Y}_t \sim I(0)$, i.e. the variables in $\mathbf{Y}_t$ are cointegrated. The cointegrating vectors are the corresponding columns in **C**, say $\mathbf{c}_1, \mathbf{c}_2, \ldots, \mathbf{c}_r$. In other words, the rank r of matrix **B** defines the number of 'cointegrating vectors', i.e. the 'cointegrating rank'. The VAR model could be formulated in terms of a 'vector error correction' (VEC) model.

The three findings above constitute the generalisation of the Granger Representation Theorem. The work of Johansen (1988), and similarly of Stock and Watson (1988), is to identify the cointegrating rank r and to provide estimates of the cointegrating and adjustment matrices, using the maximum likelihood method. The steps of the Johansen's approach may be formulated as follows (Dickey *et al.* (1994), Charemza and Deadman (1997)):

Step 1

Using unit roots tests, say ADF, find the order of integration of the variables involved, say m in number.

Step 2

Using the variables in level terms formulate a VAR model and select the VAR order, say k, by using LR, AIC, or SC, or other tests.

Step 3

Regress $\Delta \mathbf{Y}_t$ on $\Delta \mathbf{Y}_{t-1}, \Delta \mathbf{Y}_{t-2}, \ldots, \Delta \mathbf{Y}_{t-k+1}$ and save the residuals. From these residuals, construct the $m \times 1$ vector $\mathbf{R}_{0t}$ taking the tth element from the saved residuals from each one of the assumed regressions of the m variables.

Step 4

Regress $\mathbf{Y}_{t-k}$ on $\Delta \mathbf{Y}_{t-1}, \Delta \mathbf{Y}_{t-2}, \ldots, \Delta \mathbf{Y}_{t-k+1}$ and save the residuals. From these residuals, construct the $m \times 1$ vector $\mathbf{R}_{kt}$ taking the tth element from the saved residuals from each one of the assumed regressions of the m variables.

Step 5

If n is the sample size, using the following formula

$$\mathbf{S}_{ij} = \frac{1}{n} \sum_{t=1}^{n} \mathbf{R}_{it} \mathbf{R}'_{jt} \text{ for i, j} = 0, k \tag{7.148}$$

compute the four $m \times m$ matrices $\mathbf{S}_{00}$, $\mathbf{S}_{0k}$, $\mathbf{S}_{k0}$, and $\mathbf{S}_{kk}$.

Step 6

Find the squared canonical correlations which correspond to the ordered characteristic roots of the matrix

$$\mathbf{S} = \mathbf{S}_{00}^{-1/2} \mathbf{S}_{0k} \mathbf{S}_{kk}^{-1} \mathbf{S}_{k0} \mathbf{S}_{00}^{-1/2} \tag{7.149}$$

or find the characteristic roots, or eigenvalues, of the polynomial equation in μ

$$|\mu \mathbf{S}_{kk} - \mathbf{S}_{k0} \mathbf{S}_{00}^{-1} \mathbf{S}_{0k}| = 0 \tag{7.150}$$

Having m variables, m is also the maximum number of characteristic roots that can be found. Let us denote these roots, ordered in decreasing value, as $\hat{\mu}_1 > \hat{\mu}_2 > \hat{\mu}_3 > \ldots > \hat{\mu}_m$.

Step 7

Recall that if rank($\mathbf{B}$) = 0, the variables are not cointegrated, if rank($\mathbf{B}$) = m, the variables are stationary, and if rank($\mathbf{B}$) = r, where $0 < r < m$, the variables are cointegrated. Furthermore, it is known that the rank of matrix $\mathbf{B}$ is equal to the number of the characteristic roots that are significantly different from zero. Therefore, the exercise of finding the rank of matrix $\mathbf{B}$ reduces to the testing of the significance of the characteristic roots $\hat{\mu}_1 > \hat{\mu}_2 > \hat{\mu}_3 > \ldots > \hat{\mu}_m$, or of the insignificance of $1 - \hat{\mu}_j$ (for $j = 1, 2, \ldots, m$) from unity. The test is based on the following two likelihood ratio (LR) statistics:

$$\lambda_{\text{trace}}(r) = -n \sum_{j=r+1}^{m} \ln(1 - \hat{\mu}_j) \tag{7.151}$$

and/or

$$\lambda_{\text{max}}(r, r+1) = -n \cdot \ln(1 - \hat{\mu}_{r+1}) \tag{7.152}$$

For the statistic (7.151) the hypotheses to be tested are in a sequence the following:

H_0: $r = 0$ vs. H_a: $r \geq 1$ (if $\lambda_{\text{trace}}(r) >$ critical value)

H_0: $r \leq 1$ vs. H_a: $r \geq 2$ (if $\lambda_{\text{trace}}(r) >$ critical value)

$\ldots$

H_0: $r \leq m - 1$ vs. H_a: $r = m$ (if $\lambda_{\text{trace}}(r) >$ critical value) $\qquad$ (7.153)

For the statistic (7.152) the hypotheses to be tested are in a sequence the following:

H_0: $r = 0$ vs. H_a: $r = 1$ (if $\lambda_{\text{max}}(r) >$ critical value)

H_0: $r \leq 1$ vs. H_a: $r = 2$ (if $\lambda_{\text{max}}(r) >$ critical value)

$\ldots$

H_0: $r \leq m - 1$ vs. H_a: $r = m$ (if $\lambda_{\text{max}}(r) >$ critical value) $\qquad$ (7.154)

Critical values for these statistics can be found in Johansen (1988), Johansen and Juselius (1990), Osterwald-Lenum (1992) and in Enders (1995). Table 7.15 presents critical values reproduced from Enders (1995) for various specifications of the VAR model and of the cointegrating vector. However, in both tests (7.153) and (7.154) the testing of the hypotheses stops when going from top-to-bottom we encounter the first non-significant result. For this case the rank r of matrix **B** is that shown by the corresponding null hypothesis.

Step 8

To each of the characteristic roots corresponds an eigenvector, say $v_1, v_2, \ldots, v_m$, which can constitute the eigenmatrix $\mathbf{V} = [v_1 \ v_2 \ \ldots \ v_m]$. These eigenvectors can be normalised by using $\mathbf{V}'\mathbf{S}_{kk}\mathbf{V} = \mathbf{I}$. If in step 7 it has been found that r is the order of matrix **B**, then the first r eigenvectors in **V** are the r cointegrating vectors that constitute the cointegrating matrix $\mathbf{C} = [v_1 \ v_2 \ \ldots \ v_r]$. The adjustment matrix is found by $\mathbf{D} = \mathbf{S}_{0k}\mathbf{C}$. These are the ML estimators of **C** and **D**.

Table 7.15 Critical values of the λ_{max} and λ_{trace} statistics

	80%	90%	95%	97.5%	99%
λ_{max} and λ_{trace} statistics with trend drift					
m-r			λ_{max}		
1	1.699	2.816	3.962	5.332	6.936
2	10.125	12.099	14.036	15.810	17.936
3	16.324	18.697	20.778	23.002	25.521
4	22.113	24.712	27.169	29.335	31.943
5	27.889	30.774	33.178	35.546	38.341
			λ_{trace}		
1	1.699	2.816	3.962	5.332	6.936
2	11.164	13.338	15.197	17.299	19.310
3	23.868	26.791	29.509	32.313	35.397
4	40.250	43.964	47.181	50.424	53.792
5	60.215	65.063	68.905	72.140	76.955
λ_{max} and λ_{trace} statistics without trend or constant					
m-r			λ_{max}		
1	4.905	6.691	8.083	9.658	11.576
2	10.666	12.783	14.595	16.403	18.782
3	16.521	18.959	21.279	23.362	26.154
4	22.341	24.917	27.341	29.599	32.616
5	27.953	30.818	33.262	35.700	38.858
			λ_{trace}		
1	4.905	6.691	8.083	9.658	11.576
2	13.038	15.583	17.844	19.611	21.962
3	25.445	28.436	31.256	34.062	37.291
4	41.623	45.248	48.419	51.801	55.551
5	61.566	65.956	69.977	73.031	77.911
λ_{max} and λ_{trace} statistics with constant in cointegrating vector					
m-r			λ_{max}		
1	5.877	7.563	9.094	10.709	12.740
2	11.628	13.781	15.752	17.622	19.834
3	17.474	19.796	21.894	23.836	26.409
4	22.938	25.611	28.167	30.262	33.121
5	28.643	31.592	34.397	36.625	39.672
			λ_{trace}		
1	5.877	7.563	9.094	10.709	12.741
2	15.359	17.957	20.168	22.202	24.988
3	28.768	32.093	35.068	37.603	40.198
4	45.635	49.925	53.347	56.449	60.054
5	66.624	71.472	75.328	78.857	82.969

Source: Enders, W. (1995), reproduced from Johansen, S. and Juselius, K. (1990).

Example 7.13 Testing cointegration between the log of private consumption, the log personal disposable income and inflation rate for an EU member (the Johansen approach)

Suppose that we want to test cointegration between the variables of the log of private consumption ($c_t = \log C_t$), the log of personal disposable income ($y_t = \log Y_t$) and inflation ($z_t = \log P_t - \log P_{t-1}$) of an EU member, for which the original values are presented in Table 7.1.

Following the steps presented above, and using for proper comparisons the common sample 1964–1995, treating thus the initial values of the variables as presample values, we get:

Step 1

Using the ADF test we found that all $m = 3$ variables are integrated of order one, i.e. it is $c_t \sim I(1)$, $y_t \sim I(1)$ and $z_t \sim I(1)$.

Step 2

Using the variables in level terms, i.e. as c_t, y_t and z_t, we have formulated a VAR model and selected the VAR order, say k, by using the LR, AIC, and SC statistics, presented in Table 7.16.

For the LR test the degrees of freedom are $v = 9$, because going down from one lag length to an immediately lower lag length we exclude one lag on each of the three variables in each of the three equations. For a 0.05 significant level the critical value for the $\chi^2(9)$ distribution is 16.919. Because going down from $q = 3$ to $q = 1$ none of the LR values in Table 7.16 is greater that $\chi^2(9) = 16.919$, this means that none of the null hypotheses in (7.138) is rejected. Therefore, this test indicates that the proper VAR order of this model is $q = 1$. However, the other two statistics, AIC and SC, indicate an order of $q = 2$, because these statistics take their minimum value for $q = 2$.

Step 3–8

There is no need to go through one by one all the other steps presented above. Econometric packages like Microfit and Eviews include the Johansen test as a standard procedure. In what follows we made use of Eviews.

Assuming that the correct VAR order is $k = 2$, according to AIC and SC, and assuming further that the time series have means and linear trends but the cointegrating equations have only intercepts, Table 7.17 presents statistics of the Johansen tests based on the λ_{trace} LR statistic (7.151).

On the basis of the statistics in Table 7.17, or taking into account the critical values reported in Table 7.15, and for the 5% significant level, we see that the hypothesised cointegrating equations are at most one, because according to the hypotheses testing in (7.153) the LR ratio statistic cannot reject the H_0: $r \leq 1$, because this statistic is less than the corresponding critical value.

Table 7.16 Statistics for the consumption–income-inflation VAR model for an EU member

Lag length (q)	Log likelihood (ln ℓ)	AIC(q)	SC(q)	LR
1	290.9015	−23.44498	−23.26176	–
2	297.2352	−23.65333	−23.33270	12.6674
3	300.0237	−23.64011	−23.18207	2.7885

Notes: Estimates with Eviews.

Table 7.17 Test statistics for cointegration of c_t, y_t, and z_t. VAR $(k=2)$ and $n=32$ (linear deterministic trend in the data)

Eigenvalue	LR	5% critical value	1% critical value	Hypothesised No. of CE(s)
0.518647	37.04098	29.68	35.65	None
0.317653	13.64407	15.41	20.04	At most 1
0.043199	1.413110	3.76	6.65	At most 2

Notes: Estimates with Eviews.

Table 7.18 Test statistics for cointegration of c_t, y_t, and z_t. VAR $(k=1)$ and $n=32$ (linear deterministic trend in the data)

Eigenvalue	LR	5% critical value	1% critical value	Hypothesised No. of CE(s)
0.601504	46.95860	29.68	35.65	None
0.375848	17.51672	15.41	20.04	At most 1
0.073218	2.433170	3.76	6.65	At most 2

Notes: Estimates with Eviews.

The estimate of the single cointegrating vector in normalised form has been estimated with Eviews to be $[1.000000 \ -0.802239 \ 0.76507 \ -2.404530]$ and the corresponding long run relationship is given by

$$c_t = 0.802239y_t - 0.76507z_t + 2.404530$$

s.e. (0.0858)   (0.3689)   (7.155)

t   [9.3467]   [−2.1319]

Let us assume now that the correct VAR order is $k=1$, according to LR, and let us also consider as before that the time series have means and linear trends but the cointegrating equations have only intercepts. Table 7.18 presents statistics of the Johansen tests based on the λ_{trace} LR statistic (7.151).

On the basis of the statistics in Table 7.18, and for the 5% significant level, we see that the hypothesised cointegrating equations are at most two, because according to the hypotheses testing in (7.153) the LR ratio statistic cannot reject the H_0: $r \leq 2$, because this statistic is less than the corresponding critical value.

The estimates of the two cointegrating vectors in normalised forms have been estimated with Eviews to be $[1.000000 \ 0.00000 \ 1.869601 \ -12.74157]$, with the corresponding long-run relationship,

$$c_t = -1.869601z_t + 12.74157$$

s.e.   (1.7578)   (7.156)

t   [−1.0636]

and $[0.00000 \ 1.00000 \ 1.467762 \ -12.89880]$ with the corresponding long-run relationship

$$y_t = -1.467762z_t + 12.89880$$

s.e.   (1.8557)   (7.157)

t   [−0.7910]

Comparing the two results above for VAR(2) and VAR(1) we see that our conclusions changed completely with respect to the number of the long run relationships that exist between these two variables. Furthermore, if we change the assumptions referring to the data generating process (DGP), i.e. assuming linear trends or not in the data and/or intercepts or trends in the cointegrating equations, we may obtain altogether different results. Table 7.19 presents the conclusions reached with the Johansen approach assuming different initial conditions for testing cointegration between these three variables.

The contradicting results presented in Table 7.19 about the cointegration between the three variables c_t, y_t and z_t show that although cointegration tests are very valuable in distinguishing between spurious and meaningful regressions, we should not rely only on cointegration methods. We should use economic theory and all the a priori knowledge that is associated with this theory in order to decide the number and the form of the cointegrating regressions. From the three long-run equilibrium relationships estimated in (7.155), (7.156) and (7.157), it seems that the cointegrating equation (7.155) is very reasonable, taking into account that the long-run income elasticity of consumption is 0.80, a highly acceptable value, and the coefficient of the inflation variable is negative, showing thus the negative effect of rising inflation on consumption.

Finally for the case of VAR(2) and one cointegrating regression the estimated first equation of the vector error correction model is shown below (the other two equations or the other estimated terms are not shown to save space):

$$\Delta \hat{c}_t = -0.317084(c_{t-1} - 0.802239y_{t-1} + 0.786507z_{t-1} - 2.40453) + \text{other items}$$

t   [−5.0259]   [−9.3467]   [2.1319]   (7.158)

$R^2 = 0.6740$

In equation (7.158) the coefficient of the estimated equilibrium error is negative (−0.317084) and significant, showing thus, that it is using the hypothesis of the error correction mechanism.

Table 7.19 Number of cointegrating regressions by VAR order and DGP assumptions

	No linear trends in series		Linear trends in series	
	No intercept – no trend in CEs	Intercept – no trend in CEs	Intercept – no trend in CEs	Intercept – trend in CEs
VAR(1)	2	2	2	2
VAR(2)	2	1	1	1

Notes: Estimates with Eviews.

7.3 Granger causality

When in a regression equation we say that the 'explanatory' variable X_t affects the 'dependent' variable Y_t, we indirectly accept that variable X_t *causes* variable Y_t, in the sense that changes in variable X_t induce changes in variable Y_t. This is in simple terms the concept of 'causality'. With respect to the direction of causality we can distinguish the following cases:

1   *Unidirectional causality:* This is the case when X_t causes Y_t, but Y_t does not cause X_t.
2   *Bilateral*, or *feedback causality:* This is the case when variables X_t and Y_t are jointly determined.

Because in most cases the direction of causality is not known, various tests have been suggested to identify this direction. The most well-known test is the one proposed by Granger (1969). This test being based on the premise that 'the future cannot cause the present or the past', utilises the concept of the VAR models. Let us therefore consider the two variable, X_t and Y_t, VAR(k) model:

$$Y_t = \alpha_{10} + \sum_{j=1}^{k} \alpha_{1j} X_{t-j} + \sum_{j=1}^{k} \beta_{1j} Y_{t-j} + \epsilon_{1t} \tag{7.159}$$

$$X_t = \alpha_{20} + \sum_{j=1}^{k} \alpha_{2j} X_{t-j} + \sum_{j=1}^{k} \beta_{2j} Y_{t-j} + \epsilon_{2t} \tag{7.160}$$

With respect to this model we can distinguish the following cases:

1   If $\{\alpha_{11}, \alpha_{12}, \ldots, \alpha_{1k}\} \neq 0$ and $\{\beta_{21}, \beta_{22}, \ldots, \beta_{2k}\} = 0$, there exists a unidirectional causality from X_t to Y_t, denoted as $X \rightarrow Y$.
2   If $\{\alpha_{11}, \alpha_{12}, \ldots, \alpha_{1k}\} = 0$ and $\{\beta_{21}, \beta_{22}, \ldots, \beta_{2k}\} \neq 0$, there exists a unidirectional causality from Y_t to X_t, denoted as $Y \rightarrow X$.
3   If $\{\alpha_{11}, \alpha_{12}, \ldots, \alpha_{1k}\} \neq 0$, and $\{\beta_{21}, \beta_{22}, \ldots, \beta_{2k}\} \neq 0$, there exists a bilateral causality between Y_t and X_t, denoted as $X \Leftrightarrow Y$.

In order to test the hypotheses referring to the significance or not of the sets of the coefficients of the VAR model of equations (7.159) and (7.160) the usual Wald F-statistic could be utilised, which is the following:

$$F_c = \frac{(SSR_r - SSR_u)/k}{SSR_u/(n - 2k - 1)} \sim F(k, n - 2k - 1) \tag{7.161}$$

where: SSR_u = sum of squared residuals from the complete equation (unrestricted)

   SSR_r = sum of squared residuals from the equation under the assumption that a set of variables is redundant (restricted)

The hypotheses in this test may be formed as follows:

H_0: X does not Granger cause Y, i.e. $\{\alpha_{11}, \ldots, \alpha_{1k}\} = 0$, if $F_c <$ critical value of F

H_a: X does Granger cause Y, i.e. $\{\alpha_{11}, \ldots, \alpha_{1k}\} \neq 0$, if $F_c >$ critical value of F

$$(7.162)$$

and

H_0: Y does not Granger cause X, i.e. $\{\beta_{21}, \ldots, \beta_{2k}\} = 0$, if $F_c <$ critical value of F

H_a: Y does Granger cause X, i.e. $\{\beta_{21}, \ldots, \beta_{2k}\} \neq 0$, if $F_c >$ critical value of F

$$(7.163)$$

It has to be noted here that in the hypotheses (7.162) and (7.163) it is not tested if 'X causes Y', but instead it is tested if 'X causes Y according to the Granger type'. This is because the Granger test is just a statistical test based not on a specific theory of causation but based on the ability of the equation to predict better the dependent variable. Furthermore, the validity of the test depends on the order of the VAR model and on the stationarity or not of the variables. The validity of the test is reduced if the variables involved are non-stationary (Geweke, 1984). Finally, other tests have been proposed, as those of Sims (1972) and Geweke *et al.* (1983), for example, and moreover Granger (1988) extended his test to also consider the concept of cointegration.

Example 7.14 Testing Granger causality between the log of private consumption and inflation rate for an EU member (the Granger test)

Consider the following two assumptions: (1) Demand (consumption) depends on the level of prices (inflation rate); and (2) The level of prices (inflation rate) depends on demand (consumption), because it is formed by a 'pull type inflation mechanism'. Either one, or both, of these assumptions may be correct. We will use the Granger causality test between the variables of the log of private consumption $(c_t = \log C_t)$ and inflation rate $(z_t = \log P_t - \log P_{t-1})$ of an EU member, for which the original values are presented in Table 7.1. Let us consider the case of a VAR(2) model because our data are annually.

In Table 7.20 we present the calculated F_c statistics according to formula (7.161) for all the possible cases of redundant variables in the two equations of the VAR(2) model.

From the results in Table 7.20 we are interested in those referring to the testing of the set of coefficients $\{\alpha_{11}, \alpha_{12}\}$ and $\{\beta_{21}, \beta_{22}\}$. We see that in both cases the F statistics are significant (significant levels less than 0.05) meaning that $\{\alpha_{11}, \alpha_{12}\} \neq 0$ and $\{\beta_{21}, \beta_{22}\} \neq 0$, or that z Granger cause c, and c Granger cause z, respectively. In other words, the Granger causation between consumption and inflation in an EU member is of the bilateral type, i.e. $c \Leftrightarrow z$, something which was possibly expected.

Table 7.20 F statistics for testing the redundancy of variables

	Redundant: z_t, z_{t-1}	Redundant: c_t, c_{t-1}
Dependent, c_t	$\{\alpha_{11}, \alpha_{12}\} = 0$ vs. $\{\alpha_{11}, \alpha_{12}\} \neq 0$ 3.9359 (0.0316)	$\{\beta_{11}, \beta_{12}\} = 0$ vs. $\{\beta_{11}, \beta_{12}\} \neq 0$ 2767.940 (0.0000)
Dependent, z_t $\{\beta_{21}, \beta_{22}\} \neq 0$	$\{\alpha_{21}, \alpha_{22}\} = 0$ vs. $\{\alpha_{21}, \alpha_{22}\} \neq 0$ 12.5418 (0.0001)	$\{\beta_{21}, \beta_{22}\} = 0$ vs. 5.6961 (0.0086)

Note: significant levels in parentheses

DISCUSSION QUESTIONS

1 Distinguish between stationary and non-stationary time series and explain why it is important to determine the properties of economic time series in empirical analysis.
2 Explain what you understand by [a] $I \sim (0)$, [b] $I \sim (1)$ and [c] $I(2)$ time series. What are the consequence of mixing an $I \sim (0)$ time series with an $I(1)$?
3 'Cointegration methodology is basically an "inductive" method lacking a theoretical foundation'. Discuss.
4 'The basic problem with unit root tests is lack of "power"'. Discuss.

Bibliography

Bartlett, M. S. (1946) 'On the theoretical specification of sampling properties of autocorrelated time series', *Journal of the Royal Statistical Society,* Series B, 27, 27–41.

Box, G. E. P. and Piece, D. A. (1970) 'Distribution of residual autocorrelations in autoregressive integrated moving average time series models', *Journal of The American Statistical Association,* 65, 1509–1525.

Campbell, J. Y. and Perron, P. (1991) 'Pitfalls and opportunities: What macroeconomists should now about unit roots', *Technical Working Paper 100,* NBER Working Papers Series.

Charemza, W. W. and Readman, D. F. (1997) *New Directions in Econometric Practice: General to Specific Modelling, Cointegration and Vector Autoregression,* 2nd edn., Cheltenham: Edward Elgar.

Cuthbertson, K., Hall, S. G. and Taylor, M. P. (1992) *Applied Econometric Techniques,* New York: Philip Allan.

Davinson, R. and MacKinnon, J. G. (1993) *Estimation and Inference in Econometrics,* New York: Oxford University Press.

Dickey, D. A. and Fuller, W. A. (1979) 'Distributions of the estimators for autoregressive time series with a unit root', *Journal of American Statistical Association,* 74, 427–431.

Dickey, D. A. and Fuller, W. A. (1981) 'Likelihood ratio statistics for autoregressive time series with a unit root', *Econometrica,* 49, 1057–1072.

Dickey, D. A. and Pantula, S. (1987) 'Determining the order of differencing in autoregressive processes', *Journal of Business and Economic Statistics,* 15, 455–461.

Dickey, D. A., Bell, W. and Miller, R. (1986) 'Unit roots in time series models: Tests and implications', *American Statistician,* 40, 12–26.

Dickey, D. A., Hasza, D. P. and Fuller, W. A. (1984) 'Testing unit for roots in seasonal time series', *Journal of the American Statistical Association,* 79, 355–367.

Dickey, D. A., Jansen, D. W. and Thornton, D. L. (1994) 'A primer on cointegration with an application to money and income', in B. B. Rao (ed.), *Cointegration for the Applied Economist,* New York: St. Martin's Press.

Doldado, J., Jenkinson, T. and Sosvilla-Rivero, S. (1990). 'Cointegration and unit roots', *Journal of Econometric Surveys,* 4: 249–273.

Enders, W. (1995) *Applied Econometric Time Series,* New York: John Wiley.

Engle, R. F. and Granger, C. W. J. (1987) 'Cointegration and error correction: Representation, estimation and testing', *Econometrica,* 55, 251–276.

Engle, R. F. and Yoo, B. S. (1987) 'Forecasting and testing cointegrated systems', *Journal of Econometrics,* 35, 145–159.

Engle, R. F., Granger, C. W. J. and Hallman, J. J. (1989) 'Mergin short- and long-run forecasts: an application of seasonal cointegration to monthly electricity sales forecasting', *Journal of Econometrics,* 40, 45–62.

Fuller, W. (1976) *Introduction to Statistical Time Series,* New York: John Wiley.

Geweke, J. (1984) 'Inference and causality in economic time series models', in Z. Griliches and M. D. Intriligator (eds), *Handbook of Econometrics,* vol. 2, Amsterdam: North Holland.

Geweke, J., Meese, R. and Dent, W. (1983) 'Comparing alternative tests of causality in temporal systems', *Journal of Econometrics,* 77, 161–194.

Granger, C. W. J. (1969) 'Investigating causal relations by econometric models and cross-spectral models', *Econometrica,* 37, 424–438.

Granger, C. W. J. (1986) 'Developments in the study of cointegrated economic variables', *Oxford Bulletin of Economics and Statistics,* 48, 213–228.

Granger, C. W. J. (1988) 'Some recent developments in a concept of causality', *Journal of Econometrics,* 39, 199–221.

Granger, C. W. J. and Newbold, P. (1974) 'Spurious regressions in econometrics', *Journal of Econometrics,* 35, 143–159.

Greene, W. H. (1999) *Econometric Analysis,* 4th edn., New Jersey: Prentice-Hall.

Gujarati, D. N. (1995) *Basic Econometrics,* 3rd edn., New York: McGraw-Hill.

Holden, D. and Perman, R. (1994) 'Unit roots and cointegration for the economist' in Rao, B. B. (ed.), *Cointegration for the Applied Economist,* New York: St. Martin's Press.

Hylleberg, S., Engle, R. F., Granger, C. W. J. and Yoo, B. S. (1990) 'Seasonal integration and cointegration', *Journal of Econometrics,* 44, 215–238.

Jenkins, G. M. and Watts, D. G. (1968) *Spectral Analysis and Its Applications,* San Francisco: Holden-Day.

Johansen, S. (1988) 'Statistical analysis of cointegration vectors', *Journal of Economic Dynamics and Control,* 12, 213–254.

Johansen, S. and Juselius, K. (1990) 'Maximum likelihood estimation and inference on cointegration – with applications to the demand of money', *Oxford Bulletin of Economics and Statistics,* 52, 169–210.

Judge, G. G., Hill, R. C. Griffiths, W. E. Lutkepohl, H. and Lee, T-C. (1988) *Introduction to Theory and Practice of Econometrics,* 2nd edn., New York: John Wiley.

Ljung, G. M. and Box, G. P. E. (1978) 'On a measure of lack of fit in time series models', *Biometrica,* 66, 66–72.

MacKinnon, J. G. (1991) 'Critical values of cointegration tests', in R. F. Engle and C. W. J. Granger (eds), *Long-Run Econometric Relationships: Readings in Cointegration,* New York: Oxford University Press.

Maddala, G. S. (1992) *Introduction to Econometrics,* 2nd edn., New Jersey: Prentice-Hall.

Nelson, C. R. and Plosser, C. I. (1982) 'Trends and random walks in macroeconomic time series', *Journal of Monetary Economics,* 10, 139–162.

Osborn, D. R., Chui, A. P. L., Smith, J. P. and Birchenhall, C. R. (1988) 'Seasonality and the order of integration in consumption', *Oxford Bulletin of Economics and Statistics,* 50, 361–377.

Osterwald-Lenum, M. (1992) 'A note with quantiles of the asymptotic distribution of the maximum likelihood cointegration rank test statistics', *Oxford Bulletin of Economics and Statistics,* 54, 461–472.

Phillips, P. C. B. and Loretan, M. (1991) 'Estimating long-run economic equilibria', *Review of Economic Studies,* 5, 407–436.

Phillips, P. and Perron, P. (1988) 'Testing for a unit root in time series regression', *Biometrica,* 75, 335–346.

Sargan, J. D. (1964) 'Wages and prices in the United Kingdom: A study of econometric methodology', in P. E. Hart, G. Mills and J. K. Whitaker, *Econometric Analysis for National Economic Planning,* London: Butterworths.

Sargan, J. D. and Bhargava, A. (1983) 'Testing residuals from least squares regression for being generated by the Gaussian random walk', *Econometrica,* 51, 153–174.

Sims, C. (1972) 'Money, income, and causality', *American Economic Review,* 62, 540–552.

Stock, J. H. (1987) 'Asymptotic properties of least squares estimators of cointegrating vectors', *Econometrica,* 55, 1035–1056.

Stock, J. H. and Watson, M. (1988) 'Testing for common trends', *Journal of the American Statistical Association,* 83, 1097–1107.

Thomas, R. L. (1997) *Modern Econometrics: an introduction,* Harlow: Addison-Wesley.

8 Applications of cointegration

A person who understands econometrics may still not know how to utilise the knowledge to cope with empirical. This is a common problem facing students. To address this issue, this chapter aims to employ four case studies to display key practical insights in using cointegration. Readers should find these useful. They should benefit both in terms of understanding the underlying economic principles of each study and experiencing the skills required to tackle economic issues when using cointegration. The nature of the economic problems are different and need different strategies for applications. Accordingly, different conclusions and recommendations are needed. The methodological thinking behind these estimations are typical General to Specific studies. Hence, this chapter will benefit all who seek to utilise cointegration as a time series analysis methodology. The four case studies relate to topics, namely, UK brewing, advertising investment, Internet growth and aggregate expenditure. The UK brewing case evaluates government policy implications to test price-supply relationships. Secondly, since advertising is critical for brewers for market share, an analysis of advertising investment investigates the impact of advertising on sales and assesses the relative economic influences of 'main media advertising' and the 'below-the-line' advertising. A cointegration analysis of the Internet produces a forecasting system to identify the key elements influencing growth. Finally, macro-economic study of the UK import function relates elasticities to aggregate spending in terms of a cointegration study.

CASE 1 COINTEGRATION ANALYSIS IN UK BREWING

I Background

The UK brewing industry has featured vertical integration in brewing and retailing for decades. In 1989, the Monopolies and Mergers Commission (MMC) investigated so-called 'monopolised' retail linkages and imposed viable restrictions to frustrate the potential for development of 'monopoly' practices in pricing. Ironically, with continued excess capacity and changing consumption patterns, widespread re-structuring has occurred via merger and industrial concentration increased. The newly created industrial structure in UK brewing is displayed in Table 8.1.1. In terms of market shares, industrial concentration in 1998/9 contradicts the MMC's avowed policy intentions of 1989. Inspection of Table 8.1.1 shows that concentration of markets share distribution is heavily skewed in favour of large players.

Table 8.1.1 Market shares in UK brewing

	1985	1989	1991	1994	1995	1998$^{\psi}$	1999$^{\phi}$
Bass	22	23	22	22	22	23.9	24
Courage	9	20	19	19	19		
Grand Metropolitan	12						
S&N	10	10	11	11	11 (30)*		
Scottish Courage						29.7	30
Whitbread	11	11	12	13	13	14.0	11
Carlsberg			1				
Allied	13	14	13				
Carlsberg-Tetley				19	19	14.6	15
National Brewers	*77*	*78*	*77*	*84*	*84*	*82.2*	*80*
Other	*23*	*22*	*23*	*16*	*16*	*17.8*	*20*

Source: CAMRA and MMC report on the supply of beer and the Allied Carlsberg merger; $^{\psi}$*Financial Times*, p. 24, 27th November, 1998 (William, J); $^{\phi}$Estimated by an industry analyst in March 1999.

Note: ( )* denotes the new market share after S&N took over Courage in August 1995. In 1991, Grand Metropolitan sold its breweries to Courage and exited brewing. In 1992, Allied and Carlsberg merged their brewing and wholesaling operations and formed a company called Carlsberg-Tetley Ltd.

Inspection of Table 8.1.2 shows that the MMC's strictures and Orders reduced vertical linkages controlled by major brewers. However, the disposal of premises by national players may have occurred anyway as a natural process of disposal of low barrelege establishments following declining beer sales in the UK. This declining trend is reflected in the quantity released for aggregate UK consumption, and total UK production (Table 8.1.3). The declining demand for beer in the UK coincided with the present era of barrier-free European trade in beers and ales in 1993. Boot-legging between Calais and Dover has been common in recent years. This trade supplies vast quantities

Table 8.1.2 Brewers' ownership in the UK (number of premises)

	1980	1985	1990	1991	1992	1993	1994
Managed on licenses	14,000	12,900	13,500	13,000	12,000	12,000	11,700
Tenanted on licenses	35,000	33,600	29,800	25,100	19,100	18,600	18,600
All on licenses	49,000	46,500	43,300	38,100	31,100	30,600	30,300

Source: Brewers and Licensed Retailers Association (BLRA), 1995.

Table 8.1.3 Brewing consumption and production in the UK (000' barrels)

	Beer – quantity released for UK consumption	*UK beer production*	*UK import*	*% of import from Europe*	*UK export*	*% of export to Europe*
1988	39,732	37,813	2,658	95.3	753	54.7
1989	39,902	37,965	2,770	95.8	816	51.7
1990	39,829	37,751	3,096	94.0	992	53.2
1991	38,518	36,388	3,256	90.7	1,126	56.8
1992	37,256	35,206	3,382	93.0	1,273	59.8
1993*	35,901	34,673	2,800	92.6	1,283	57.0
1994	36,515	35,643	3,160	93.8	1,936	67.0

Source: Brewers and Licensed Retailers Association, 1995

Note: *According to CSO, the Introduction of the Single Market in January 1993, the method of recording imports to the EU changed and the related data may not be comparable with earlier years.
Europe here refers to all countries from the European Continent.

Table 8.1.4 General retail price index and beer price index

Year	$^\phi$All Items (index)	All Items (% change)	$^\phi$In-sales (index)	In-sales (% change)	$^\phi$Off-sales (index)	Off-sales (% change)
1987	101.9		103.7		103.7	
1988	106.9	4.91	110.9	6.94	108	4.15
1989	115.2	7.76	117.5	5.95	112.2	3.89
1990	126.1	9.46	132.4	12.68	121.5	8.29
1991	133.5	5.87	149.4	12.84	134.2	10.45
1992	138.5	3.75	157.4	5.35	139.0	3.58
1993	140.7	1.59	165.4	5.08	143.7	3.38
1994	144.2	2.49	166.7	0.79	142.0	−1.18

Source: $^\phi$Employment Gazette Jan 1987–December 1994, Employment Department UK.

of cheap beers to domestic black markets. This large influx of foreign beer also made UK brewers aware of the strong competition from foreign traders, particularly brewers from European Union who pay relatively low tax rates on beers.

In Table 8.1.4, the beer price index for in-sales has been growing faster than the retail price index (RPI) since the Beer Orders (1989). In 1989, the rate of growth for the All Item General Price Index was 7.76%, which was higher than that for in-sales (5.95%). However, since 1990, the situation changed, whereby the growth rates of the in-sales index took the lead. For instance, the growth of the in-sales index peaked in 1990 (12.68%), whereas the General Price Index peaked in same year with a growth rate of 9.46%. Moreover, the divergence between the two indices has widened steadily in recent years. In contrast, beer prices for off-sales have been increasing at a slower rate, lagging behind the RPI with rare exceptions in 1991 and 1993.

So, if the question is raised as to what extent the price of beers have fluctuated around a steady state equilibrium in the UK, no unequivocal conclusion can be advanced. In order to understand the structural stability of UK prices, it is necessary to investigate the nature of the equilibrium conditions existing between beer prices and other key supply-side variables. By using cointegration, long-run findings are used to develop an error correction model (ECM), to track down short-run price variations in the UK market. Moreover, ex-ante forecasting testing is used to evaluate the structural impact of the MMC's 1989 recommendations on pricing/supply relationships within the UK brewing industry.

II The model

Stage 1

To demonstrate supply-side response relationships in the UK market, the following model specification was constructed:

$$P_t = \alpha C_t + \beta M_t + e_t \qquad (8.1.1)$$

where;

α and β: estimated coefficients

t: time

e: residual term

Hence the beer price index in the UK (P_t) is assumed to be associated with the total volume of beer released for consumption in the UK (C_t) and the quantity of net imports of beer (M_t). M_t is obtained by measuring the difference between C_t and UK total beer production. It is worth noting that not all beers brewed in the UK are available for domestic consumption and C_t includes imported beers. Utilisation of C_t as the only explanatory variable ignores the increasing significance of imported beers. Thus a more appropriate model specification for estimating trends in price indices would be to include an explanatory variable which captures imported volumes. However, since 1993, official methods of recording the relevant data changed and so these data are no longer reliable and comparable with previous series. The second best alternative is to use M_t from reliable official data sources and consider M_t as the proxy capturing the gross influence of imported beer volumes. Accordingly using Johansen (1988) multivariate cointegrating procedures, the model was estimated over the sample period, 1983 Q1 to 1994 Q4.

Stage 2

The estimation begins with an Augmented Dickey-Fuller (ADF) test to confirm that all variables of the model are suitable to be included for cointegration. Results shown in Table 8.1.A1 indicate that all variables are of I(1) and their first differences are stationary. Hence, all variables are suitable for estimation by the long-run cointegration technique. The next step is to choose a specific lag length of VAR. Thereafter suitable trace tests and maximum eigenvalue tests can be undertaken to identify the number of cointegrating vectors in the model. The results from the hypothesis testing shown in Table 8.1.A2 reveal that there is only one cointegrating vector for the long-run estimation. The normalised coefficients for this model generated from the cointegrating vector are thus given as follows:

$$P_t = -16.6812C_t + 1.0227M_t \tag{8.1.2}$$

The sign of coefficient C_t is correct. This negative sign implies that as total volumes released increases, there is a tendency in the long-run to push aggregate beer prices downwards. However, the magnitude of the coefficient is rather large, but it should be remembered that C_t is of a much larger absolute scale, such that a small percentage change in C_t is strong enough to induce rather larger movements in the beer price index. For instance, if C_t increased by 1%, the beer index would fall by 16.68%. This means that the price index in the long-run is very sensitive to changes in the total volumes released for consumption. Given the fact that the demand for beer in the UK had decreased significantly since 1980 (Table 8.1.5), then as more imported beers enter the UK market the divergence between aggregate demand and aggregate supply for beer inevitably widens. A prolonged period of excess supply is likely to generate long-run elastic responses to quantity changes.

Generally, if concentration increases, then real market prices can be expected to rise because of the increased market power. However, in the UK excess brewing capacity has been a critical factor hindering the practise of tacit collusion in beer pricing. Mergers in brewing in recent years has obviously increased industrial concentration. However, this process should be viewed as a long-run rationalisation tendency keeping surviving brewers potentially more efficient. Hence, the rising

Table 8.1.5 The consumption of alcoholic drinks in the UK

	Beer (pints/head)	Cider (litres/head)	Wine (litres/head)	Spirit – 100% alcohol (litres/head)
1980	263.6	4.8	10.2	2.2
1988	247.7	6.1	15.3	2.2
1990	246.0	7.3	15.7	2.1
1992	229.2	8.2	16.4	1.8
1994	223.7	9.5	17.9	1.9

Source: Brewers and Licensed Retailers Association, 1995

Note: Conversion Factor: One litre = 1.75975 pints

trends in industrial concentration in UK brewing have not led to sustained real price escalation. Table 8.1.4 shows that inflation in beer price index for in-sales and off-sales compared to the RPI, slowed down markedly since 1992. Moreover, in 1994 both beer price indices showed a relatively lower rate of increase compared to that of the RPI. In equation (8.1.2), M_t displays a positive and approximately unitary response. Moreover, it well documented that in recent years the premium market segment grew rapidly and since higher prices are applied to imported beers, it is to be expected that a positive long-run influence is experienced.

Stage 3

Further investigation of the supply-price relationship can be obtained using a short-run ECM. This is constructed by employing the residuals saved from the long-run estimation as the error correction variable (ecv). In estimating the short-run model using the General to Specific modelling procedure, the following short-run coefficients are recorded in equation (8.1.3):

$$\Delta P_t = -1.3731 - 0.28845 \Delta C_{t-1} - 0.34408 \Delta C_{t-2} - 0.21957 \Delta C_{t-3} - 0.11498 \Delta C_{t-4}$$
$$(-2.6039)\,(-3.3031) \quad\quad (-4.7629) \quad\quad (-4.123) \quad\quad (-2.8634)$$

$$+ 0.008811 \Delta M_{t-1} - 0.012914 D3 + 0.013134 ecm_{t-1}$$
$$(2.3721) \quad\quad (-3.0395) \quad\quad (2.6215) \quad\quad\quad\quad\quad (8.1.3)$$

$$R^2 = 0.5492 \; DW = 2.0957 \quad\quad \text{Serial Correlation: } \chi^2(4) = 3.3970$$

$$\text{Functional Form: } \chi^2(1) = 3.3147 \quad \text{Normality: } \chi^2(2) = 0.99288$$

$$\text{Heteroscedasticity: } \chi^2(1) = 0.39781 \; \text{Chow: } F(8, 27) = 1.1897)$$

According to the t-statistics, all coefficients of model (8.1.3) are statistically significant at the 5% level. All C_t variables are of the same sign as is shown in the long-run model, but their short-run responses are inelastic. The response shown by M_t is inelastic too, however, this is positively related with P_t. Moreover, since the demand for beer is assumed to follow dominant seasonal trends, three different seasonal dummies were inserted in the short-run estimation and eventually, the third quarter seasonal dummy was found significant and produced a negative sign. This scenario is likely in the summer 'season' since more varieties of beer and higher aggregate supply volumes are available in retail outlets to match random increases in summer demand. All the results from the diagnostics tests generated from the short-run model indicate that the ECM behaves with a white noise process. For example, Figure 8.1.1 indicates that the estimated values from model 3 match the actual values reasonably well.

Given the changes in nominal structural supply relationships created by the Beer Orders 1989, it is of paramount importance to investigate possible structural changes in aggregate supply-price relationships in UK beer markets. Hence, a CUSUM test was utilised to evaluate the stability of parameters in model (3). The results derived are depicted in Figure 8.1.2 and indicate that at 5% significance level, the parameters in model (3) are statistically stable over the study period between 1983 Q1 and 1994 Q4. Another method for confirming the structural stability of the model for the study period was to run an ex-ante forecasting procedure for the post 1989 period based on model (3). The forecasting diagnostic tests reported in Table 8.1.6 confirm the accuracy of the model in predicting the short-run fluctuations in the UK demand for beer. The forecasting results, depicted in Figure 8.1.3 indicate

Table 8.1.6 Forecast diagnostics generated from short-run error correction model

Mean Prediction Error = 0.0012358	Mean Sum Absolute Prediction Errors = 0.0035072
Sum of Squared Prediction Errors = 0.0000188	Root Mean Sum Square Prediction Errors = 0.0043354
Predictive Failure Test: $F(21, 14) = 0.55469$	Structural Stability Test $F(8, 27) = 1.1897$

Note: 22 observations used in estimation of the model from 84 Q2 to 89 Q3.

Table 8.1.A1 Unit root tests

Variable	AD test (level)	ADF test (first difference)
P_t	$-1.9470*$ (0)	-6.9395 (0)
M_t	-1.637 (1)	-18.6164 (0)
C_t	$-0.84125*$ (3)	-9.9104 (0)

Note: the null hypothesis in each case is that the variable in question is I(1); the 95% critical value are given in brackets derived from Fuller (1976). In the majority of cases the DF and ADF values are greater than the corresponding critical values thus confirming the existence of a unit root.

Table 8.1.A2 Maximal eigenvalue tests for cointegrating vectors

Null hypothesis: the number of cointegrating vectors, r.	Likelihood ratio statistic LR (n)	5% critical value	10% critical value
$r \leq 0$	66.9349	21.0740	18.904
$r \leq 1$	5.3147	14.9000	12.912
$r \leq 2$	1.4553	8.1760	6.503

Note: The LR(n) statistics are asymptotically $\chi^2(n)$ variates under the null hypothesis. The likelihood ratio test tests that the number of cointegrating vectors is at most equal to r. The sequential testing stops when H_0 can not be rejected. H_0: $r \leq 2$ is reported for additional information.

Trace test

Null	Alternative	L-R statistic LR (n)	5% critical value	10% critical value
$r = 0$	$r = 1$	73.7049	31.525	28.709
$r \leq 1$	$r = 2$	6.77	17.953	15.663
$r \leq 2$	$r = 3$	1.4553	8.176	6.503

Note: Test statistics are asymptotically $\chi^2(n)$ variates under H_0. The sequential testing stops when H_0 can not be rejected. H_0: $r \leq 2$ is reported as additional information.

that the forecast values track the actual values well and capture most of the upswings and downswings in prices.

All in all, this empirical study utilises cointegration to produce findings to suggest that there has been no significant change in aggregate supply relationships with reference to beer price indices, following changes in supply structures in UK brewing industry induced by government intervention.

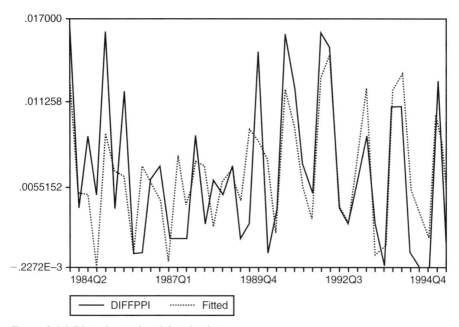

Figure 8.1.1 Plot of actual and fitted values

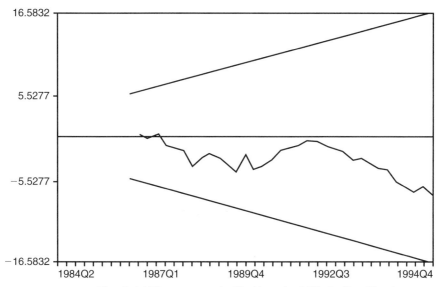

The straight lines represent critical bounds at 5% significant level

Figure 8.1.2 Plot of cumulative sum of recursive residuals

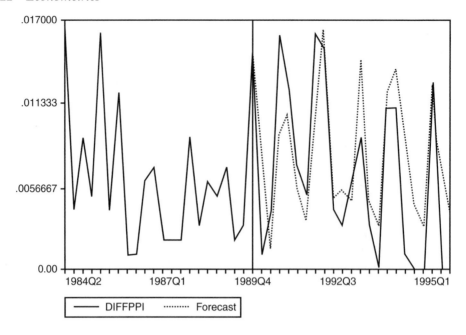

Figure 8.1.3 Plot of actual and static forecast(s)

CASE 2 CONTINUOUS ADVERTISING INVESTMENT IN UK BREWING

I Background

In this analysis we test this hypothesis with respect to main forms of advertising available to brewers – 'main-media' and 'below the line' advertising. The incumbent firms' long-run advertising policy can influence the strategic behaviour of potential entrants. Following the Arrow-Nerlove (1962) approach we adopt the view that market structure is, to a significant degree, endogenous in all but the short run. Strategic conduct by incumbents is critical in affecting future profitability. It is shown from the evidence that given a choice of advertising techniques, brewers will normally aim to undertake continuous 'main-media' advertising investment to improve potential market share. This is a long-run strategic decision. The traditional static approach typically ignores this key phenomenon, whereas in our approach continuous advertising investment is the central theme of the analysis.

In 1993 UK national brewers spent more than £70 million on advertising (Table 8.2.1). The advertising/turnover ratio for national brewers varies from 0.3% (Carlsberg and Tetley) to 1.5% (Courage). Table 8.2.2 illustrates the effectiveness of using advertising to stimulate lager sales. Despite the fact that sales performance by leading brewers solely attributed to advertising investment, the data presented suggests that the efficient market leader is Bass PLC since it has the lowest advertising sales ratio (1.11), followed by Carlsberg-Tetley (1.96). Whitbread however appears to be less efficient in utilising advertising investment. Both Tables 8.2.1 and 8.2.2 indicate that the beer market is very competitive. These ratios suggest that advertising has a powerful intrinsic commercial value, otherwise brewers would not be

Table 8.2.1 Advertising/turnover ratio, 1993

Brewer	Turnover (£m)	Advertising (£000)	Advertising/turnover ratio (%)
Bass	4451	14654	0.3292
Carlsberg & Tetley	5266	16209	0.3078
Courage	1326	20129	1.5180
S&N	1514	7880	0.5205
Whitbread	2346	18282	0.7793

Source: Register-Meal.

Table 8.2.2 Advertising-sales ratio for lager products, 1993

Brewer	Lager sales (£m)	Lager volume share in the UK (%)	Advertising on lager products (£000)	Advertising sales (%)
Bass	755	25	8403	1.1130
Carlsberg Tetley	635	21	12456	1.9616
Courage	745	25	17114	2.2972
S&N	220	7	4889	2.2223
Whitbread	390	13	10155	2.6039

Source: Register-Meal.

willing to put such a large proportion of their sales revenue into long-run advertising investments.

Theory suggests advertising expenditure has distinctive features, since it can be seen as a durable good with positive long-term effects on brands and individual firms, through building up stocks of goodwill. Moreover, advertising produces scales economies and generates entry barriers which discourage potential entrants. Brand building is currently a key strategy in UK brewing. In this paper, investment used for long term brand promotion is treated as an asset instead of a pure suck cost with no salvage value. Tables 8.2.3 and 8.2.4 indicate that brewers are willing to allocate far larger budgets to 'main-media' advertising than 'below-the-line' advertising investment. Accordingly, this paper seeks to isolate the rationale for this long-run advertising strategy.

Table 8.2.3 Main-media advertising budgets

	Advertising (ale and stout)	Advertising (lager) (£m)	Total advertising (£m)
1987	29.2	68.9	98.1
1988	31.4	70.4	101.8
1999	29.7	72.7	102.4
1990	28.6	73.7	102.3
1991	33.4	73.0	106.4
1992	41.8	85.0	126.8
1993	45.1	68.8	113.9
1994	44.7	60.0	104.7

Source: Register-Meal.

Table 8.2.4 Below-the-line advertising in the UK (1987 prices)

Year	Advertising rate card prices (£m)	Index
1983	2.0	133.85
1984	2.4	160
1985	1.4	93.85
1986	1.5	100
1987	1.5	100
1988	1.5	92
1989	1.1	68
1990	1.5	92
1991	2.9	181

Source: Register-Meal.

II The model and data

Stage 1

Our analysis assumes that a typical large brewer maintains a continuous flow of real advertising investment resources over time. This means that such investments are not concentrated on given periods of the year, but are distributed equally in each quarter of a year. Consequently, systematic advertising investment implies the stock of 'goodwill' for any given firm takes time to accumulate. Hence, continuous advertising investment can be conveniently considered as:

$$A = (\text{Annual real advertising Investment/RPI}) \div 4 \qquad (8.2.1)$$

Moreover, we utilise a standard linear model of the demand for beer (see Lynk, 1985; Penm, 1988).

The model can be written as follows:

$$Q_D = f(Y, P_B, P_S, A) \qquad (8.2.2)$$

where:

Q_D = a proxy of the demand for beer in real terms (total barrelage)

Y = real UK personal disposable income

P_B = the ratio of an index of beer prices to the retail price index

P_S = the ratio of a price index of substitute alcohol products to the retail price index

A = real advertising investment; (A_M = 'main-media' advertising; A_B = 'below-the-line' advertising)

Economic theory would predict that the income elasticity of demand for beer should be positive over time. Moreover, whereas the advertising elasticity could be expected to be positive, the own price elasticity of demand should be negative and the cross-price elasticity should be positive. However, short-run prices discounting, 'tit-for-tat' marketing strategies and intensive advertising strategies in the short-run could produce perverse effects. This would not be unusual in any given short-run analysis of oligopoly. Moreover, given the long-run decline in the demand for beer, competitive long-run advertising investment expenditure by leading brewers could counteract short-run advantages attained by 'first movers' in such games. Indeed, over the long-term, the overall impact of both types of advertising strategy by leading brewers may not have any significant impact on the aggregate demand for beer, or its long-term decline, but have serious consequences for any individual brewer which does not implement a continuous advertising investment approach.

Stage 2

Testing and estimation

In order to test for the long-run impact of 'main-media' and 'below-the-line' advertising on aggregate UK beer sales, the Johansen multivariate cointegratiion procedure was used. To implement this method, it is necessary to: (a) test for the order of integration of each variable in the model; and (b) use likelihood ratio hypothesis testing to identify the number of distinct cointegrating vectors. Given the problems of data availability, specifically for the advertising variables, it was only possible to estimate equation (8.2.1) with 'main-media' advertising over the period 1987 Q4 to 1994 Q1 and 'below-the-line' advertising over the period 1983 Q4 to 1990 Q4. This lack of observations may slightly impair our empirical analysis since all the tests are only asymptotically valid.

Table 8.2.A1 shows the ADF tests, which indicate that the levels of each variable are integrated or order one, and the first differences are stationary. Given these results, each variables satisfies the requirements to be included in the long-run cointegration model. To implement the Johansen procedure it was necessary to specify the lag length of the VAR at one time period, which in each case ensured that the error term was white noise. Given the seasonal fluctuations in the barrelage sales over each year, seasonal dummy variables were also included in the long run cointegration equation. The Johansen procedure produces Trace and Maximum Eigenvalue Tests, from which the number of cointegrating vectors can be identified. The results of the hypothesis testing are shown in Tables 8.2.A2 and 8.2.A3. The results from Table 8.2.A2 clearly reject the null hypothesis of zero cointegrating vectors against the alternative, at both 5 and 10% significance levels. Each test indicates the presence of at most one cointegrating vector in respect to the variables specified in

Table 8.2.A1 ADF tests

	1987 Q4–1994 Q1 Main-media advertising		1983 Q4–1990 Q4 Below-the-line advertising	
	ADF test (levels)	*ADF test (1st difference)*	*ADF test (levels)*	*ADF tests (1st difference)*
Q_D	−2.6784 (1)*	−8.9377 (0)	−3.4134 (2)*	−6.9335 (0)
Y	−1.7602 (1)	−10.4326 (0)	−2.7793 (1)*	−13.2213 (1)
P_B	−1.7898 (0)*	−4.0243 (0)	−2.1142 (1)	−7.8139 (0)
P_S	−2.9023 (0)*	−6.9391 (0)	−2.6054 (0)*	−6.3145 (0)
A_M	−1.1163 (0)*	−4.9169 (0)	–	–
A_B	–	–	−2.2636 (0)	−5.2902 (0)

Note: * denotes a significant time trend. The lag length of the ADF regression is specified in parentheses.

Table 8.2.A2 Maximum eigenvalue and trace tests – 1987 Q4–1994 Q4
Main-media advertising

No. of cointegrating vectors	Likelihood ratio Statistics LR(n)	95% critical value	90% critical value
Maximum engenvalue test			
r = 0	70.1365	34.4000	31.6640
r = 1	15.4378	28.11380	25.5590
r = 2	8.1760	22.0020	19.7660
Trace test			
r ≤ 0	101.1010	76.0690	71.8620
r ≤ 1	30.9645	53.1160	49.6480
r ≤ 2	15.5267	34.9100	32.0030

Table 8.2.A3 Maximum eigenvalue and trace tests – 1983 Q4–1990 Q4
Below-the-line advertising

No. of cointegrating vectors	Likelihood ratio statistics LR(n)	95% critical value	90% critical value
Maximum engenvalue test			
r = 0	28.9126	34.4000	31.6640
r = 1	19.2930	28.11380	25.5590
r = 2	13.8464	22.0020	19.7660
Trace test			
r ≤ 0	69.4845	76.0690	71.8620
r ≤ 1	40.5719	53.1160	49.6480
r ≤ 2	21.2789	34.9100	32.0030

the model. The results from Table 8.2.A3 indicate that when the 'below-the-line' advertising variable is included, the model does not cointegrate at both the 5 and 10% significance levels. This is an important finding, since the Johansen procedure indicates which particular form of advertising is more important in influencing aggregate beer sales.

The Johansen procedure also produces the normalised coefficients associated with the unique cointegrating vector, obtained from the results of Table 8.2.A2, which are given as follows:

$$Q_D = 10.0311 + 0.030203Y - 0.36169P_B + 0.39551P_S - 0.13809A_M \qquad (8.2.3)$$

The income elasticity is positive in this case although the magnitude of the coefficient is far smaller than one would normally expect. Both the coefficients of the P_B and P_S variables have appropriate signs and magnitudes. The demand for beer is own price inelastic and the cross elasticity is positive, indicating that beer is a substitute for other alcoholic products, over the period examined. Surprisingly, the coefficient of the A_M variable is negative and has a smaller magnitude than expected. A likelihood ratio test, which imposed a zero restriction on this particular coefficient, yielded a test statistic of 0.95156, which for a $\chi^2(1)$ of 3.84146 at the 5% significance level, indicates that 'main-media' advertising had no significant influence on aggregate beer sales over the period examined. A plot of the residuals from the cointegrating vector in equation (8.2.3) indicates an absence of serial correlation (see Figure

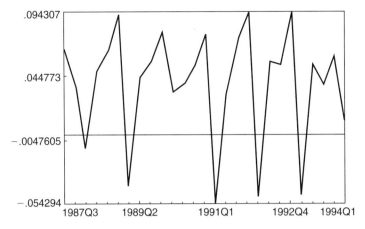

Figure 8.2.1 Plot of estimated residuals from cointegration regression

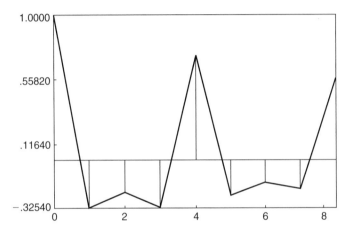

Figure 8.2.2 Autocorrelation function of estimated residuals from cointegration regression

8.2.1). This conclusion is supported from the plot of the estimated residuals auto correlation function, shown in Figure 8.2.2. An ADF test for the residuals was esti-mated to be -5.3641, which given the 95% critical value of -4.5449 indicates that the estimated residuals are I(0) and thus the cointegration regression specified is a valid long run equilibrium relationship, since the variables have a tendency to move together over time.

The residuals from the cointegration equation were then incorporated into a short-run error correction model, with a view to investigating the short-run move-ments in aggregate beer sales:

$$\Delta Q_D = \alpha + \sum_{i=0}^{5} \beta_i \Delta Y_{t-i} + \sum_{i=0}^{5} \delta_i \Delta P_{B,t-i} + \sum_{i=0}^{5} \gamma_i \Delta P_{S,t-i} + \sum_{i=0}^{5} \phi_i \Delta A_{M,t-i}$$

$$+ \sum_{i=0}^{5} \lambda_i \Delta Q_{D,t-i} + \sum_{i=1}^{3} v_i DUM_i = \pi ECM_{t-1}$$

(8.2.4)

Given the quarterly frequency of the data, up to five lags were included in the ECM. Hypothesis testing was then used to derive a parsimonious equation, using the general to specific procedure, the results of which are given as follows:

$$\Delta Q_{D,t} = 0.023714 + 1.7886\Delta Y_t + 13.98047\Delta P_{B,t} + 2.4114\Delta P_{s,t} - 0.45114\Delta A_{M,t}$$
$$\quad\quad (2.1465)\quad\quad (2.5834)\quad\quad\quad (2.2718)\quad\quad\quad\quad (1.6705)\quad\quad\quad\quad (-1.2991)$$

$$\quad - 1.1716 ECM_{t-1}$$
$$\quad\quad (-6.1450)$$

$R^2 = 0.82110, DW = 1.9806$ $\quad\quad\quad\quad\quad\quad\quad\quad\quad\quad\quad\quad\quad$ (8.2.5)

Serial Correlation, $\chi^2(4) = 8.6018$

Normality, $\chi^2(2) = 1.1.167$

Heteroscedasticity, $\chi^2(1) = 0.00084$

The diagnostic test indicates that the residuals approximate a white noise process. The plot of the actual and fitted values in Figure 8.2.3 suggest that the short-run model tracks the data well. Moreover, a CUSUM test of parameter stability (Figure 8.2.4) indicates that the estimated parameters of the model are stable over the sample period examined. All of the short-run elasticities have magnitudes which are far larger than anticipated. Strangely, the own price elasticity is positive, the coefficient of the A_M variable is also incorrectly signed and insignificant. In support of the conclusions from the cointegration equation, the results indicate 'main-media' advertising has no significant influence on short-run movements in beer sales, as well as having no long-run impact. All in all, the econometric analysis shows that in the case of 'main-media' advertising, this variable should be included in the cointegrating vector of beer demand, despite the fact that its impact was shown to be insignificant. Our evidence implies that both forms of advertising had no significant impact on beer sales in the long run. These are startling findings given that advertising investment by brewers has increased dramatically in the 1990s compared with similar expenditure in the 1980s. Ironically though, total beer sales in the UK have

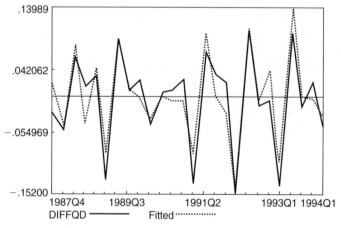

Figure 8.2.3 Plot of actual and fitted values from short run error correction model

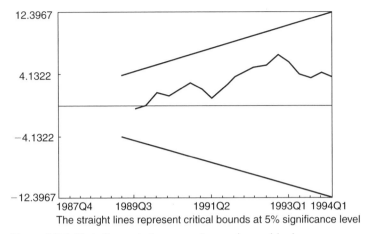

The straight lines represent critical bounds at 5% significance level

Figure 8.2.4 Plot of cumulative sum of recursive residuals

steadily declined. However, it cannot be concluded that between the 1980s and 1990s advertising produced a zero or negative impact on the demand for beer, since advertising investment may have arrested the rate of decline in beer sales. There are also other exogenous factors such as demographic changes, leisure interests which have led to the decline in beer demand.

CASE 3 INTERNET GROWTH AND ADVERTISING

I Background

This example seeks to develop a forecasting system for Internet growth. The example uses fundamental economic principles to guide a general to specific modelling procedure to isolate the key explanatory variables determining future Internet growth.

The subsequent model estimation seeks to evaluate the direct influence of advertising and other key variables affecting of the development of Internet hosts. Internet advertising is expected to increase at a dramatic rate in the near future as the number of consumers taking on-line services increase. The positive interactions between the development of the Internet and Internet advertising will fundamentally determine the communications environment for years to come. Although, the future scope of services provided by the Internet cannot be currently accurately assessed or anticipated, the potential scale of the impact of the Internet can be estimated.

Forecasts about the growth of Internet have been undertaken by several scholars and commercial researchers. Rai *et al.* (1998) argued that the growth of Internet hosts is expected to reach exponential rates in time. The results developed from their exponential model are satisfactory as compared to others which use either Logistic and Gompertz modelling techniques. Based on the foundation of the modelling specification between the number of Internet hosts and time, both short-run and long-run influences are estimated. In order to satisfy criteria for testing procedures, the following estimation is divided into three stages; these are: the ADF/VAR tests, the long-run analysis followed by the short-run estimation.

II The data

As the number of internet networks is difficult to identify and since there are no reliable statistics for estimating the global number of networks for the Internet, the best alternative is to select the number of hosts as the dependent variable for estimating the future expansion of the Internet. The analysis below employs quarterly data for Internet hosts published in the *Internet Domain Survey* produced by Network Wizards (1999). In economic terms, the growth of Internet hosts can be regarded as the diffusion rate of internet use. The rate of diffusion is supposed to generate endogenous effects on the number of Internet hosts. To facilitate the estimation of the rate of diffusion of Internet hosts, two different growth functions; namely the normal growth curve and the exponential growth curve, were used to create growth rate variables. Similarly, an ordinary time trend and an exponential time scale were derived to evaluate the relative predictive powers. Moreover, a dummy variable was utilised to capture the inconsistent influences created by the change in method of compiling the data. Furthermore, another key variable is Internet advertising revenues. These data were obtained from the Internet Advertising Bureau's Report (1999). The influence of Internet advertising relies on the sheer size of the potential audience. The expansion in Internet service consumption is therefore believed to generate a positive impact on Internet advertising growth as well as the growth of associated activities, such as e-commerce and advanced telecommunications services. The list of key variables used in this estimation is shown under Table 8.3.1.

III Test 1

Stage 1

Stage 1 relates to testing for the presence of I(1) series. At this stage, an augmented Dickey-Fuller test is carried out to identify whether the chosen variables are suitable for the use of cointegration by examining the possibility of the existence of a long-run relationship. Table 8.3.1 indicates that all variables are of I(0) for the normal levels and first differences are stationary (i.e. I(1)). By applying an Ordinary

Table 8.3.1 Augmented Dickey–Fuller test

	Levels	*First differences*
Xh	−0.7451 (0)	−5.1683 (0)
X2	−3.2943 (2)	−8.7830 (0)
X3	−2.5808 (0)*	−4.4655 (0)
X4	−3.4109 (4)*	−13.4164 (0)
X5	−3.3185 (2)*	−9.5552 (0)
X6	−3.3474 (2)	−9.3387 (0)

Note:
Xh = Number of Internet hosts
X2 = Normal growth rates of Internet hosts (%)
X3 = Internet advertising (US$m)
X4 = Exponential time trend
X5 = Exponential growth factor
X6 = A multiple of X4 and X5
The asterisks indicate test statistics with strong trends. The number in brackets refer to the lag length for the ADF test.

Least Squares (OLS) estimation on each variable given the model specification [1], the nature of stationarity, either trend stationarity or difference stationarity can be classified, which is of fundamental importance for choosing the correct class of estimation procedure at Stages 2 and 3.

$$Xi = a + (1 - p)\text{time} + pXi_{t-1} + e \qquad (8.3.1)$$

where:

Xi_{t-1} = A variable

time = Normal time scale

e = Residual term

p = Coefficient for testing trend/difference stationarity

Stage 2

Stage 2 concerns the estimation of long run models. First, the model specification is required. Based on Rai's *et al.* (1998) version of the specification, the following structural specification is used. The coefficients of explanatory variables are expected to bear a positive sign.

$$Xh = a + Xi + e \qquad (8.3.2)$$

where:

Xh = Internet host numbers

X = A chosen explanatory variable from Table 8.3.1

a = Constant term

e = Residuals

An unrestricted VAR estimation is applied to correlate the dependent variable Xh with each of the chosen independent variables, namely X2, X3, X4 and X5, to determine the possibility of running a long-run estimation and to identify the level of VAR. Deterministic and exogenous variables were added to the estimation to increase the efficiency of the model. The inclusion of these deterministic/exogenous variables is subject to passing a specific statistical test. The resulting statistics for the deletion test shown in Table 8.3.2 indicate that inclusion of these deterministic variables are significant at 95% or 90% levels. Secondly, the level of r needs to be identified before generating any long-run model. The criteria for choosing the level of r is determined by the trace test and the maximum eigenvalue test. Apart from equation (d), a level of r was determined by the test statistics in each equation. The corresponding test statistics are shown in Table 8.3.3. The non-compliance for testing criteria for equation (d) results in failure for the derivation of a long run forecasting model. The long-run model coefficients for equations (a), (b) and (c) are displayed in Table 8.3.2.

Table 8.3.2 Deletion test results

	Long-run model coefficients	VAR order	r	Deterministic/exogenous variable and test for deletion (χ^2)
a	Xh = −53.1560 + 116.4362X2	3	1	X8(−1), D, 10.3773*
b	Xh = 1.2056 + 0.6838X3	2	1	X10(−1), D, 8.843*
c	Xh = 53.8626 + 0.1988X4	2	1	X9(−1), D, 5.1774**
d	None for Xh = a + X5	2	–	X11(−1), D, 5.713**

Note: Equation (d1) was unable to be established as its variables fail to pass the trace and the maximum eigenvalue test for cointegration.
where:
X8 = first difference of X4   X9 = first difference of X5   X10 = first difference of X6
X11 = first difference for a nominal Time Trend   D = a dummy variable

Table 8.3.3 Long run model coefficients

Null	Alternative	Statistic	95% Critical value	90% Critical value
Equation a				
Maximal eigenvalue				
r = 0	r = 1	48.0582	15.8700	13.8100
r < = 1	r = 2	4.7755	9.1600	7.5300
Trace Test				
r = 0	r > = 1	52.8337	20.1800	17.8800
r < = 1	r = 2	4.7755	9.1600	7.5300
Equation b				
Maximal eigenvalue				
r = 0	r = 1	27.7748	15.8700	13.8100
r < = 1	r = 2	4.9233	9.1600	7.5300
Trace Test				
r = 0	r > = 1	32.6981	20.1800	17.8800
r < = 1	r = 2	4.9233	9.1600	7.5300
Equation c				
Maximal eigenvalue				
r = 0	r = 1	80.4159	15.8700	13.8100
r < = 1	r = 2	2.5295	9.1600	7.5300
Trace Test				
r = 0	r > = 1	82.9455	20.1800	17.8800
r < = 1	r = 2	2.5292	9.1600	7.5300
Equation d				
Maximal eigenvalue				
r = 0	r = 1	77.4867	15.8700	13.8100
r < = 1	r = 2	14.8618	9.1600	7.5300
Trace Test				
r = 0	r > = 1	92.3485	20.1800	17.8800
r < = 1	r = 2	14.8618	9.1600	7.5300
Equation e				
Maximal eigenvalue				
r = 0	r = 1	140.4942	15.8700	13.8100
r < = 1	r = 2	5.8678	9.1600	7.5300
Trace Test				
r = 0	r > = 1	146.3619	20.1800	17.8800
r < = 1	r = 2	5.8678	9.1600	7.5300

* refers to a significant value at 95% level and ** refers to a significant value at 90% level

Inspection of the sign of these long-run coefficients shows that all independent variables (X2, X3 and X4) are positively correlated to Xh. This means that the growth rate of hosts, Internet advertising revenue and the exponential time factor produce a positive impact on the future expansion of hosts. Moreover, the magnitude of coefficient for each explanatory variable indicates the responsiveness of change in Xh induced by the variables on the right-hand side. Model (a) has a very large coefficient of 116.4362. This implies that a 1% increase in X2 will result in 116.4% increase in Xh. Though this long-run growth rate seems to be high intuitively, it may nevertheless suggest a rapid exponential growth influence on the absolute number of hosts as time goes by at this stage in the product cycle. However, the most worrying issue derived from this equation is the negative intercept term. This is a non-feasible result.

The coefficients of X3 and X4 are less than 1. This means that the impact on Xh induced by each unit of change in each independent variable is inelastic. By comparing the magnitude of these variables, it is seen that Internet advertising created the greatest effects on Xh in the long run. The only weakness of this comparison is the time duration for each estimated equation. Despite the short history, however, it seems that advertising variable creates the strongest positive impact on Xh. Given the short data ran for internet advertising, the maximum length of estimation for equation (b) is restricted to 95 Q1 and 98 Q4. However, the other four equations were run from 81 Q4 to 98 Q4.

Stage 3

Stage 3 aims to generate a short-run model for each equation successfully passing the Trace and the Maximum Eigenvalue Tests. After identifying the long-run relationships, the Microfit software can also insert error-correction variables to model short-run dynamic influences for each specified equation. The results for each short-run model, including diagnostic test statistics are summarised in Table 8.3.4. In general, all these short-run models produce statistically satisfactory fitted values which are shown by the R^2 figures, ranging from 0.5711 to 0.7478. All short-run models in Table 8.3.4 passed the diagnostic tests for serial correlation and functional form. Among the three equations shown in Table 8.3.4, only equation (b1) satisfies all the diagnostic tests at 95% significance level, together with the highest R^2 value. Apart from the overall modelling efficiency, the evaluation of the coefficients of independent variables is equally important. In equation (a1), change in X2 creates a negative impact on Xh, which is not the expected sign. The change in X4 with one lag also demonstrates a negative influence on change in Xh for equation (c1). Inspection of equation (b1), shows that a change in Internet advertising produces a positive effect on ΔXh.

Resulting fitted values from equations (a1), (b1) and (c1) were listed in Table 8.3.5. Forecast values from these equations match the actual values quite well. For instance, the forecast for 96 Q3 and Q4 in equation (a1) are very near to the actual number of Internet hosts. Generally, forecasts from dynamic short-run estimations produce better forecast outcomes than the figures estimated by Rai (ibid) as shown in Table 8.3.6. Adjusted figures obtained from the Internet Domain Survey indicate that equations (b1) and (c1) generate better predictive power than the rest. This is

Table 8.3.4 Dynamic error correction models based on cointegrating VAR estimated by OLS

Equation a1

$\Delta Xh = 0.3433\Delta Xh_{t-1} - 0.0018\Delta X2_{t-1}$ (0.1768) (0.0413) $+ 0.1778\Delta Xh_{t-2} - 0.0211\Delta x2_{t-2}$ (0.2097) (0.0200) $+ 0.1084ecm_{t-1} + 0.1167D + 0.0032x8_{t-1}$ (0.0420) (0.0165) (0.0445)	Estimation: 82 Q4–96 Q4 VAR = 3 Diagnostic tests: Serial Correlation: $\chi^2(4) = 5.7036*$ Function form: $\chi^2(1) = 0.05757*$	R2 = 0.57113 DW Statistic: 2.2712 Normality: $\chi^2(2) = 7.6274$ Heteroscedasticity: $\chi^2(1) = 10.3331$

Equation b1

$\Delta Xh = 0.5562\Delta Xh_{t-1} + 0.0572\Delta X3_{t-1}$ (0.1650) (0.0764) $-0.0046ecm_{t-1} + 0.0940D - 0.0022X10_{t-1}$ (0.0083) (0.0159) (0.0014)	Estimation: 95 Q3–98 Q4 VAR = 2 Diagnostic tests: Serial correlation: $\chi^2(4) = 3.7020*$ Function Form: $\chi^2(1) = 3.4862*$	R2 = 0.7478 DW Statistic: 1.5435 Normality: $\chi^2(2) = 3.3001*$ Heteroscedasticity: $\chi^2(1) = 0.3363*$

Equation c1

$\Delta Xh = 0.3210\Delta Xh_{t-1} - 0.1716\Delta x4_{t-1}$ (0.1028) (0.0513) $-0.00005825ecm_{t-1} + 0.1096D$ (0.00001012) (0.0162) $-0.0015X9_{t-1}$ (0.00786)	Estimation: 82 Q2–96 Q4 VAR = 2 Diagnostic tests: Serial correlation: $\chi^2(4) = 4.4169*$ Function form: $\chi^2(1) = 0.4864*$	R2 = 0.6011 DW Statistic: 2.1754 Normality: $\chi^2(2) = 13.5418$ Heteroscedasticity: $\chi^2(1) = 13.7735$

Table 8.3.5 Resulting fitted values

	Actual	Equation (a1)	Equation (b1)	Equation (c1)	Adjusted values
96Q3	12,881,000	12,930,044	13,115,985	13,182,567	
96Q4	14,513,500	14,679,119	15,392,175	15,534,597	
97Q1	16,146,000	16,504,410	18,063,421	18,310,491	21,819,000*
97Q2	17,843,000	18,314,708	21,198,249	21,577,444	
97Q3	19,540,000	20,104,811	24,887,114	25,433,141	26,053,000*

* Values obtained from Network Wizards, *Internet Domain Survey* (1999).

Table 8.3.6 Estimated figures by Rai *et al.*

	Actual	Logistic	Gompertz	Exponential
96 Q3	12,881,000	4,755,203	6,832,009	10,144,100
96 Q4	14,513,500	4,931,211	7,514,849	11,768,258
97 Q1	16,146,000	5,088,032	8,245,301	13,652,457
97 Q2	17,843,000	5,226,436	9,024,760	15,838,332
97 Q3	19,540,000	5,347,564	9,854,518	18,374,185

Source: Figures are adopted from Rai *et al.* (1998).

evident given the superior predicted values for 1997 Q1 and Q3, whereas the predictions using other models suffer serious underestimation as seen in Table 8.3.6. Forecasts for the change in the number of hosts from equations (a1) and (c1) are shown in Figures 8.3.1 and 8.3.2 respectively. Moreover, the actual and fitted values for equation (b1) are illustrated in Figure 8.3.3. Inspection of these figures indicate

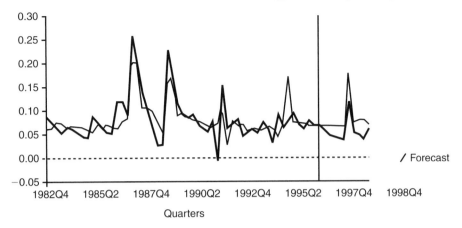

Figure 8.3.1 Multivariate dynamic forecasts

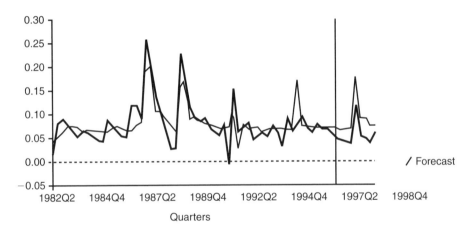

Figure 8.3.2 Multivariate dynamic forecasts

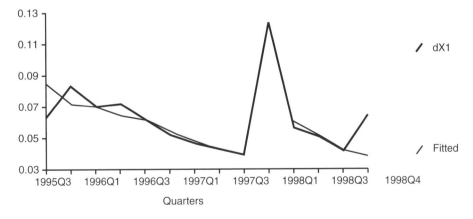

Figure 8.3.3 Plot of actual and fitted values

that forecast values pick up some essential turning points and track the data fairly well during the forecasting periods.

IV Test 2

Interpretation of the results shown in Tables 8.3.2 and 8.3.4 suggest that advertising is statistically significant insofar as exerting an influence on Internet growth in both the short and long run. Additional analysis and closer examination of the parametric impact of the advertising variables means that modification of model specification are needed.

$$Xh = a + X3 + Xi + e \tag{8.3.3}$$

Despite the fact that advertising data only has been available for a short-time duration, there is a possibility of combining the advertising variable with other explanatory variables, namely x2, x4 and x5, to improve forecast efficiency. The following

Table 8.3.7 Long-run and short-run models

	Long run model coefficients	VAR order	r	Deterministic/ exogenous variable and test statistic for restriction deletion (χ^2)
a2	Xh = −651.8533 − 6.3362X2 + 0.7321X3	2	1	X8(−1), D, $\chi^2(3) = 7.6202$**
d2	Xh = −9.9461 − 0.0586X3 − 0.6759X5 Xh = −7.3496 + 0.1305X3 − 0.3633X5	2	2	XH1(−1), D, $\chi^2(3) = 7.2905$**

* refers to a significant value at the 95% level and ** refers to a significant value at the 90% level.

Table 8.3.8 Trace and maximum eigenvalue test statistic

Null	Alternative	Statistic	95% critical value	90% critical value
Equation a2				
Maximal eigenvalue				
r = 0	r = 1	36.4293	22.0400	19.8600
r <= 1	r = 2	15.4285	15.8700	13.8100
r <= 2	r = 3	1.2149	9.1600	7.5300
Trace Test				
r = 0	r >= 1	53.0726	34.8700	31.9300
r <= 1	r = 2	16.6433	20.1800	17.8800
r <= 2	r = 3	1.2149	9.1600	7.5300
Equation d2				
Maximal eigenvalue				
r = 0	r = 1	35.5427	22.0400	19.8600
r <= 1	r = 2	27.2000	15.8700	13.8100
r <= 3	r = 3	0.8249	9.1600	7.5300
Trace Test				
r = 0	r >= 1	63.5675	34.8700	31.9300
r <= 1	r = 2	28.0249	20.1800	17.8800
r <= 3	r = 3	0.8249	9.1600	7.5300

Table 8.3.9 Long-run and short-run models

Equation a3

$\Delta Xh = -0.4589\Delta Xh_{t-1} + 0.0560\Delta X3_{t-1}$	Estimation: 95Q3–98Q4	R2 = 0.8134
(1.9664) (0.0572)	VAR = 2	DW Statistic: 1.7319
$-0.0204\Delta X2_{t-1} - 0.0016ecm_{t-1} + 0.0867d$	Diagnostic tests:	
(0.0425) (0.0039) (0.0149)	Serial correlation:	Normality:
$-18.1503 X8_{t-1}$	$\chi^2(4) = 2.3335*$	$\chi^2(2) = 1.7955*$
(43.9650)	Function form:	Heteroscedasticity:
	$\chi^2(1) = 0.2582*$	$\chi^2(1) = 0.7445*$

Equation d3

$\Delta Xh = 1.5646\Delta Xh_{t-1} + 0.0097\Delta X3_{t-1}$	Estimation: 95Q3–98Q4	R2 = 0.8319
(10.0199) (0.0734)	VAR = 2	DW Statistic: 1.7592
$+ 0.00011\Delta X5_{t-1} - 0.0266ecm_{t-1} - 0.0118ecm_{t-1}$	Diagnostic tests:	
(0.0020) (0.0272) (0.0835)	Serial correlation:	Normality:
$+ 0.0769d + 18.1503 X11_{t-1}$	$\chi^2(4) = 3.6112*$	$\chi^2(2) = 0.9049*$
(0.0173) (0.3221)	Function form:	Heteroscedasticity:
	$\chi^2(1) = 0.7309*$	$\chi^2(1) = 0.3598*$

* refers to a diagnostic test statistic of 95% significant level.

estimations utilise model specification (3) to trace the significance of the advertising variable influence on growth. By repeating stages 1 to 3, results for long-run and short-run models were generated and are summarised in Tables 8.3.7 and 8.3.9. The results for the long run estimation indicate that when X3 is added to equations (a), (c) and (d), only two long run equations can be produced, namely (a2) and (d2) (Table 8.3.7). The model utilising X3 and X4 as independent variables suffers from collinearity, hence no long- or short-run model can be estimated.

Inspection of the results found in Table 8.3.7 indicates that two out of three of the Internet advertising variables generate positive influences on Xh. Based on the trace and maximum eigenvalue test statistics (Table 8.3.8), the number of long-run models (i.e. r) was determined. Equation (d2) generates two sets of long-run series and one of the advertising coefficients shows a negative inelastic response. While equation (a2) generates a unique cointegration correlation, and the advertising influence is closer to unity as compared to both corresponding values from equation (d2). Obviously, in comparing the long-run estimations summarised in Tables 8.3.2 and 8.3.7, long-run advertising influences introduce three major changes. First, its resulting effects increase from 0.6838 (equation (b)) to 0.7321 (equation (a2)) Secondly, advertising influences tend to turn the coefficients of X2 to become negative when X3 is added to equation (a). Finally, Internet advertising variables help improve modelling for equation (d), which originally cannot be estimated before X3 is inserted.

Long-run models are regarded as a summary for all the short-run interactions, therefore it is reasonable to examine the short-run impacts created by the chosen independent variables. In using the dynamic short-run estimation, two short-run models based on equation (a2) and (d2) were produced. Both short-run models (a3) and (d3) generate statistically significant outcomes as illustrated by all the diagnostic tests, each passing the criteria for the 95% significant level. With model specification (3), the short-run advertising influence improves the 'goodness of fit' from 0.5711 (equation (a1)) to 0.8134 (equation (a3)). Consistent with the long run equations (a2) and (d2), advertising variables from equations (a3) and (d3) produced positive effects. However, the lower absolute value displayed in the short-run models indicate that the influences are relatively weaker than that obtained for long-run models. Empirically, this is consistent with the expectation of industry experts on the growth of e-commerce and Internet advertising. At the moment, advertising revenue on the Internet is relatively small. However, as the sheer scale of Internet activities gathers momentum, the advertising revenue can be expected to reach multi-billion dollar levels in the medium and long-run. Eventually, this electronic advertising medium will attract more customers and enhance global diffusion of the Internet by dissipating ever increasing benefits to world society.

In equation (d3), X5 bears the expected positive sign. While in the long-run model (d2) it is negatively correlated with Xh. This means that in short term, exponential growth in host numbers can generate positive influences on consumers currently not on-line. Supposing that there exists a maximum level of Internet connections, or when the number of hosts reaches a large level during the mature stage of life cycle, then an exponential growth is virtually impossible. The negative responses, therefore, correspond to the limited exponential effects in the long-run. Hence, negative responses (-0.6759 and -0.3633) in equation (d2) are recorded. In contrast, an exponential growth effect on host size has been illustrated by the recent explosive developments on Internet. Indeed, the absolute value of X5 is small. Considering its implicit exponential character, a 1% change can generate a more than proportionate change in Internet hosts.

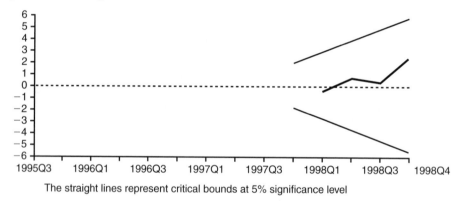

The straight lines represent critical bounds at 5% significance level

Figure 8.3.4 Plot of cumulative sum of recursive residuals

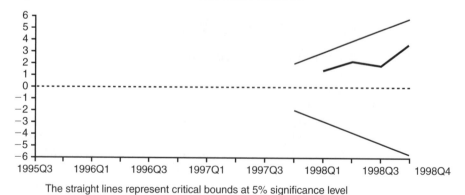

The straight lines represent critical bounds at 5% significance level

Figure 8.3.5 Plot of cumulative sum of recursive residuals

In order to investigate the stability of the parameters for the model specification, two CUSUM tests were run for equations (a3) and (d3). The results are shown in Figures 8.3.4 and 8.3.5, which imply that the parameters used in these two equations, following model specification (3) are significant at 95% significance level.

V Test 3

Once again repeating stages 1 to 3, and utilising a new model specification (4), yet another set of short run and long run models were generated.

$$Xh = a + X2 + X3 + X5 + e \qquad (8.3.4)$$

In addition to the major estimation parameter X3, this model specification consists of two growth variables, one is a log variable (X2), whilst the other is a multiplier by an exponential factor, namely X5. After carrying out a trace and a maximum eigenvalue test, the value of r was unable to be exactly determined. By looking at the test statistics in Table 8.3.10, the maximum eigenvalue test suggests $r = 3$, while the trace test recommends $r = 2$. Intuitively, there may exist three long-run relationships between Xh, X2, X3 and X5. Simply because each of the dependent variables may form a long-run relationships with Xh which is depicted in Table 8.3.2. By using the criteria of SBC and AIC statistical tests, $r = 3$ is chosen (Table 8.3.11).

Setting VAR order at 1, three long-run relationships were summarised as in Table 8.3.12. The long-run coefficients for the Internet advertising are positive but inelastic in all three equations. The growth rate of hosts (X2) also displays positive influence on Xh in all long-run estimations. However, the magnitudes are very dif-

Table 8.3.10 Test statistic for eigenvalue

Null	Alternative	Statistic	95% critical value	90% critical value
Equation a4				
Maximal eigenvalue				
$r = 0$	$r = 1$	202.5830	28.2700	25.8000
$r <= 1$	$r = 2$	41.5767	22.0400	19.8600
$r <= 2$	$r = 3$	19.9962	15.8700	13.8100
$r <= 3$	$r = 4$	0.0691	9.1600	7.5300
Trace Test				
$r = 0$	$r >= 1$	264.2249	53.4800	49.9500
$r <= 1$	$r = 2$	61.6420	34.8700	31.9300
$r <= 2$	$r = 3$	20.0652	20.1800	17.8800
$r <= 3$	$r = 4$	0.0691	9.1600	7.5300

Table 8.3.11 Choice of the number of cointegrating relations using model selection criteria

Rank	Maximised LL	AIC	SBC	HQC
Equation a4				
$r = 0$	64.9756	56.9756	54.1434	57.0057
$r = 1$	166.2671	150.2671	144.6027	150.3274
$r = 2$	187.0554	165.0554	157.2669	165.1384
$r = 3$	197.0535	171.0535	161.8488	171.1515
$r = 4$	197.0880	169.0880	156.1753	169.1936

Table 8.3.12 Long-run relationships

	Long-run model coefficients	VAR order	r	Deterministic/ exogenous variable and test statistic for restriction deletion (χ^2)
a4i	Xh = −12.0412 + 7.4638X2 + 0.0328X3 + 2.6947X5			X11(−1), D,
a4ii	Xh = −9.6133 + 1.3447X2 + 0.0069X3 − 0.0688X5	1	3	$\chi^2(4) = 11.0945^*$
a4iii	Xh = −8.5814 + 0.7919X2 + 0.0631X3 − 0.0362X5			

* refers to diagnostic test statistic of 95% significant level.

ferent in each case. With respect to percentage influences on Xh, X2 displays relatively greater momentum among the three independent variables in each estimation. Similar to the long-run model (d2) as shown in Table 8.3.5, the exponential growth variable (X5) creates a negative, inelastic influence on Xh in two of the long-run models. Comparing the negative influences of X5, (−0.6759 and −0.0688; −0.3633 and −0.03602) in Tables 8.3.7 and 8.3.12 respectively, the long-run impact of X5 from equation (a4ii) and (a4iii) are approximately ten times lower. Only the variable of X5 in equation (a4i) shows a positive elastic effect on Xh. Comparing the parametric responses from equation (a2) and all equations in Table 8.3.12, when X5 is inserted into equation (a2), then the negative elastic long-run effects introduced by X2 are changed to positive long-run influences. This means that the exponential growth is important to create a balancing effect to track the normal growth rate of The Internet. The positive response is of expected sign and follows general economic principles. By contrast, advertising influences remain positive in all cases.

Inspection of the short-run equation (a5) shows that three error correction variables with different lags were estimated. These variables help correct the unnecessary disturbances originated from serial correlation problems. By checking the DW statistics, there exists no serial correlation in the model. As compared with the R^2 in equations (a3) and (d3), equation (a5) produced a very reasonable level of good fit. Moreover, all diagnostic tests were satisfied at the 95% significant level. Moreover, by measuring the 'goodness of fit' (R^2), values increase from 0.8134 (equation (a3)) to 0.8252 (equation (a5)) when X5 is inserted. This means that the inclusion of X5 based on model specification (4) helps establish better modelling power both intuitively and statistically. A plot of the fitted and actual values is shown in Figure 8.3.6. As the ratio for the number of samples to the number of estimated parameters is small for this model testing, hence CUSUM Test is not reliable and is not produced. Instead, a multivariate dynamic forecast for the level of Xh is presented in Figure 8.3.7, which clearly suggests the presence of forecast overlapping the actual values along the upward trend for Internet host numbers.

$$\Delta Xh = 0.0642 ecm1_{t-1} - 0.0249 ecm2_{t-1} - 0.0024 ecm3_{t-1} + 0.0800d + 0.0086 X11_{t-1}$$
$$\quad (0.0626) \qquad\quad (0.0162) \qquad\quad (0.0036) \qquad\quad (0.0134) \quad (0.3900)$$

Estimation: 95 Q3–98 Q4

VAR = 1

$R^2 = 0.8252$

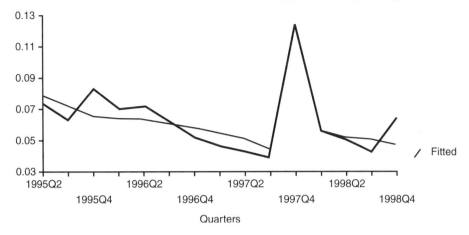

Figure 8.3.6 Plot of actual and fitted values

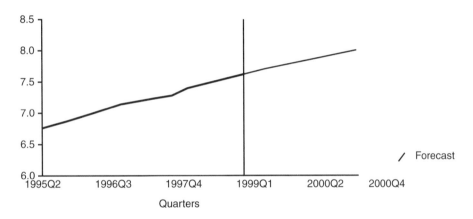

Figure 8.3.7 Multivariate dynamic forecasts

DW Statistic: 1.6955

Diagnostic Tests:

Serial Correlation: $\chi^2(4) = 1.3033$ Normality: $\chi^2(2) = 2.1961$

Function Form: $\chi^2(1) = 3.6512$ Heteroscedasticity: $\chi^2(1) = 0.5943$ (a5)

The empirical case study demonstrates that there exists a range of potential variables which affect the growth of Internet hosts. Simply utilising an exponential time scale as a model regressor, as in Rai's estimation, is insufficient to trace the potent dynamic linkages. The advertising variables utilised in this analysis show that significant potential exists to track the growth of Internet hosts. However, limited historical data for the Internet advertising restricts forecasting power. In future, further testing based on the model specification used in the foregoing analysis would certainly provide more key insight into the exact specification of the linkages between the estimated parameters. More particularly, when more comprehensive survey data on e-commerce and advertising investment for Internet become available, further time series analysis should be able to provide more accurate predictions on Internet diffusion rates.

CASE 4 IMPORTS AND THE MACROECONOMIC COMPONENTS OF FINAL EXPENDITURE

I Background

The case study examines the long-run relationship between aggregate imports and the macroeconomic components of final expenditure, using multivariate cointegration analysis. The study focuses exclusively upon the empirical examination of UK aggregate imports in order to: (a) examine the long-run relationship between aggregate imports and the macro components of final expenditure; (b) to estimate the import contents of the different macro components of final expenditure using appropriate econometric techniques; and (c) to propose a short-run model for forecasting UK aggregate imports.

II The model and data

Economic theory identifies three major factors determining a country's demand for competitive imports. First, the level of final expenditure. The composition of expenditure will also be important to the extent that the import content of the different components of expenditure differs. In the UK as in other EC Member States, the evidence available from input-output data suggests that each macro component of final expenditure corresponds to different aggregate propensities to import (Eurostat, 1991). Indeed, if the composition of final demand changes, the aggregate import propensity will change, even if the disaggregate marginal propensities remain unchanged. In this study, we have distinguished between three board categories of final expenditure: consumption expenditure by the private and public sector, investment expenditure and expenditure on exports. The underlying assumptions being that each macro component of final expenditure will have a different impact on imports.

The second factor identified by economic theory is the price of imports relative to the price of domestic substitutes. In particular, a rise in the relative price level is expected to lead to a fall in demand for imports. We have assumed that: (a) the supply elasticities are infinite and therefore import prices can be viewed as exogenously given; and (b) the domestic prices are flexible and would change to eliminate excess demand at home. The third factor is the capacity of the country to produce and supply the goods itself. However, the capacity factor is essentially a short run phenomenon and is relevant, only if excess demand at home cannot be eliminated by a change in the domestic level. Given the above assumptions, the long-run import demand function is specified as follows:

$$M = f(CG, INVT, EXPT, P_M/P_D) \tag{8.4.1}$$

where M is the volume of imports, CG is the sum of private and public consumption expenditure, INVT is expenditure on investment goods including gross domestic fixed capital formation and stock building and EXPT is expenditure on exports, all measured in constant prices. P_M is the import price deflator defined as a ratio of imports in current prices over imports in constant prices and P_D is an index of domestic prices measured by the GDP deflator. All variables are in natural logarithmic form. The data used in this study are quarterly form 1970 Q1 to 1990 Q4. The period 1990 Q1 to 1992 Q2 is used for post sample forecasting.

III Cointegration analysis

The first step is to test for the order of integration of each variable included in the model. We have listed the relevant variables in Table 8.4.1. It is clear on the basis of the DF and ADF test statistics all variables are integrated of order one, i.e. I(1) and therefore constitute potential variables for inclusion in a cointegrating vector.

We are now in the position to carry out the cointegration tests proposed by the Johansen technique. In order to carry out these tests the first step is to specify a lag length for the VAR. We set this at three periods on the basis of the results of diagnostic tests including the likelihood ratio test proposed by Sims (1980). Under this specification, Table 8.4.2 presents the results of the trace test derived from the Johansen maximum likelihood procedure. To carry out this test we proceeded sequentially by first testing for Ho:r ≤ 0, where r is the number of cointegrating vectors. If Ho: was rejected, we then tested for r ≤ 1 and so on, until the null hypothesis could not be rejected. At both 5% and 10% significancy levels, the null hypothesis of zero cointegrating vector appears to be strongly rejected, while the hypothesis of one or more cointegrating vectors is not. According to the trace test there appears to exist, at most one cointegrating vector in respect of variables specified in the model. Similarly, according to the results of maximal eigenvalue test, it appears that there is at most one statistically significant cointegrating vector in respect of variable identified in the model. Based on the above testing procedures,

Table 8.4.1 DF and ADF test statistic variables

Variable	DF test	ADF test
M	−1.1577 (−2.8939)	−0.91035 (−2.8955)
CG	−0.93213 (−2.8939)	−0.97152 (−2.8955)
INVT	−0.66324 (−2.8939)	−0.40978 (−2.8955)
EXPT	−1.2213 (−2.8939)	−1.3565 (−2.8955)
P_M/P_D	0.30527 (−2.8939)	−0.09983 (−2.8955)

Note: the null hypothesis in each case is that the variable in question is I(1); the 95% critical value are given in brackets derived from Fuller (1976). In all cases the DF and ADF values are greater than the corresponding critical values thus confirming the existence of a unit root.

Table 8.4.2 Trace tests for cointegrating vectors

Null hypothesis: the number of cointegrating vectors, r	Likelihood ratio statistic LR (n)	5% critical value	10% critical value
r ≤ 0	72.5906	68.5240	64.8430
r ≤ 1	29.4068	47.210	43.9490
r ≤ 2	15.6075	29.680	26.7850

Table 8.4.3 Test results of maximal eigenvalue test

Null	Alternative	LR statistic LR(n)	5% critical value	10% critical value
$r \leq 0$	$r = 1$	43.1838	33.4610	30.90
$r \leq 1$	$r = 2$	13.7998	27.0607	24.7340
$r \leq 2$	$r = 3$	10.2973	20.9670	18.5980

Note: Test statistics are asymptotically $\chi^2(n)$ variates under Ho. The sequential test stops when Ho cannot be rejected. Ho: $r \leq 2$ is reported as additional information.

equation 8.4.2 represents the long-run relationship between variables identified in the model.

$$M = 1.6371CG + 0.150INVT - 0.0202EXP - 0.0339(P_M/P_D) \tag{8.4.2}$$

The equilibrium relationship indicates clearly that the major determinant of the UK aggregate imports in the long run, is consumption expenditure. This category of expenditure appears to dominate the influence of other macro components of final expenditure in the long run. Moreover, there appears to exist significant differences between the partial elasticities of demand to import with respect to consumption expenditure, investment expenditure and expenditure on exports with estimated elasticities of 1.63, 0.15 and -0.020 respectively. This latter observation was tested statistically by imposing two linear restrictions on the parameters of the cointegrating vector. The null hypothesis was specified as follows: the long-run coefficients on consumption expenditure, investment expenditure and expenditure on exports are all equal when normalised on aggregate imports. The restricted estimates are given as equation 8.4.3 below:

$$M = 0.4394CG + 0.4394INVT + 0.4394EXPT - 0.36457(P_M/P_D) \tag{8.4.3}$$

The likelihood ratio test statistic on the restrictions in $LR(2) = 30.615$. The restriction can thus be rejected at 5% critical value of 5.99. The long-run price elasticity of demand for imports is only -0.034 indicating that the UK aggregate imports are not price respective. This observation was tested statistically by imposing an exclusion restriction on the relative price term. The likelihood ratio statistic listed alongside the restricted equation (8.4.4) below, shows that the hypothesis is easily accepted at the 1% significant level.

Exclusion restriction:

$$M = 1.730CG + 0.143INVT - 0.0670EXPT$$

$$LR(1) = 0.44651 \tag{8.4.4}$$

The cointegration analysis therefore suggests that (a) the major long run determinants of aggregate imports in the UK is aggregate consumption expenditure; (b) there are significant differences between the elasticities of import with respect to each macro component of final expenditure; and (c) change in the prices of imports relative to the price of domestically produced goods have no significant impact on the UK aggregate imports in the long run.

IV The short-run behaviour of UK aggregate imports

In order to examine (a) the short-run behaviour of UK aggregate imports and (b) to derive a short-run forecasting equation, a dynamic error correction model was estimated. For this purpose the lagged residual error derived from the cointegrating vector was incorporated into a highly general error correction model with four lags. The general model was then tested downwards to arrive at a specific short-run data consistent equation. The final equation obtained from the General to Specific modelling procedures is reported below.

$$\Delta M_t = \alpha_0 + \alpha_1 \Delta CG_t + \alpha_2 \Delta INVT_t + \alpha_3 \Delta EXP_t + \alpha_4 \Delta EXP_{t-3} + \alpha_5 \Delta(P_M/P_D)_t$$
$$+ \alpha_6 ECM_{t-1} + \text{error term}$$

$$\Delta M_t = -4.7381 + 0.54209 \Delta CG_t + 0.25913 \Delta INVT_t + 0.39435 \Delta EXP_t$$
$$\quad (-5.5788) \quad (2.0763) \qquad\quad (5.1908) \qquad\qquad (4.8179)$$

$$+ 0.12906 \Delta EXP_{t-3} - 0.079216 \Delta(P_M/P_D)_t - 0.47464 ECM_{t-1}$$
$$\quad (1.8233) \qquad\qquad (-0.97017) \qquad\qquad (-5.5801)$$

R = 0.58799

DW = 2.1190

Serial Correlation: $\chi^2(4) = 3.0969$

Serial Correlation: $\chi^2(12) = 16.00$

Predictive Failure $\chi^2(6) = 3.414$

ARCH $\chi^2(4) = 5.783$

ARCH $\chi^2(12) = 13.599$

Functional Form: $\chi^2(1) = 1.2024$

Normality $\chi^2(1) = 0.78138$

Heteroscedasticity $\chi^2(1) = 0.006189$

According to the results of the diagnostic test, the short-run model appeared to be well behaved with a white noise error term. Moreover the CUSUM test of parameter stability indicates that the estimated parameters of the model has remained stable over the sample period. The short-run variation in imports appear to be mainly determined by the variations in macro components of final expenditure. In particular, changes in consumption expenditure seems to be the major determinants of variations in aggregate imports in the short-run. In order to investigate the ex ante forecasting performance of the model, six additional observations covering period 91 Q1 to 92 Q2 were added to the original sample period. The forecast results are presented in Table 8.4.4. The model appears to track the data well, picking up all turning points in the data. The forecasting diagnostic tests reported in Table 8.4.4 confirm the accuracy of the model in predicting the short-run fluctuations in the UK aggregate imports. All in all, it is found that there are significant differences between the long-run elasticities import demand with respect to the different components of final expenditure. An error correction model is proposed for short-run forecasting of the UK aggregate imports. The short-run model appears to track the data well.

Table 8.4.4 Static forecasts generated from short-run ECM

Observation	Actual	Forecast	Error	SD of Error
91Q1	−0.025872	0.0042402	−0.30112	0.024776
91Q2	0.014615	0.023989	−0.0092833	0.024988
91Q3	0.0068286	0.016635	−0.0098062	0.024615
91Q4	0.0055178	0.0041176	0.0014002	0.024659
92Q1	0.028887	0.0083097	0.020577	0.024676
92Q2	0.022696	−0.0009175	0.023614	0.024779

Mean prediction error = −0.0006017	Mean sum absolute prediction Error = 0.0157
Sum of squared prediction errors = 0.0003453	Root mean sum square Prediction errors = 0.018583
Predictive failure test: $F(6,72) = 0.56911$	

Note: 79 observations used in estimation of the model from 71 Q2 to 90 Q4.

Bibliography

Brewers and Licensed Retailers Association (1995) *Statistical Handbook: A Compilation of Drinks Industry Statistics*, Thruman, C. and Witheridge, J. (eds.), BLRA, London.

Brewers and Licensed Retailers Association (1995) 'Beer and Pub Facts 1994', London: BLRA.

Chow, G. C. (1960) 'Tests of equality between sets of coefficients in two linear regressions', *Econometrica*, 28, 591–605.

Durbin, J. and Watson, J. S. (1971) 'Testing for serial correlation in least squares regression III', *Biometrika*, 58, 1–42.

Engle, R. F. and Grainger, C. W. J. (1987) 'Cointegration and error correction: representation, estimation and testing', *Econometrica*, 55, 251–276.

Fuller, W. A. (1976) 'Introduction to Statistical Time Series', New York: Wiley.

IAB (1999) *Internet Advertising Bureau Report*, http://www.iab.net/.

Lynk, W. (1985) 'The price and output of beer revisited', *Journal of Business*, 58, 433–437.

Jarque, C. M. and Bera, A. K. (1987) 'A test for normality of observations and regression residuals', *International Statistical Review*, 55, 163–172.

Johansen, S. (1988) 'Statistical analysis of cointegrating vectors', *Journal of Economic Dynamics and Control*, 17, 231–254.

Johansen, S. and Juselius, K. (1990) 'Maximum likelihood estimation and inference on cointegration – with applications to the demand for money', *Oxford Bulletin of Economics and Statistics*, 52, 2, 169–210.

Monopolies and Mergers Commission (1989) *The Supply of Beer: A Report on the Supply of Beer for Retail Sale in the United Kingdom*, Cm 651, March (HMSO).

Nerlove, M. and Arrow, K. (1962) 'Optimal advertising policy under dynamic conditions', *Econometrica*, 29, 129–142.

Network Wizards (1999) *Internet Domain Survey*, http://www.nw.com/zone/www/report.html.

Nua survey (1999) http://www.nua.ie/surveys, 28 January 1999.

Penm, J. (1988) 'An econometric study of the demand for bottled, canned and bulk beer', *Economic Record*, 64, 268–274.

Pesaran, M. H. and Pesaran, B. (1991) *Microfit 3: An Interactive Econometrics Package*, Oxford: Oxford University Press.

Rai, A., Ravichandran, T. and Samaddar, S. (1998) 'How to anticipate the Internet's global diffusion', *Communication of the ACM*, 41, 10 (October 1998), 97–106.

Ramsey, J. B. (1969) 'Test for specification errors in classical linear least squares regression analysis', *Journal of Royal Statistical Society, Series B*, Vol. 31, pp. 350–371.

Sims, C. (1980) 'Macroeconomics and Reality', *Econometrica*, 48, 1–48.

Appendix 1 The Internet and data for economists

1 Information sources and trading

The World Wide Web (WWW) provides practically all of the information which is needed from the INTERNET, particularly by those who are not technically inclined. The WWW is obtained using browsers such as *Netscape.*

One of the longest established information gateways is Bill Goffe's *Resources for Economists on the Internet.* It is located at:

http://econwpa.wustl.edu/EconFAQ/EconFAQ.html

Note that Web addresses must be typed precisely; lower and upper case distinctions must be recognised.

http://wueconb.wustl.edu/EconFAQ/sc.html

provides a 'shortcut to all resources'

Included will be found:

US Macro and Regional Data, Other US Data, World and non-US Data, Finance and Financial Markets, Economic Consulting and Forecasting Services, Working Papers, Bibliographical Databases and Information Working Papers, Publication, and Report Notification Services, Information About Conferences etc.

http://www.helsinki.fi/WebEcl

categorises free economics information on the WWW by subject area.

The information above relates to a fraction of the material assembled by Goffe at his site. It is based in the USA and Finland. Its availability partly depends upon the time of the day access is sought, partly upon the quality of the user's computer resources, etc. If difficulty is experienced, it is preferable first to try bizIed, as detailed in Section 2.

The Web page for the computer training initiative in economics is at:

http://sosig.ac.uk/ctiecon

2 Economics education on the Internet

In this chapter, particular reference is made to company data. However, there is one site which provides the location for, perhaps, 95% of all useful information on the Internet, both microeconomic and macroeconomic. *Business Education on the Internet* (bizIed), at the location:

http://bizIed.ac.uk/

is a business education gateway. The project director is Richard Young. He is based at the University of Bristol (Richard.Young@bristol.ac.uk).

A part of the site map is reproduced below (by 'clicking' on the phrase underlined in blue, the computer user jumps to the site of his or her choice).

A hard copy of the information available from this site is provided by Young (1996). BizIed is such an excellent site, however, that it is best to find out what is available by using the Web itself.

The main features of BizIed are:

The data server

It is easy to down load both macro- and microeconomic (mainly company based) data sets. Most users generally down load first to a spreadsheet file and

Company Facts	Data	Listings
AMW	Office for National Statistics	COM9MY sites
The Body Shop		Economics data site S.
BMW	World Penn Date Table	
BP		Business & finance data sites
	US Census Bureau	
Euroster(UK) Ltd		Government, politics and special interest groups
	Economics data	
The Financial Times		Media sites

Figure A.1.1 Part of the site map for bizIed

then, perhaps, to Microfit. The data set made available by the Office for National Statistics is most impressive. UK and overseas data are available.

Company facts

Data on particular companies is provided; their field of operation, company history, etc., are detailed.

BizIed listings

This provides users with an up to date list of resources.

Study skills

Advice is provided on the presentation and interpretation of data, essay skills, revision and note taking. Students are encouraged to pose questions and to consider the adequacy of explanations often provided in traditional texts. *The Financial Times*, for instance, is used as a study tool and as an education resource.

Contacts

An online directory of national and regional contacts

Tutor support

Provides classroom material to further the use of the Internet for teaching.

3 The search for information using key phrases

Alta Vista, *Info Seek, Yahoo!,* etc provide general search facilities. Alta Vista is excellent, easy to use and enables Advanced Queries by means of AND, OR expressions.

It is advisable to use *general* search engines such as Alta Vista in order to obtain references to major sites (e.g. HM government). You can then use the *dedicated* search facility of the provider. For instance, the UK central government's search facility is at the location:

http://www.open.gov.uk/

You are recommended to use this site for information on, say, monetary policy, rather than to use Alta Vista, Info Seek, etc.

Of course, there are numerous invaluable sites. Listed below are just five sites.

Current exchange rates between any two currencies can be found by consulting:

http://www.dna.lth.se/cgi-binlkurt/rates?

The Institute of Fiscal Policy's site at:

http:U~.ifs.org.uk/

may be referred to for articles and files held by the IFS.

The Ontario Health Care Evaluation Network:

http://HIRU.MCMASTER.CA/OHCEN/DEFAULT.HTM#Projects

provides a search facility for the health economist.

The National Bureau of Economic Research:

www.nber.org

provides on line macro and industry based data for many countries. International comparisons are facilitated.

The Social Science Information Gateway (SOSIG):

http://sosig.ac.uk/

is ESRC funded, provides an excellent search facility, and attempts to filter out material which is inappropriate, out of date, etc.

4 Web sites for econometrics: a modern and practical approach

4.1 e-mail enquiries

The location for e-mail requests concerning questions raised in this book is:

www.sunderland.ac.uk/~bsOprilemetrics.htin

4.2 Analysis of UK share prices and the market for equities

Readers may wish to refer to:

www.sunderland.ac.uk/~bsOpri/javas.ht[n

for an interactive page which allows them to test their knowledge of the stock market, the analysis of risk in portfolio holdings, etc.

The java script code used may be of particular interest for those intending to develop their own pages on any subject.

An extract from the site is shown below:

Econometrics, A Modern and Practical Approach (Interactive Pages)
Consider each of the questions below:
1. What is the main index of share prices in Germany: Dax, Topix, CAC, etc.?

> [   ]

> [Check answer]

2. Which of the following is most likely to cause UK gift yields to fall?
   Click on *the arrow and* select option

> [Allowing the Bank of England to control the base rate]

3. You wish to invest in two companies from the FTSE. You aim to choose companies with negatively correlated earnings characteristics. Which of the following would achieve this aim?

Air Tours and Lucas Varity
Dixons and Argos
NatWest and Barclays
ICI and RMC

Figure A.1.2 Contents of interactivelm

Figure A.1.2 displays the content of javas.htm at the time of writing. This information will be updated from time to time to take account of issues of current importance. An extract on the feedback provided to users on question 3 is shown in below.

Your answer is incorrect
Air Tours is a company which supplies UK residents with tours abroad. When the pound is strong it is cheaper to travel abroad. Lucas Varity is a major UK manufacturing company. It exports a large proportion of its products and is disadvantaged by competition from abroad when the pound is strong...

Figure A.1.3 Sample response to a user's answer

Readers could consider the information reproduced in Figure A.1.4, below, the interactive page, and the city pages published daily in the national broadsheets. They could consider how risk is assessed and compare their notes with the view expressed in mainstream econometrics texts.

4.3 FTSE100 (leading UK company) information

By accessing the following site:

http://www.ftse.com/ftse-100.html

readers will be able to obtain information on the leading UK 'blue chip' companies which make up the FTSE100. Companies which belong to the FTSE100 are revised from time to time. For instance, when the Halifax building society became a PLC it joined the index. Other sites of interest can be obained by using Alta Vista's search engine.

Real time news on stock price, information on selected company report data, etc., is available at:

www.sunderland.ac.uk/~bsOpri/bus/44.html

Reference

Young, R. (1966) 'The Internet and Economics Education', *Computers in Higher Education Research* (CHEER), 10, 3.

Appendix 2 A Basic Guide to Microfit 4.0

1 Introduction

This guide provides an introduction to using Microfit version 4.0, an econometric modelling package specifically designed for analysing time series data. Microfit 4.0 provides a wide range of econometric modelling and analysis features for:

- Data entry and editing
- Graphics display
- Estimation
- Hypothesis testing
- Univariate modelling and forecasting
- Multivariate modelling and forecasting

The software contains sufficient features for experienced users to carry out sophisticated univariate and multivariate modelling studies with full featured diagnostics tests. However the simple windowing environment, whilst not closely following standard windows formats, is sufficiently simple to make the software accessible to beginners. The purpose of this guide is to get you started using Microfit version 4.0. It will not cover all the features of Microfit (for example GARCH modelling is only covered in outline and multivariate analysis not at all) and it is certainly not intended to be a course in time series analysis. Where theoretical results are required to explain the use of a Microfit feature they are simply quoted and a reference made to the relevant chapter for more information. A detailed discussion of all the features of Microfit 4.0 can be found in the book which accompanies the software[1].

2 Starting Microfit

Assuming that the software has been correctly installed when the package was purchased Microfit is started in the same way as any other windows based package by clicking on the desktop icon or selecting the package via the program menu option.

For a few seconds the copyright screen is displayed followed by the opening menu options as shown in Figure A.2.1.

You are now ready to begin working with Microfit by entering time series data.

Figure A.2.1 Microfit opening screen

3 Format of Microfit data

Microfit inherently assumes that data sets represent either time series data or cross sectional data. Data frequency can be selected from the following options:

Undated   Annual   Half Yearly   Quarterly   Monthly

The undated option is used for cross section data or time periods other than the standard, for example Weekly. It is important to remember that Microfit assumes that undated data is unordered.

3.1 Variable names

Each data column has an associated variable name. If you do not specify a name Microfit will assume the name sequence X1, X2,..., etc. You can devise your own data names according to the following rules:

The name is alphanumeric and must begin with a letter.
The name can be a maximum of 9 characters.
The underscore character '_' can appear at any point.
The name is not case sensitive.

The following are all allowed variable names:

GDP   Unit_Cost   Sel_Price   SEL_PRICE

The last two examples will be treated as the same variable.
  The following are examples of illegal variable names:

Selling_Price: Too many characters
25year_G: Numeric start character
M2$:   Invalid symbol ($)

The 9 characters limitation is now anachronistic but at least Microfit allows the user to attach an 80 word description to each variable.

3.2 Size limitations

For data Microfit imposes the following limitations:

The total number of variables NV must be = 200
The total number of observations NO must be = 3,100

A further restriction is imposed on the combination of variables and number of observations as follows:

NV*NO = 124,000

Other size limitations exist with the Microfit package (e.g. at the estimating and forecasting stages) and these will be discussed in the appropriate sections.

3.3 Data formats

Microfit recognises 3 data formats in addition to its own '.FIT' format:

Plain ASCII
Plain ASCII text is expected to contain only numeric values. The data is expected to be in one of two basic formats; *Free format* or *Fixed format*. *Free format* data has individual data items separated by a space, comma or an end of line/carriage return. *Fixed format* follows a packed data format defined for example using a FORTRAN statement. In the majority of cases you will probably encounter *Free format* data.

Having defined the format there is still one further decision to be made with regard to a table of data namely whether the variables are organised by rows or columns. For example consider the following simple example:

Year	1993	1994	1995	1996
Cost (£)	30	32	36	38
Price (£)	53	50	45	40

Microfit would recognise this table of data to be *organised by variable* if the numbers appeared in the text file as:

30 32 36 38
53 50 45 40

In other words each row contains all the values for a single variable. If on the other hand the data is organised with all the variables at a particular time before the data for subsequent time periods then Microfit refers to this as *organised by observation* as follows:

30 53
32 50
36 45
38 40

Comma delimited CSV
This is a common data format understood by spreadsheets. It is an ASCII text file with items comma delimited. This format saves only the text and values as they are displayed in cells of the active worksheet. All rows and all characters in each cell are

saved. Columns of data are separated by commas, and each row of data ends in a carriage return. If a cell contains a comma, the cell contents are enclosed in double quotation marks.

You should be aware that if cells display formulas instead of formula values, the formulas are converted as text.

AREMOS (TSD) files
This is a special time series data format generated by the AREMOS package.

4 Data entry

Data can be entered into Microfit either from the keyboard, from a data file stored on disc or from data placed on the windows clipboard. Keyboard entry is only used for entering very small quantities of data and clipboard data is limited to 32,585 characters.

4.1 Keyboard data entry

Before you begin entering data from the keyboard it is a good idea to have the data listed in front of you. This will make it simpler to use the *'New Data Set'* dialog box.

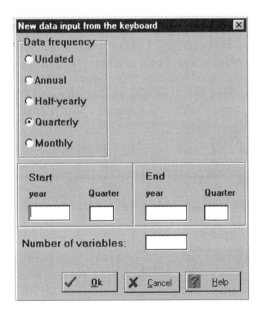

Figure A.2.2 Keyboard data entry

Begin by choosing *'New'* from the *'File'* Menu which will open up the *'New Data Set'* dialog box shown in Figure A.2.2. Fill in the data requested and select *'OK'* If you choose the option *'Undated'* the data requested changes from start and end dates to number of observations. Having clicked *'OK'* you are presented with the variables dialog box listing the chosen number of variables with the default names (Figure A.2.3). At this stage you may change the variable names and add descriptions but remember to follow the naming conventions. Selecting 'Go' will open up the data entry screen with the variables listed and the entry 'NONE' for each field. Now you are free to enter the data values. When you have completed the data entry it is a good idea to save your data because at the entry or editing stage the data is only held in memory.

4.2 Loading data sets from a file

Data can be loaded from any of the four data formats recognised by Microfit.

To load any existing data file select *'Open'* from the *'Files'* menu. If the data is in one of the three standard formats (CSV, TSD or FIT) it will be loaded automatically.

Note: If the file is in CSV format Microfit expects:

- Columns to be variables and rows to be observations.
- The first row must contain variable names and additionally the second row may contain a description of the variables.
- The first column must contain dates or an observation number.
- If the file contains data which uses the comma (e.g. 3,798) the whole data item is enclosed in quotes.

A TSD file may only contain the date periods recognised by Microfit (Year, Quarter and Month)

4.3 Data entry via the clip board

For small quantities of data (maximum of 32 Kbytes) the clipboard provides a simple means of transferring data from a spreadsheet to Microfit. The data copied to the clipboard from the spreadsheet has to be arranged in columns by variable. It may optionally contain the variable names (maximum of 9 characters) in the first row and a variable description (maximum 80 characters) in the second row.

To read in clipboard data first select *'Paste Data'* from the *'Edit'* menu. You are then requested to supply information describing the structure of the data on two forms (Figure A.2.3). Having entered the correct information data import follows. Having viewed the imported data accept it by 'clicking' on *'Go'*.

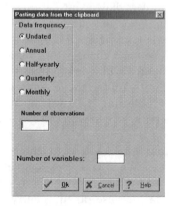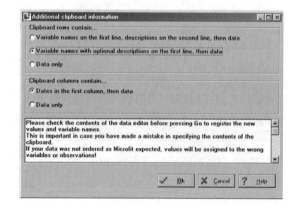

Figure A.2.3 Clipboard data entry

5 Appending data files

There are many occasions when it is necessary to add data to an existing data set. For example data may become available covering an extended time period or better values may become available to replace existing data. Microfit allows the easy incorporation of new data into an existing table or extending an existing table with new variables. The two files must both be in the Microfit data format and clearly if the new data is not in this format the first task is to convert the data by first loading the data as described in section 4 and then saving as a Microfit data file.

To combine two Microfit files load the first file in the normal way. Now select '*Add*' from the '*Edit*' menu. An open dialog box now allows you to select the file to be added. provided the two files have the same set of variables and the same data interval (years, quarters, months) the combination will take place.

If the second file contains variables not available in the first the combination will still take place with the list of variables extended in the combined table to cover both the original data files.

If the data is incompatible between the two files (e.g. one file has quarterly data and the second annual data) then an error message is generated.

5.1 *Appending data to an existing table with the same variables*

This covers the case of using additional data to extend the range of a table or to introduce more up to date information into a table. If both tables being combined contain different data values for an overlapping time period then the data in the second table (the one being 'added') over rides the data in the initial table. Figure A.2.4 shows the general concept.

Obs	C	S
...	...	...
1985	238.5	379.6
1986	321.0	405.7
1987	332.8	429.2
1988	340.9	452.1

+

Obs	C	S
1986	355.2	473.6
1987	398.2	447.2
1988	407.0	488.3
1989	415.6	531.6
...	...	...

=

Obs	C	S
...	...	...
1985	238.5	379.6
1986	321.0	405.7
1987	398.2	447.2
1988	407.0	488.3
1989	415.6	531.6
...	...	...

Figure A.2.4 Appending data

If data gaps exist in the combined table (e.g. the first table ends in 1988 and the second commences in 1990) then the unassigned values are given a null indicator in the form of the string *NONE*.

5.2 *Appending with a different variable list*

The list of variables associated with the first data file are simply augmented with the new variables from the second file. Clearly both data files must use the same data frequency but they need not cover the same time period.

Note: The variables in the two data sets *must* be mutually exclusive.

5.3 *Saving data files*

Microfit can save data files in any of the supported formats:

- Microfit file format
- Plain ASCII
- Comma delimited CSV
- AREMOS (TSD) files

When you wish to save your data click on the 'Save As' item in the 'File' menu or click on the standard windows save icon. You are then given the choice of selecting the desired format.

All the options are quite straightforward with the possible exception of the plain ASCII format. A number of choices have to be made corresponding to the data formats outlined in section 3.3 (plain ASCII). A form will guide you through the choice of fixed or free format and save by variable or observation.

6 Basic data analysis

Microfit contains a range of built in functions and commands which enable you to generate variables which satisfy specific functional requirements for use in your analysis. Appendices 2 and 3 contain a complete listing of the available functions and commands respectively. To support econometric analysis it is often useful to add constant terms, trend lines or seasonal dummies to our data files. Let us see how we achieve this with Microfit.

6.1 *Creating constant term and time trend*

As an example we will take our modified Microfit file UKCON2.FIT. To add a constant term representing an intercept for regression analysis we will add the constant INCP. There are two ways to do this. First of all we can select 'Const' (Intercept) term from the main 'Edit' menu option or selecting the 'Constant' button in the 'Process Window'. In the dialog box which this option opens up you are asked to supply a name for the term. When you supply the name and click OK a constant term is created with all the values set to a default of 1. Alternatively if you wish to add a constant term with any other values you can use the 'Process Window' to enter the constant name and the required value. For example if you wish to add a constant 'Baseload' with a value of 250 you would enter BASELOAD = 250; and click on 'Go'.

Trend lines can be added with equal ease either by selecting 'Time Trend' from the 'Edit' menu or by selecting the 'Time Trend' button in the 'Process Window'. In either case a dialog box opens up requesting a variable name for the time trend. After you have supplied this and accepted it Microfit creates the trend variable setting the first value to one and adding an increment of 1 to each succeeding term.

6.2 *Creating seasonal dummies*

It is very often the case when building models for time series data that seasonal effects have to be accounted for. This is normally done by adding dummy variables known as *'seasonal dummies'*. For example consider a variable D_2 defined as follows:

$D_2 = 1$ in quarter 2. $D_2 = 0$ for all other quarters.

This is a seasonal dummy used to model the seasonal effects in quarter two of quarterly data. We can define D_3 and D_4 in an analogous manner. However we would not normally define a D_1 for two reasons.

1   Normally we use the first quarter intercept as the 'benchmark' from which we measure changes in other quarters.
2   If we did introduce D_1 we would then have the relationship:

$$D_1 + D_2 + D_3 + D_4 = 1$$

This exact linearity leads to perfect multi co-linearity under which circumstances analysis is not possible.

If we add quarterly seasonal dummies to the classical linear regression model:

$$y(t) = \beta_0 + \sum_{i=1}^{k} \beta_i x_i(t) + u(t) \qquad t = 1, \ldots, n$$

we obtain

$$y(t) = \beta_0 + \delta_2 D_2 + \delta_3 D_3 + \delta_3 D_3 + \sum_{i=1}^{k} \beta_i x_i(t) + u(t) \qquad t = 1, \ldots, n$$

From this we can see immediately that the 'constant' term for the first quarter is β_0 and for the second and subsequent quarters $\beta_0 + \delta_2$, $\beta_0 + \delta_3$, $\beta_0 + \delta_4$ respectively.

   Seasonal dummies can of course be defined over other time periods with biannual and monthly being the most common alternatives.

   It is also possible to use the concept of dummy variables to represent effects other than seasonal variations. For example suppose we are looking at the changing demand for a soft drink over time in the UK market. Clearly price P will be a factor and perhaps the average disposable income E for the population. However sometimes the manufacturers run an advertising campaign and at other times there is no campaign. We can use a dummy variable A where $A = 1$ when a campaign is running and 0 when it is not. The pattern of A (e.g. 0011010010011...) may be quite irregular but the variable can be used in the same way as a seasonal dummy. The demand model will then be the following:

$$Q = \beta_0 + \alpha A + \beta_1 E + \beta_2 P + \epsilon$$

Q is the demand for the soft drink and in effect the above equation is saying that the residual demand for the drink is β_0 when there is no advertising and $\alpha + \beta_0$ when a campaign is running.

   Within Microfit there are three forms of seasonal dummies accessed either from the 'Edit' – Seasonal dummies sub menu or from the buttons in the 'Process Window':

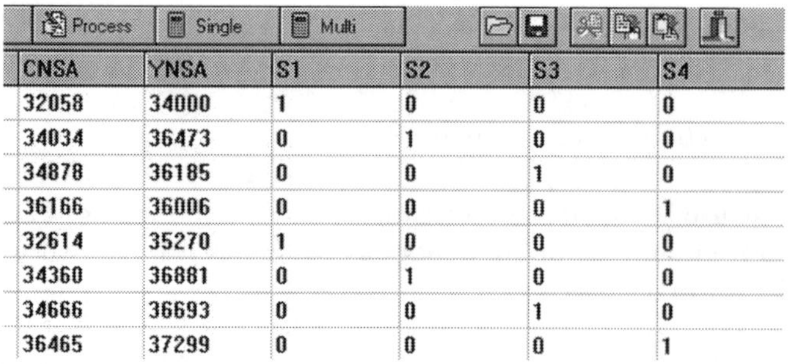

CNSA	YNSA	S1	S2	S3	S4
32058	34000	1	0	0	0
34034	36473	0	1	0	0
34878	36185	0	0	1	0
36166	36006	0	0	0	1
32614	35270	1	0	0	0
34360	36881	0	1	0	0
34666	36693	0	0	1	0
36465	37299	0	0	0	1

Figure A.2.5 Seasonal dummies

Seasonal(1) standard

When this option is selected the standard seasonal dummy pattern as outlined above is set up. If the data in the Microfit file is seasonal then the appropriate 0, 1 pattern of values is set up. When the option is selected you are given the choice of accepting the default names (S1,..., S4 for quarterly data S1,..., S12 for monthly data etc.) or using your own names. Figure A.2.5 shows the result of adding standard seasonal dummies to the UKCON2.FIT file. You will notice that four seasonal dummies are created and so it is *imperative* to avoid also adding a constant term as well for the reasons outlined above. If the Microfit data is undated then you can choose any period for the Seasonal(1) dummy.

Seasonal(2) centred

When this option is selected you will obtain a seasonal dummy pattern centred on zero as shown in the following example using quarterly data

Period	*SC1*	*SC2*	*SC3*	*SC4*
Q1	0.75	−0.25	−0.25	−0.25
Q2	−0.25	0.75	−0.25	−0.25
Q3	−0.25	−0.25	0.75	−0.25
Q4	−0.25	−0.25	−0.25	0.75

SC1 ... SC4 are the default variable names but you can change these if you wish.

Seasonal(3) relative

This option creates seasonal dummies relative to the last period of the preceding season. It is a precondition of seasonal(3) dummies that you have already created a seasonal(1) dummy pattern. To take a specific example if you have created a monthly seasonal(1) dummy pattern then the equivalent seasonal(3) pattern will compute SR1 = S1 − S12, SR2 = S2 − S12,..., SR11 = S11 − S12. Relative seasonal dummies can be incorporated into a regression equation in the normal way.

7 Using Microfit's standard functions

Microfit has a range of standard functions to enable you to build formulae to support your analysis. The list of available functions can be viewed from the drop

Figure A.2.6

down list box on the right hand side of the 'Process Window' (Figure A.2.6). You can either type the function name and any required parameters directly in the process window when you want to use it or you can highlight the function in the drop down list box and then press *'enter'* which will place the function at the current cursor location in the process window.

The functions provided by Microfit can be divided into two groups:

- ***Mathematical functions*** such as COS(X) which computes the cosine of the angle X and EXP(X) which computes e^X. The range of mathematical functions available is rather limited. A complete list of the mathematical functions with a description is given in appendix 2.

- ***Specialised statistical functions*** such as PTTEST(X, Y) which returns the Pesaran Timmermann test statistic or HPF(X, λ) which runs the parameter X through the Hoderick Prestcott filter. These are also described in appendix 1.

Usually the built in functions are used as part of formulae developed in the 'Process' Window'.

7.1 *Writing formulae in Microfit*

Microfit gives you the capability of generating complicated multi line formulae which can be applied to the data sets to generate new variables or new values for existing variables. Formulae are constructed using the standard arithmetic operators '+', '−', '*', '/' and the built in functions. Complex formulae can be generated on a single line by the use of parenthesise but it is generally better to split a complicated formula of several lines. Formulae remain in the 'Process Edit Window' after execution and so are available for editing if required. Each line is ended with a semicolon. No support is provided for the logical operators (e.g. AND, OR, NOT) but they can be generated using dummy variables if required.

To illustrate the process let us take a simple example. We will use the SIN function which calculates the sine of an angle but rather than calculate a single sine wave between 0° and 360° we will compound two sine terms using the following formula:

$$F = \sin(x) + \frac{1}{3}\sin(3x)$$

where x is the angle. One slight complication that we meet is that, in line with most computer languages, the trigonometric functions (SIN and COS) assume that the angle is in radians. The conversion from degrees is quite simple if we remember that 180° is equivalent to π radians. Unfortunately Microfit does not give us a value for π and so we will use the approximation 3.1416. First we create a new Microfit project and add a variable 'Angle' with 36 values 10 ... 360 as described in section 6.1. A plot of our function can now be created by adding the following formulae to the 'Process Edit Window'.

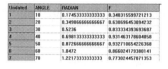

radian = 3.1416*angle/180;
F = SIN(radian) + SIN(3*Radian)/3;
scatter F angle

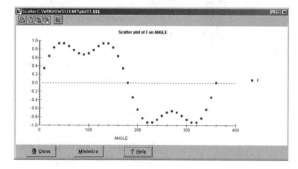

Figure A.2.7

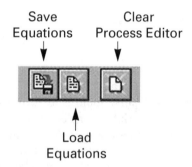

Figure A.2.8

When you click on 'Go' this will generate two new variables 'radian' and 'F', compute the values and then produce a scatter plot of F against angle (Figure A.2.7). For details on using the scatter command see section 8.2.

In the top left hand corner of the process window editor panel are three buttons (Figure A.2.8) which enable you to save and reload an equation set or to clear the 'Process Edit Window'. Equations are saved with an extension '.EQU' and can only be read by a subsequent Microfit session.

8   Using Microfit commands in batch mode

In addition to the functions the 'Process Window' also gives access to the Microfit commands via a second drop down list box (Figure A.2.6). We have already used some of the commands (e.g. SCATTER and DELETE) and others are described in subsequent sections. Appendix 2 gives a complete list of all the available commands.

It is quite a common requirement to apply the same set of formulae and commands to different data sets. A batch file provides a convenient mechanism for bundling together an appropriate set of formulae and commands. Microfit does not provide a tool for creating batch files but you can use any text editor or word processor to create the file. If you use a word processor remember to save the file as TEXT ONLY and always use a file extension '.BAT' which makes it easier for Microfit to find the file.

As an example let us convert the sine formulae example above to a batch file. First of all use the 'process editor' to combine the formulae and commands together to generate the desired result and then when you are satisfied that all is working correctly copy the editor panel to the clipboard. Open your text editor and paste in the formulae and commands. Add any comment lines which you feel will make it easier to understand the function of the batch file and save the result with a '.BAT' extension. Comment lines begin with the '$' character. The resulting batch file for the sine wave example used earlier might look something like the following.

```
$ First two terms of a Fourier
series radian = 3.1416*angle/180;
F = SIN(radian) + SIN(3*Radian)/3;
$ The end
```

Notice that the 'SCATTER' command has not been included in the file. Unfortunately there is a set of commands (including 'SCATTER') which cannot be included in batch files. The full list of such commands is given in appendix 2.

To use the batch file you can type BATCH <filename> in the 'Process Editor Window' followed by 'Go' and adding any further commands you may require. Alternatively you can click on the 'BATCH' button in the 'Process window' which will bring up a file dialog box.

9 Microfit graphics

9.1 Basic graphics with the PLOT command

The PLOT command provides access to a wide range of graphical display options. Let us begin by creating a basic plot of a simple data set. We will use the 'OIL.FIT' data set which displays oil production in millions of barrels from 1971 to 1989.

Once the data file is loaded go to the 'Command Editor' and type: **PLOT OP**. This will produce a very basic plot with default options as shown in Figure A.2.9.

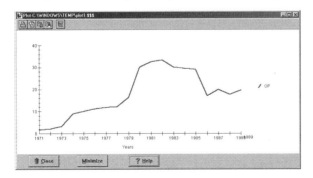

Figure A.2.9

The parameter 'OP' selects the variable to be plotted. Microfit allows up two three variables to be plotted simultaneously and has a limit of 6000 points in total. For example if you issue the command **PLOT X Y Z** you can plot up to 2000 points each for the three variables X, Y and Z.

For the type of data displayed the default is clearly not appropriate and we might decide that a two dimensional bar chart is more appropriate. The first task is to select the graphics toolbar in the PLOT window which gives us access to the graphics customisation features Figure A.2.10(a).

From the 2D Gallery we can now select the 'Bar' option. This gives us a bar chart. We can add titles to our graph by picking the 'Titles' tab and entering a graph title (e.g. Oil production) and a left title (e.g. Production (Million Barrels)) and selecting the print orientation to be 'up'. You may find the psychedelic colour display with each bar a different colour rather distracting! We can improve this by selecting the option 'Markers' when the cursor changes to an arrow and we can select any particular column for a colour change. Changing the colour of any one column initially changes the colour of all other columns to black. To change the colour of all other columns to our selected colour is, I'm afraid, a rather tedious task

Graphics toolbar

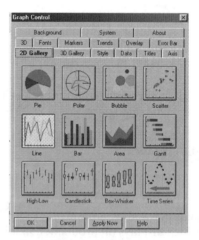

(a)

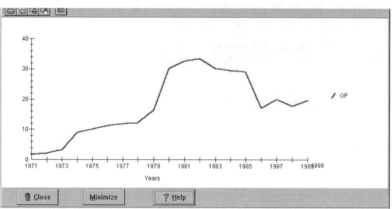

Figure A.2.10

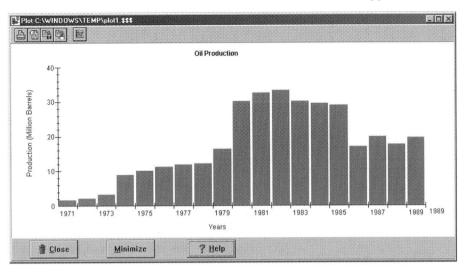

Figure A.2.11

of editing the colour of each column in turn. Figure A.2.11 shows the results of these changes.

The Graphics toolbar contains many other 2D and 3D graphical functions which you should explore if you wish to enhance your graphics displays.

9.2 Using the SCATTER command

The SCATTER command gives us a very convenient way to visualise a relationship between two variables. We can explore this by loading the Microfit data file UKCON.FIT which lists real disposable income (RDI) and consumers expenditure (CE). We obtain a scatter plot by issuing the command: SCATTER CE RDI which produces the scatter plot shown in Figure A.2.12. Notice that the variable plotted along the X axis is defined second. The limit of a total of 6000 data points applies also to scatter plots. The quality of the graphics can be enhanced by using the Graphics toolbar.

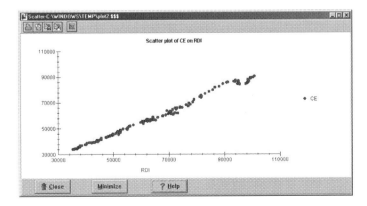

Figure A.2.12

9.3 Saving and printing graphs

Once you have generated a graph and arranged the presentation to suite your requirements the graph can be saved to a file in order to use it in other applications (e.g. Word Documents) or to read it back into a Microfit session. Three basic formats are available (Figure A.2.13) as follows.

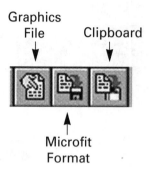

Figure A.2.13

- *Graphics file.* The graph is saved either as a Windows bitmap or Metafile. Once converted to this format the graph simply becomes a picture for importing into other applications. It can no longer be edited as a graph.
- *Microfit format.* This is a special Microfit graphics format with the extension .MFW. When saved in this format the graph can be loaded into another Microfit session and the graph edited.
- *Clipboard.* The graph can be copied to the Clipboard for subsequent use in Microfit or other applications.

10 Single equation estimation

The classical linear regression model for a dependent variable y(t) as a function of a set of regressors $x_1(t), x_2(t), \ldots, x_n(t)$ is given by:

$$y(t) = \beta_0 + \sum_{i=1}^{k} \beta_i x_i(t) + u(t) \qquad t = 1, \ldots, n \tag{A.2.1}$$

β_i are the regression coefficients to be estimated (β_0 the constant term) and u(t) are the error terms subject to the following limitations:

1 **Zero means** $E(u(t)) = 0 \, \forall \, n$

2 **Homoskedasticity** $Var[u(t) | x_1(t), x_2(t), \ldots, x_k(t)] = \sigma^2$

3 **Non autocorrelated errors** $Cov[u(t), u(s)] = E[u(t), u(s)] = 0$

4 **Orthogonality** $E[u(t) | x_1(t), x_2(t), \ldots, x_k(t)] = 0$

5 **Normality** $Norm(0, \sigma)$

As is well known the critical assumptions are 3 and 4 above and in particular 3 the failure of which can lead to spurious regression. This is particularly true when lagged terms (y(t − 1) etc.) are included in the regression equation.

Let us see how we can use Microfit to perform a classical linear regression analysis.

10.1 Specifying a linear regression equation

Let us start with a very simple example. First of all select the data file 'UKCON.FIT'. We will use the '*Variables*' form to rename the two key variables.

The real personal disposable income we will rename 'RINC' and the consumers' expenditure 'CEXP'. An examination of the scatter plot for these two variables suggests that there is an approximate linear relationship between them of the form:

$$CEXP = INCP + \beta.RINC \tag{A.2.2}$$

A closer examination of the scatter plot suggests that this may not be the best model but we will return to this later. INCP is a constant term representing the intercept and β is the regression coefficient. These are the two parameters we need to estimate. We can add the constant term to our data file either by using the *'CONSTANT'* button in the *'Command Editor'* or by typing INCP = 1 in the *'Process Window'*. Either method will add INCP to the data items with the value set to one for all rows.

Suppose we wish to use the method of ordinary least squares. From the main menu select *'Univariate'* and then from the sub menus *'linear regression menu'* and finally *'Ordinary least squares'* We are now ready to use the Ordinary Least Squares (OLS) option.

10.2 Using the OLS option

After the OLS estimation method has been selected the OLS window opens as shown in Figure A.2.14.

The first task is to use the two data entry boxes to select the time span for the OLS estimation. The next step is to enter the dependent variable followed by the regressors in the order specified in equation (A.2.2). For this example we enter the following: CEXP INCP RINC. When the 'Start' button is selected Microfit performs the required calculation followed by a summary display of the results as shown below. This is suggesting that our model should be.

CEXP = 1981 + 0.875*RINC

Microfit also provides a range of test statistics to help us evaluate the quality of our model. Using a single statistic to evaluate a model is usually inadequate. For example the R^2 parameter (0.992) suggests that the model is quite a good fit. However the low value of the DW statistic (0.405) shows that the model is fundamentally incorrect since the value is far from the 'best fit' value of 2.

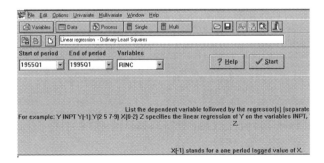

Figure A.2.14 Data entry point

```
                    Ordinary Least Squares Estimation
*********************************************************************************
Dependent variable is CEXP
160 observations used for estimation from 1955Q1 to 1994Q4
*********************************************************************************
Regressor        Coefficient      Standard Error      T-Ratio[Prob]
INCP             1980.9           406.9290            4.8680[.000]
RINC             .87502           .0060946            143.5741[.000]
*********************************************************************************
R-Squared                      .99239    R-Bar-Squared                .99235
S.E. of Regression             1476.8    F-stat.  F( 1, 158)          20613.5[.000]
Mean of Dependent Variable     57949.1   S.D. of Dependent Variable   16879.4
Residual Sum of Squares        3.45E+08  Equation Log-likelihood      -1393.6
Akaike Info. Criterion         -1395.6   Schwarz Bayesian Criterion   -1398.7
DW-statistic                   .40489
*********************************************************************************
```

On closing the results display the post regression menu is displayed. We will follow this up in section 11 below.

10.3 Testing the model assumptions

As outlined in the introduction to section 9 above the classical regression model is based on five assumptions of which 3 and 4 in the list are particularly critical. Any model that we develop should be tested against these assumptions to check for violations. Of course it goes without saying that passing this evaluation does not verify the truth of our model (several different models may pass the evaluation) it only checks that it does not fail these basic assumptions. Similarly failing one of the assumptions does not identify a unique cause for the failure since several different causes can lead to a violation of a given assumption. Microfit provides the researcher with an impressive set of tests to apply to the regression equation and full details can be found in the manual which accompanies the software. Using our former OLS example the summary display of results also includes a set of diagnostic statistics as shown below.

```
                              Diagnostic Tests
***************************************************************************
*    Test Statistics  *      LM Version        *        F Version        *
***************************************************************************
*                     *                        *                         *
* A:Serial Correlation*CHSQ(  4)= 108.3511[.000]*F(  4, 154)=  80.7668[.000]*
*                     *                        *                         *
* B:Functional Form   *CHSQ(  1)=  15.8264[.000]*F(  1, 157)=  17.2344[.000]*
*                     *                        *                         *
* C:Normality         *CHSQ(  2)=   8.1751[.017]*   Not applicable       *
*                     *                        *                         *
* D:Heteroscedasticity*CHSQ(  1)=  28.9265[.000]*F(  1, 158)=  34.8689[.000]*
***************************************************************************
    A:Lagrange multiplier test of residual serial correlation
    B:Ramsey's RESET test using the square of the fitted values
    C:Based on a test of skewness and kurtosis of residuals
    D:Based on the regression of squared residuals on squared fitted values
```

For example the parameter 'A' displays the Lagrange Multiplier test of residual serial correlation. This parameter is interpreted as follows. Under the assumption of homoskedasticity ($Var(u_t)$ constant) we expect the residuals to show the same distribution for the complete range of the dependent variable. A more formal way to test for this is to regress the residuals (e_i) onto the regressor $x_1(t)$.

$$e_t = \alpha_1 + \alpha_2 x_1(t) + u_t$$

This a simple form of White's equation and an equivalent expression exists when more than two explanatory variables are used. This equation with a change of explanatory variable can also be written:

$$e_t = \alpha^* + \beta^*[E(Y_t)]^2 + u_i$$

Now the variance of the disturbances is given by

$$Var(\epsilon_t) = E(\epsilon_t^2) = \beta_0 + \beta_1[E(Y_t)]^2$$

However since $E(Y_i)$ is unknowable then we can replace $E(Y_t)$ by Y_{bt} the predicted values from the OLS estimation.

$$e_t = \alpha^* + \beta^* Y_{bt} + u_i \tag{A.2.3}$$

Now it can be shown that under the assumption of homoskedasticity ($\beta^* = 0$) that nR^2 (n the number of observations and R^2 the coefficient of determination) obtained from the above equation has a χ^2 distribution with the number of degrees of freedom given by the number of non constant explanatory variables. In the example quoted at the 5% level of significance the critical χ^2 value is 3.841. The parameter 'A' considerably exceeds this indicating that homoskedasticity does not hold.

10.4 The post regression Menu

When you have finished examining the summary report and closed the form you are taken directly to the post regression menu. Here you are provided with a wide range of options to allow you to explore the validity of the model that you have developed. Your post regression analysis may be to amplify some of the conclusions formed from the diagnostics tests in the summary report.

Figure A.2.15 shows the first post regression menu options.

It is not possible in this brief introduction to do justice to all the features available from this basic menu. Instead, by way of illustration, we will follow through our conclusion in 9.5 that the residuals indicate that homoskedasticity does not hold. First we will select item 3 on the menu to bring up the residuals sub menu. From the residuals sub menu we can again pick item three to plot the residuals as a function of time (Figure A.2.16). This certainly lends support to the conclusion that the variance is changing (in this case increasing) over time. However a far better illustration of heteroskedastic disturbance is to plot the square of residuals against the regressor variable. To do this select item 6 (*save residuals*) from the residuals sub menu in

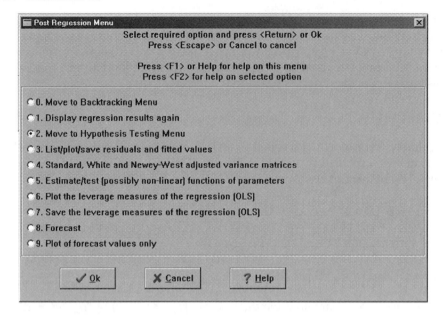

Figure A.2.15

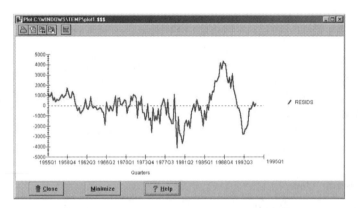

Figure A.2.16

order to save the residuals as part of the data file. You are asked to supply a name for the residuals (e.g. RESIDS) and an optional description. Having saved the residuals we can use the process window to calculate the square of the residuals (e.g. SQRESID = RESID*RESID) and then call for a scatter plot of SQRESID on RINC.(SCATTER SQRESID RINC). The result is shown in Figure A.2.17 clearly demonstrating the heteroskedastic behaviour.

It is of course unwise to rely on a single test for heteroskedasticity and the other diagnostic parameters should be used to confirm your conclusions.

If we suspect that the disturbances are subject to auto correlation up to the pth order i.e.

$$e_t = \rho_1\epsilon_{t-1} + \rho_2\epsilon_{t-2} + \rho_3\epsilon_{t-3}, \ldots, \rho_p\epsilon_{t-p} + u_t$$

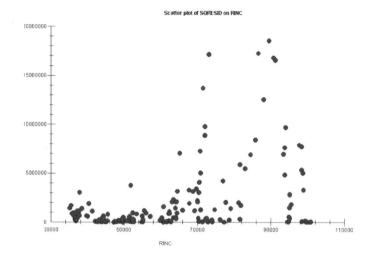

Figure A.2.17

Then we can test the null hypothesis H_0: *there is no auto correlation* $(\rho_1 = \rho_2 = \ldots = \rho_p = 0)$ by replacing the ϵ_i by e_1 and performing the Lagrange Multiplier test of order p. In this case equation (A.2.3) is modified by the addition of the auto correlation terms giving:

$$e_t = \alpha^* + \beta^* Y_{bt} + \rho_1 \epsilon_{t-1} + \ldots + \rho_p \epsilon_{t-p} + u_i$$

In fact the 'A' parameter in the diagnostics of the summary report is the χ^2 with $p = 4$. Item 2 of the post regression menu (*Move to hypothesis testing menu*) allows you to explore this further by choosing different values for p (item 1 *LM tests for serial correlation* on the sub menu).

It is of course unwise to rely on a single test for heteroskedasticity and the other diagnostic parameters should be used to confirm your conclusions.

11 Co-integration

As a final example of using Microfit we will take a simple example of examining a data set for co-integration. Recall that a basic two variable Error Correction Model (ECM) is given by:

$$y_t = \beta_0 + \beta_1 x_t$$

This model defines a possible long running relationship between the two time series y_t and x_t. The associated disequilibrium error is given by $u_t = y_t - \beta_0 - \beta_1 x_t$ and if the long-term relationship underlying the ECM model really exists then the u_t should be centred on zero. It can be shown more formerly that a long running relationship exists between the variables if the following two conditions hold

- Both y_t and x_t are I(1). In other words they become stationary on first differencing.
- A linear combination of y_t and x_t exists (the ECM) that is stationary or I(0).

If both conditions hold the two time series y_t and x_t are said to be co-integrated. In recent years co-integration has become a key econometric analysis tool.

As usual testing for co-integration is hampered by the fact that the parameters β_0 and β_1 are unknown and hence the disequilibrium errors are unknown. We attempt to overcome this problem by first using OLS to estimate the two parameters and then take the residuals as an approximation to the disequilibrium errors. We test the residuals for stationarity by using the Dickey–Fuller (DF) and Augmented Dickey–Fuller tests. The DF and ADF tests for unit roots and their limitations are discussed in chapters 6, 7 and 8 and will not be repeated here.

Let us apply these tests to the 1st data set supplied with the Microfit package. This data set records the expenditure on food (in $billion) and the total expenditure (in $billion) for the years 1963 to 1992 for the USA. Of the four data columns provided we are particularly interested in the two normalised columns, RF expenditure on food in constant prices and RTE the total expenditure in constant prices. We will assume that we have already demonstrated that both time series are I(1)

Let us seek a relationship between log(RF) and log(RTE). Using the process window we can soon compute LINRF = LOG(RF) and LINRTE = LOG(RTE). producing a scatter plot of these two new variables (Figure A.2.18) suggests indeed that an approximate linear relationship does exist.

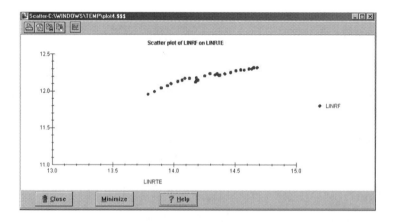

Figure A.2.18

Using the method described in section 9.1 we can find estimates for β_0 and β_1 from the linear regression equation which gives:

LINRF = 7.1138 + 0.35537*LINRTE $R^2 = 0.935$

Whilst R^2 is not a good diagnostic statistic with a small sample size (25) the high value is at least reassuring. We now save the residuals in the usual way into a variable RESIDS. Producing a plot of RESIDS does indicate that they are small and fairly well distributed about zero (Figure A.2.19). We can now apply the augmented Dickey–Fuller test to the residuals by typing in the process window

ADF RESIDS(4)

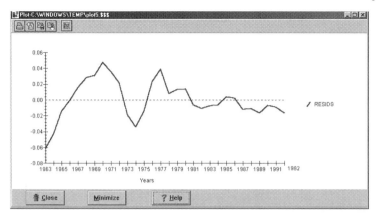

Figure A.2.19

The factor 4 indicates that up to the 4th order process is included in the ADF equation.

The summary page produced by Microfit is illustrated below

```
                    Unit root tests for variable RESIDS
          The Dickey-Fuller regressions include an intercept but not a trend
**********************************************************************************
     25 observations used in the estimation of all ADF regressions.
     Sample period from 1968 to 1992
**********************************************************************************
             Test Statistic      LL         AIC        SBC        HQC
     DF         -1.9584       69.3670     67.3670    66.1482    67.0290
     ADF(1)     -3.1395       72.9751     69.9751    68.1468    69.4680
     ADF(2)     -2.4466       72.9776     68.9776    66.5399    68.3015
     ADF(3)     -2.3804       73.1931     68.1931    65.1459    67.3480
     ADF(4)     -2.2897       73.3071     67.3071    63.6504    66.2929
**********************************************************************************
     95% critical value for the augmented Dickey-Fuller statistic =  -2.9850
     LL  = Maximized log-likelihood     AIC = Akaike Information Criterion
     SBC = Schwarz Bayesian Criterion   HQC = Hannan-Quinn Criterion


                    Unit root tests for variable RESIDS
          The Dickey-Fuller regressions include an intercept and a linear trend
**********************************************************************************
     25 observations used in the estimation of all ADF regressions.
     Sample period from 1968 to 1992
**********************************************************************************
             Test Statistic      LL         AIC        SBC        HQC
     DF         -2.6219       70.8608     67.8608    66.0325    67.3537
     ADF(1)     -4.4974       76.7848     72.7848    70.3471    72.1087
     ADF(2)     -4.0315       77.3259     72.3259    69.2787    71.4808
     ADF(3)     -4.0186       77.8977     71.8977    68.2411    70.8835
     ADF(4)     -3.8573       78.0437     71.0437    66.7777    69.8605
**********************************************************************************
     95% critical value for the augmented Dickey-Fuller statistic =  -3.6027
     LL  = Maximized log-likelihood     AIC = Akaike Information Criterion
     SBC = Schwarz Bayesian Criterion   HQC = Hannan-Quinn Criterion
```

The fact that the ADF test statistic is not more negative than the critical value for the case of an intercept but no trend suggests that LINRF and LINRTE are not co-integrated. However when the linear trend is included for ADF(1) ... ADF(3) the test statistic is more negative than the critical value strongly suggesting the two series are co-integrated.

APPENDIX 1 – LIST OF MICROFIT FUNCTIONS

(The following information is taken from the Microfit help file)

ABS()

This function replaces X by its absolute value.

$$\text{ABS}(X) = X \text{ if } X > 0, \text{ and } \text{ABS}(X) = -X \text{ if } X < 0$$

Clearly, $\text{ABS}(0) = 0$.

COS()
The function takes cosine of the expression that follows in the brackets.
   Note the expression represents the angle in radians

CPHI()
This is the cumulative standard normal function so that CPHI(X) represents the integral between minus infinity to X of the standard normal distributions.
   Examples:

$$Y = \text{CPHI}(0.0)$$

returns the value of 0.5 for Y.

$$Y = \text{CPHI}(w + 2*(z1/z2))$$

first computes the expression inside the brackets, and then returns the values of the integral of the standard normal distribution from minus infinity and $w + 2*(z1/z2)$.

CSUM()
This function, when applied to a variable X, calculates the cumulative sum of X. For example, if $X = (6, 2, -1, 3, 1)$, then typing the formula

$$Y = \text{CSUM}(X)$$

will result in $Y = (6, 8, 7, 10, 11)$.
   The argument X can itself be a function of other variables as in the following example:

$$Y = \text{csum}((z - (\text{sum}(z)/\text{sum}(1))^2)$$

EXP()
The function takes the exponential of the expression that follows in the brackets.
   Example:

$$Y = 2 + 3.4*\text{EXP}(1.5 + 4.4*X)$$

GDL()

This is the geometric distributed lag function and has the following form:

$$Y = GDL(X, lambda)$$

and computes Y as

$$SIM\ Y = lambda*Y(-1) + X$$

over the sample period which is in effect. Lambda is the parameter of the distributed lag function, and SIM is the SIM command. The initial value of Y is set equal to the value of X at the start of the selected sample period.

For example, suppose you wish to compute geometric distributed lag of X over the period 1950–1980, with lambda = 0.8. You need to type

$$Sample\ 1950\ 1980;\ Y = GDL(X, 0.8)$$

The value of Y in 1950 will be set equal to the value of X in 1950.

HPF()

This function has the form

$$Y = HPF(X, lambda)$$

and runs the variable X through a Hodrick-Prescott filter using the parameter lambda as the smoothing coefficient. In this function X is a vector, and lambda is a non-negative scaler (i.e. a vector with all its elements equal to lambda $>=0$). This filter is used extensively in the Real Business Cycle literature as a de-trending procedure.

The choice of lambda depends on the frequency of the time series, X. For quarterly observations Hodrick and Prescott set lambda = 1600. Often the value of lambda = 7 for annual observations, and lambda = 126,400 for monthly observations is made but, in general, the optimal choice of lambda will depend on the particular time series under consideration.

The argument X could also be specified to be a function of other variables on the workspace.

Example: Suppose USGNP contains quarterly observations on US aggregate output. Then the trend series (in logarithms) are given by:

$$YT = HPF(log(USGNP), 1600)$$

To compute the filtered, or de-trended, series you need to type

$$YD = log(USGNP) - YT$$

INVNORM()

This function computes the inverse of the cumulative standard normal distribution, so that for a given probability P, Y = INVNORM(P) computes Y such that the area under the standard normal curve between minus infinite to Y is equal to P.

$$Y = INVNORM(0.975)$$

In this example, Y will be set to 1.9600, the 95 per cent critical value of the standard normal distribution. Note that

$$0 = < P < = 1$$

LOG()
The function takes logarithm to the base e (natural logarithm) of the expression that follows in brackets. Example:

$$Y = 2.4 + 3.5*LOG(X + 3)$$

For negative values of X, this function returns the missing values, *NONE*.

MAV()
This function has the form Y = MAV(X, p) and places the pth order moving-average of the variable X in Y, namely

$$Y(t) = [X(t) + X(t-1) + \ldots + X(t-p+1)]/p$$

Variable X could be any of the variables on the workspace or a function of them. If p is not an integer Microfit chooses the nearest integer to p in order to carry out its computations. If p is negative, this function returns the missing value, *NONE*.

MAX()
This function has the form Y = MAX(X, Z) and places the maximum of X and Z in Y. For example, if X = (1, 7, 2, 3, 6) and Z = (6, 2, −1, 3, 1), then Y will be set to (6, 7, 2, 3, 6).

The arguments X and Z themselves could be functions of other variables. As in the following example

$$Y = MIN((G1/G2) + 1, (H1/H2) - 1)$$

MEAN()
This function, when applied to a variable X, calculates the mean of X over the specified sample period. For example,

$$Sample\ 1970\ 1995;\ Y = (X - MEAN(X))$$

generates deviations of variable X from its mean computed over the sample period 1970 to 1995, inclusive.

Note that the value of MEAN(X) will be set to *NONE* (i.e. missing) if one or more values of X are missing over the specified sample period.

MIN()

This function has the form $Y = MIN(X, Z)$ and places the minimum of X and Z in Y. Clearly, X and Z could themselves be functions of other variables, as in

$$Y = MIN(2 + x1/x2, 3/z1)$$

NORMAL()

This function can be used to generate independent standardised normal variates (i.e. with zero means and unit variances). The function should be used in the form of NORMAL(j) within a formula, where j represents the 'seed' for the quasi-random numbers generated, and it must be an integer in the range $0 < j < 32000$. By changing the value of j, different quasi-random number series can be generated. Examples of the use of this function are:

$$X = NORMAL(1)$$

$$Y = 2 + 3.5*NORMAL(124) + Z$$

ORDER()

This function has the form $Y = ORDER\ (X, Z)$ and orders X according to the sorting implied by Z, where Z is sorted in an ascending order. The results are placed in Y. For example, if $X = (1, 7, 2, 3, 6)$ and $Z = (6, 2, -1, 3, 1)$, then Y will be set to $(2, 6, 7, 3, 1)$.

Clearly, as in the case of other functions, the arguments of the function, namely X and Z, could themselves be functions of other variables.

PHI()

This function gives the ordinate of the standard normal distribution for the expression that follows in brackets. Examples:

$$Y = .5*(1/PHI(0))^2$$

$$Y = PHI(1)$$

$$Z = PHI(1 + 0.5*W)$$

The general formula for the PHI function is given by

$$PHI(x) = ((2*pi)^{(-.5)})*EXP(-.5*x^2),$$

where EXP(.) is the exponential function and $pi = 3.14159\ldots$

PTTEST()

This function has the form $T = PTTEST(Y, X)$ and returns the Pesaran–Timmermann statistic for a non-parametric test of association between Y and X. Under the null hypothesis that Y and X are distributed independently, the PTTEST statistic is approximately distributed as a standard normal.

RANK()

This function returns the value of *NONE* if the observations in Y and/or X do not change signs over the specified sample period.

This function when applied to a variable X gives the ranks associated with the elements of X, when X is sorted in an ascending order. For example, if $X = (6, 2, -1, 3, 1)$ then typing the formula:

Y = RANK(X) will give $Y = (5, 3, 1, 4, 2)$.

RATE()

This function enables you to generate per cent change in a variable. For example $PI = RATE(P)$ generates percentage change of the variable P, and stores the result in PI. More specifically, PI will be computed as:

$PI = 100*(P - P(-1))/P(-1)$.

An alternative approximate method of computing rate of change in a variable would be to use changes in logarithms, namely

$PIX = 100*(\log(P/P(-1))$

It is easily seen that both approximations are reasonably close to one another for values of PI and PIX around 20 per cent or less. The argument of this function, namely P, can itself be a function of other variables as in the following example:

$Y = Rate(w + u/v)$

SIGN()

This function when applied to a variable X returns the value of 1 when X is positive, and 0 when X is zero or negative. For example, if $X = (3, -4, 2, 0, 1.5)$, then typing

Y = SIGN(X) will set $Y = (1, 0, 1, 0, 1)$. You can also have
$Y = 2 + Z * SIGN(DY - 0.05)$ etc.

SIN()

This function takes sine of the expression that follows in the brackets. Example:

$Y = 2 + 3*SIN(5 + 7*X)$

Note the argument of the function is the angle in radians.

SORT()

This function, when applied to a variable X will sort X in an ascending order. For example, if $X = (6, 2, -1, 3, 1)$ then typing Y = SORT(X) will set $Y = (-1, 1, 2, 3, 6)$, while:

$Z = -SORT(-X)$ will sort X in descending order in Z so that $Z = (6, 3, 2, 1, -1)$.

In the example $Y = Sort(2 + w/z)$ the expression $2 + w/z$ will be first computed and the resultant expression will be sorted in Y as above.

SQRT()
This function takes square-roots of the expression that follows in the brackets. Example:

$Y = 3 + 5*SQRT(X)$

For negative values of X, this function returns the missing values, *NONE*.

STD()
This function, when applied to a variable X, calculates the standard deviation of X over the specified sample period. For example

SAMPLE 1970 1995; $Z = (X - MEAN(X))/STD(X)$

places the standardised values of X over the period 1970–1995 (inclusive) in the variable Z.

Note that the value of STD(X) will be set to *NONE* (i.e. missing) if one or more values of X are missing over the specified sample period.

SUM()
This function first calculates the expression specified within closed brackets immediately following it, and then computes the sum of the elements of the result over the relevant sample period. Examples of the use of this function are:

SAMPLE 1960 1970; $XBAR = SUM(X)/SUM(1)$; $XD = X - SUM(X)/SUM(1)$;

$YD = Y - SUM(Y)/SUM(1)$; $BYX = SUM(XD*YD)/SUM(XD^2)$

In the above examples, SUM(X) is a variable with all its elements equal to the sum of the elements of X, over the period 1960–1970. SUM(INPT) is equal to the number of observations in the sample period (i.e. 11). XBAR is, therefore, equal to the arithmetic mean of X, computed over the specified sample period. XD and YD are deviations of X and Y from their respective means and BYX is the ordinary least squares (OLS) estimates of the coefficients of X in the regression of Y on X (including an intercept term).

UNIFORM()
This function can be used to generate independent random numbers from a uniform distribution within the range 0 and 1. The function should be used in the form of UNIFORM(j) within a formula, where j represents the 'seed' for the quasi-random numbers generated, and must be an integer in the range $0 < j < 32000$. By changing the value of j, different quasi-random number series can be generated.

Examples of the use of this function are:

$X = UNIFORM(1)$

$Y = 2 + 3.5*UNIFORM(124) + Z$

APPENDIX 2 – LIST OF MICROFIT COMMANDS

(The following information is taken from the Microfit help file)

<NB> following the name of a command indicates that it cannot be part of a batch file.

ADD <NB>
This command enables you to add a new variable to the list in the Variables window.

The form of this command is

ADD XNEW where XNEW is the name of the new variable.

The variable is added to the Variables window and to the Data Editor with all its values set to *NONE*. To insert the values for your variable, click the DATA button to move to the Data Editor, then click on each of the cells in turn and type in the value you want.

ADF <NB>
This command, when followed by a variable name, displays the Dickey–Fuller (DF) and the Augmented Dickey–Fuller (ADF) statistic for testing the unit root hypothesis together with the associated critical values. For example, SAMPLE 75Q1 87Q1; ADF X computes the DF and the ADF test statistics of up to order 4 (i.e. the periodicity of the data) for the variable X, and displays the statistics together with their 95 per cent critical values on the screen. The pth order ADF test statistic is given by the t-ratio of a2 in the ADF regression

$$DX(t) = a0 + a1*t + a2*X(t-1) + b1*DX(t-1) + bp*DX(t-p)$$

where $DX(t) = X(t) - X(t-1)$. Microfit computes the ADF statistic both with and without a time trend. The 95 per cent critical values for the test are given in brackets.

The ADF command can also be used to compute the Augmented Dickey–Fuller test statistics, up to an order specified by the user. The desired order should be specified in parentheses immediately after the variable name. For example,

ADF X(12)

gives the ADF test statistics for the variable X up to the order 12, assuming, of course, that there are enough observations. 12 is the maximum that can be used.

BATCH
This command has the form

BATCH   or BATCH <filename>

Section 7.5 discusses the writing of batch files.

COR <NB>

This command has different effects depending on whether it is followed by one variable or more. When only one variable is specified after COR, as in the example

SAMPLE 1 20; COR X>

summary statistics for X (i.e. mean, standard deviation, coefficient of variation, skewness, kurtosis, minimum, and maximum values) and its auto-correlation coefficients of up to the order of a third of the number of specified observations will be shown on the screen. The plot of the auto-correlation function will also be displayed.

The COR command can also be used to compute auto-correlation coefficients up to an order specified by the user. The desired order should be specified in parentheses immediately after the variable. For example,

COR X(12)

gives the auto-correlation coefficients for the variable X up to the order of 12 (assuming, of course, that there are enough observations). When the COR command is followed by two or more variables, as in the example

COR X Y Z

then summary statistics and the correlation coefficients for these variables, over the specified sample period, will be provided.

DELETE

This command enables you to delete one or more variables from the list of your existing variables on your work-space. The names of the variables to be deleted should follow the command, separated by spaces. For example,

DELETE X Y Z deletes the variables X, Y, and Z from the list of your existing variables. If you wish to delete a single variable, you can also type

X =

This operation has the effect of deleting variable X.

ENTITLE

This command allows you to enter or change the description of one or more of the variables on your work space. For example, if you type

ENTITLE X Y Z

you will be asked for the titles (or descriptions) of the variables X, Y, and Z, in that order. Note that the description of a variable can be at most 80 characters. If you type in a title which is more than 80 characters long, only the first 80 characters will be saved. If you enter the command ENTITLE on its own, namely

ENTITLE

you will be asked to give titles for those variables in your work space which do not have titles.

When a new variable is generated using the data transformation facilities, the first 80 characters after the quality sign will be automatically used as the title of the generated variable. Also, when a variable XNEW is created by the formula

XNEW = XOLD

the title of the variable XOLD, if any, will be passed on to the new variable, XNEW.

HIST <NB>
When followed by a variable name, this command displays the histogram of the variable. For example,

SAMPLE 1 20;

HIST X

The number of bands is automatically chosen between 6 and 15 according to the formula

Min 15, Max(N/10, 6)

where N is the total number of observations. This command can also be used to plot the histograms for any numbers of intervals chosen by the user. The desired number of classes should be specified in parentheses immediately after the variable. For example,

HIST X(12)

KEEP
This command deletes all the variables in the work space except those specified. For example, suppose you have 10 variables named X1, X2,..., X10 in your work space, and you wish to keep only the variables X1 and X2. Then type

KEEP X1 X2

LIST <NB>
This command allows you to inspect your data on screen and/or to save them in a file to be printed out later. If the command LIST is typed on its own followed by GO, then the values of all the variables will be displayed over the current sample period set by the SAMPLE command. If the command LIST is followed by one or more variable names, then only the values of the specified variables will be listed.

For example,

SAMPLE 1940 1980;

LIST

displays all the existing variables over the period 1940–1980.

SAMPLE 80Q1 85Q2;

LIST X Y Z

displays the observations on the variables X, Y, Z over the period from the first quarter of 1980 to the second quarter of 1985, inclusive.

PLOT <NB>

This command produces a line graph of up to a maximum of three variables against time. You must specify at least one variable name. For example

SAMPLE 1950 1970;

PLOT X produces a plot of variable X against time, over the period 1950–1970.

PLOT X Y Z produces a plot of the three variables X, Y, and Z against time.

REORDER

This command enables a complete reordering of the observations on the workspace according to the ordering of the variable that follows the command. For example,

REORDER X

produces a reordering of observations according to the ordering of the observations in variable X. This command is particularly useful when analysing cross-sectional observations where the investigator wishes to carry out regression analysis on a subset of the observations. The exact nature of the particular sub-set of interest is defined by the ordering of the observations in the variable X.

As an example, suppose that the undated observations in the variables list refer to both male and female indexed by 0 and 1, respectively, stored in the variable SEX. Issuing the command

REORDER SEX

reorders the observations in the variables list so that observations referring to females appear first. The number of such observations is equal to SUM(SEX).

RESTORE

This command should be used after the REORDER command, and restores the ordering of the observations to their original state, before their ordering was changed by the use of the REORDER command.

SAMPLE

This command can be used to change the sample period for subsequent data analysis in the data processing stage, but does not carry over to the other parts of the program. Examples include

SAMPLE 3 26 for undated observations

SAMPLE 1972 1986 for annual observations

SAMPLE 50H2 72H1 for half-yearly data

SAMPLE 75Q1 78Q2 for quarterly data

SAMPLE 70M1 80M11 for monthly data

Note: The default century for the half-yearly, quarterly, and monthly observations is taken to be the 20th century. This can be overwritten in Microfit 4.0. When specifying sample periods that include dates from the 21st century you need to give the dates in full. Examples are:

SAMPLE 1996 2005

SAMPLE 1998H1 2003H2

SAMPLE 1999Q1 2010Q4

SAMPLE 1998M1 2020M12

SCATTER <NB>
This command can be used to produce scatter diagram of a variable against another. When issuing this command, you must specify exactly two variable names. For example,

SCATTER X Y

produces a scatter plot of the variable X against the variable Y. Graphs may be saved, printed, and annotated.

SIM
This is a stimulation command, and enables you to solve numerically any general univariate linear or non-linear difference equation. For example, to solve the non-linear difference equation

$$X(t) = 0.2X(t-1) + 0.7Log(X(t-2)) + Z$$

for $t = 3, 4, \ldots, 20$, with initial values $X(1) = 0.05$ and $X(2) = 0.10$, you need to issue the following commands

SAMPLE 1 1; X = 0.05;

SAMPLE 2 2; X = 0.10;

SAMPLE 3 20; SIM X = .2*X(−1) + .7*LOG(X (−2)) + Z;

SAMPLE 1 20;

PLOT X

The first four commands in the above example set the initial values for X, which are used to simulate the values of X for observations 3, 4,..., 20.

The following points should be kept in mind when using the SIM command.

- When the SIM command is used, the values of the simulated variable will be overwritten. To avoid this problem, one possibility would be to create a new variable called, say XNEW, which contains the appropriate values, but is other wise undefined for other periods. The SIM command can then be applied to XNEW over the sample period for which XNEW is undefined. A typical example of this procedure would be (assuming that the specified sample period runs from 1950 to 1980)

    SAMPLE 1950 1950;

    XNEW = 0.05;

    SAMPLE 1951 1980;

    SIM XNEW = 4*XNEW(−1)*(1 − XNEW(−1))

The above will solve the well-known chaotic bifurcation equation.

$$X(t) = 4 X(t − 1)(1 − X(t − 1))$$

starting with the initial value X1950 = 0.05 over the period 1951–1980.

- Choose your sample period carefully and make sure that well defined initial values exist for the simulation. Otherwise all the values of the variable being simulated will be set equal to *NONE*.
- In the case of unstable difference equations, the use of the SIM command may cause an overflow. When the value of the simulated variable exceeds 10 to the power 50, to prevent the program from crashing, all the subsequent values will be set equal to *NONE*.

SIMB

This is a simulation command which allows you to solve numerically any general univariate linear or non-linear difference equation involving lead (not lagged) values of the left-hand side variable. The difference equation is solved backwards over the specified sample period. For example, to solve the linear difference equation:

$$X(t) = 1.2*X(T + 1) + Z(t), \text{ for } t = 20, 21,..., 1$$

with a terminal value of 30.5 at observation 20, the following commands should be issued:

    SAMPLE 20 20; X = 30.5;

    SAMPLE 1 19;

SIMB X = 1.2*X(+1) + Z;

SAMPLE 1 20;

PLOT X

SPECTRUM <NB>

This command, when followed by a variable name, displays the estimates of the standardised spectral density function of the variable and their estimated standard errors using Bartlett, Tukey, and Parzen lag windows as in the following example:

SAMPLE 1 120;

SPECTRUM X

The window size will be taken to be twice the square root of the number of specified observations. If you have a graphics adaptor, the plot of the different spectral density functions and the associated standard error bands will also be displayed.

The SPECTRUM command can also be used to estimate the spectrum for a window size specified by the user. The desired window size should be specified in parentheses after the variable. For example,

SPECTRUM X(12)

TITLE <NB>

This command generates a list of the names of your variables, together with their description, if any. If you type

TITLE

the variable names and the descriptions (if any) of all your variables will be displayed. To see the description of some of your variables you can either type TITLE followed by the list of the variables you are interested in, or simply type the list of variables and press <GO>. For example,

TITLE X Y Z W

has the same effect as

X Y Z W

and generates a list and description (if any) of the variables X, Y, Z, and W.

XPLOT <NB>

This command can be used to plot up to a maximum of three variables against another variable. When issuing this command you must specify at least two variable names. For example,

XPLOT X Y produces a plot of the variable X against the variable Y.

XPLOT X1 X2 X3 Y produces a plot of the variables X1, X2, and X3 against the variable Y.

The graph may be saved, printed, or its display adjusted.

&

This command enables you to continue the list of variables appearing after the commands DELETE, KEEP, TITLE, ENTITLE, LIST, or COR on a new line. To continue the list of variables on a new line, type & (the ampersand) and click <GO>. For example, the command

KEEP X Y Z & <GO>

W D <GO>

applies the KEEP command to all the variables X, Y, Z, W, and D. Similarly, the command

COR A B C D & <GO>

X W <GO>

applies the COR command to all the variables A, B, C, D, X, and W, and computes the correlation matrix for the variables A, B, C, D, X, and W.

Reference

Pearson, M. H. and Pearson, B. (1997) *Working with Microfit 4.0, Interactive Econometric Analysis*, Oxford: Oxford University Press.

Index